FOOD PRODUCT OPTIMIZATION FOR QUALITY AND SAFETY CONTROL

Process, Monitoring, and Standards

FOOD PRODUCT OPTIMIZATION FOR QUALITY AND SAFETY CONTROL

Process, Monitoring, and Standards

Edited by

Juan Carlos Contreras-Esquivel, PhD

Laxmikant S. Badwaik, PhD

Porteen Kannan, PhD

A. K. Haghi, PhD

AAP | APPLE ACADEMIC PRESS

First edition published 2021

Apple Academic Press Inc.
1265 Goldenrod Circle, NE,
Palm Bay, FL 32905 USA
4164 Lakeshore Road, Burlington,
ON, L7L 1A4 Canada

CRC Press
6000 Broken Sound Parkway NW,
Suite 300, Boca Raton, FL 33487-2742 USA
2 Park Square, Milton Park,
Abingdon, Oxon, OX14 4RN UK

Library and Archives Canada Cataloguing in Publication

Title: Food product optimization for quality and safety control : process, monitoring, and standards / edited by Juan Carlos Contreras-Esquivel, PhD, Laxmikant S. Badwaik, PhD, Porteen Kannan, PhD, A.K. Haghi, PhD.

Names: Contreras-Esquivel, Juan Carlos, editor. | Badwaik, Laxmikant S., editor. | Kannan, Porteen, editor. | Haghi, A. K., editor.

Description: Includes bibliographical references and index.

Identifiers: Canadiana (print) 20200292242 | Canadiana (ebook) 20200292382 | ISBN 9781771888790 (hardcover) | ISBN 9781003003144 (PDF)

Subjects: Canadiana (print) 20200292242 | Canadiana (ebook) 20200292382 | ISBN 9781771888790 (hardcover) | ISBN 9781003003144 (PDF)

Classification: LCC TP373.5 .F68 2020 | DDC 363.19/26—dc23

Library of Congress Cataloging-in-Publication Data

CIP data on file with US Library of Congress

ISBN: 978-1-77188-879-0 (hbk)
ISBN: 978-1-00300-314-4 (ebk)

About the Editors

Juan Carlos Contreras-Esquivel, PhD

coyotefoods@hotmail.com

*Department of Food Research, Faculty of Chemistry,
Universidad Autonoma de Coahuila, Blvd. V. Carranza,
Colonia Republica Oriente, Saltillo 25280, Coahuila, Mexico*

Juan Carlos Contreras-Esquivel, PhD, is a full Professor and Head of the Laboratory of Applied Glycobiotechnology at the Food Research Department, School of Chemistry, the University of Coahuila, Mexico. He is involved in teaching undergraduate and graduate students with a special focus in food science and food bioscience. His research interests are mainly applied glycoltechnology for polysaccharide extraction by using ecofriendly technologies (enzymatic, microwave, pulsed electric field, ultrasonic), and oligosaccharide production by degradative or biosynthetic routes. Dr. Contreras-Esquivel earned a bachelor's degree in Biological Chemistry and holds an MSc degree in Food Science and Technology, both from the Universidad Autonoma de Chihuahua, Mexico. He received his PhD from the Universidad Nacional de La Plata, Argentina.

Laxmikant S. Badwaik, PhD

laxmikantbadwaik@gmail.com

*Associate Professor, Department of Food Engineering and Technology,
Tezpur University, Tezpur, Assam, India*

Laxmikant S. Badwaik, PhD, is Associate Professor of the Department of Food Engineering and Technology at Tezpur University, Assam, India. He is also serving as Deputy Director of the National Accreditation Board for Testing and Calibration Laboratories-accredited Food Quality Control Laboratory of Tezpur University, Tezpur. He has been involved in teaching and research for the last 15 years in the area of food engineering and technology. His main areas of research are food quality and safety, food packaging, and food processing waste utilization. He has published more than 25 research articles and chapters in various journals and books and also edited one book,

Innovations in Food Processing Technology. He has handled various funded research projects related to food processing and food packaging. He holds BTech in Food Technology from the University Department of Chemical Technology (UDCT), S.G.B. Amravati University, Amravati, Maharashtra, India; MTech in Food Engineering and Technology from the Sant Longowal Institute of Engineering and Technology (SLIET), Longowal, Punjab, India; and a PhD in Food Engineering and Technology from Tezpur University, India.

Porteen Kannan, PhD

rajavet2002@gmail.com

Assistant Professor, Department of Veterinary Public Health, Madras Veterinary College, Tamil Nadu Veterinary and Animal Sciences University, India

Porteen Kannan, PhD, is an Assistant Professor in the Department of Veterinary Public Health at Madras Veterinary College, Tamil Nadu Veterinary and Animal Sciences University, India. The research activities of Dr. Kannan include food safety and antimicrobial resistance. He performed his postdoctoral studies from the US Department of Agriculture, Maryland, USA, with a specialization on foodborne pathogens. He has published his work in both national and international journals. He is actively involved in mentoring both MVSc and PhD students.

A. K. Haghi, PhD

AKHaghi@Gmail.com

Former Editor-in-Chief, International Journal of Chemoinformatics and Chemical Engineering and Polymers Research Journal; Member, Canadian Research and Development Center of Sciences and Cultures (CRDCSC), Canada

A. K. Haghi, PhD, is the author and editor of over 165 books, as well as 1000 published papers in various journals and conference proceedings. Dr. Haghi has received several grants, consulted for a number of major corporations, and is a frequent speaker to national and international audiences. Since 1983, he served as a professor at several universities. He formerly served as the Editor-in-Chief of the *International Journal of Chemoinformatics and Chemical Engineering* and *Polymers Research Journal* and is on the editorial boards of many international journals. He is also a member of the Canadian Research and Development Center of Sciences and Cultures (CRDCSC), Montreal, Quebec, Canada. He holds a BSc in urban and environmental engineering from the University of North Carolina (USA), an MSc in mechanical engineering from North Carolina A&T State University (USA), a DEA in applied mechanics, acoustics and materials from the Université de Technologie de Compiègne, France, and a PhD in engineering sciences from Université de Franche-Comté, France.

Contents

Contributors

Rafail A. Afanas'ev
Pryanishnikov All-Russian Scientific Research Institute of Agrochemistry, d. 31A, Pryanishnikova Street, Moscow 127550, Russia

Juan José Gaytán Andrade
Research Group of Chemist-Pharmacist-Biologist, School of Chemistry, Autonomous University of Coahuila, Blvd. Venustiano Carranza, Col. República Oriente, Saltillo, Coahuila 25280, Mexico

Madesh Angappan
Division of Animal Health, ICAR Research Complex for NEH Region, Umiam, Meghalaya 793103, India

Laxmikant S. Badwaik
Department of Food Engineering and Technology, School of Engineering, Tezpur University, Napaam, Assam 784028, India

Sonia Yesenia Silva Belmares
Research Group of Chemist-Pharmacist-Biologist, School of Chemistry, Autonomous University of Coahuila, Blvd. Venustiano Carranza, Col. República Oriente, Saltillo, Coahuila 25280, Mexico

Girija Brahma
Department of Food Engineering and Technology, Tezpur University, Napaam, Sonitpur, Assam 784028, India

Sebastian Fernando Cavalitto
Research and Development Center for Industrial Fermentations, CINDEFI (CONICET, La Plata, UNLP), Calle 47 y 115 B1900ASH, La Plata, Argentina

Patricia Isabel Torres-Chávez
Departamento de Investigación y Posgrado en Alimentos, Universidad de Sonora, Hermosillo, Sonora 83000, Mexico

Agustin Rascón-Chu
Centro de Investigación en Alimentación y Desarrollo, CIAD, A. C. Carretera Gustavo Enrique Astiazarán Rosas No. 46, Col. La Victoria, Hermosillo, Sonora 83304, Mexico

Alicia Couto
Universidad de Buenos Aires, FCEN, Departamento de Química Orgánica—CONICET, CIHIDECAR, Intendente Güiraldes 2160, C1428GA, Ciudad Universitaria, Buenos Aires, Argentina

Amit Baron Das
Department of Food Engineering and Technology, Tezpur University, Napaam, Sonitpur, Assam 784028, India

Pranjal Pratim Das
Department of Chemical Engineering, IIT Guwahati, Assam, India

Jorge Márquez-Escalante
Centro de Investigación en Alimentación y Desarrollo, CIAD, A. C. Carretera Gustavo Enrique Astiazarán Rosas No. 46, Col. La Victoria, Hermosillo, Sonora 83304, Mexico

Juan Carlos Contreras-Esquivel
Food Research Department, School of Chemistry, Universidad Autonoma de Coahuila, Saltillo City 25250, Coahuila State, Mexico

María Irene Silvas-García
Universidad Estatal de Sonora, Hermosillo, Sonora, Mexico

Sandeep Ghatak
Division of Animal Health, ICAR Research Complex for NEH Region, Umiam, Meghalaya 793103, India

Carlos Neftalí Cano-González
Food Research Department, School of Chemistry, Universidad Autonoma de Coahuila, Saltillo City 25250, Coahuila State, Mexico

Tabli Ghosh
Department of Chemical Engineering, Indian Institute of Technology Guwahati, North Guwahati 781 039, Assam, India

Andrea Carolina Ramirez-Gutierrez
Research and Development Center for Industrial Fermentations, CINDEFI (CONICET, La Plata, UNLP), Calle 47 y 115 B1900ASH, La Plata, Argentina

Manuj Kumar Hazarika
Department of Food Engineering and Technology, Tezpur University, Assam, India

A. K. Haghi
Professor Emeritus of Engineering Sciences, Former Editor-in-Chief, *International Journal of Chemoinformatics and Chemical Engineering and Polymers Research Journal*; Member, Canadian Research and Development Center of Sciences and Culture
E-mail: akhaghi@yahoo.com

Raúl Rodríguez-Herrera
Food Research Department, School of Chemistry, Universidad Autonoma de Coahuila, Saltillo City 25250, Coahuila State, Mexico

Dipankar Kalita
Department of Food Engineering and Technology, School of Engineering, Tezpur University, Assam, India

Porteen Kannan
Department of Veterinary Public Health and Epidemiology, Madras Veterinary College, Chennai, Tamil Nadu 600007, India

Vimal Katiyar
Department of Chemical Engineering, Indian Institute of Technology Guwahati, North Guwahati 781 039, Assam, India

Faustino Lara
CISEF, Saltillo, Coahuila, Mexico

Beatriz Montaño-Leyva
Departamento de Investigación y Posgrado en Alimentos, Universidad de Sonora, Hermosillo, Sonora 83000, Mexico

Merab Magaly Rios-Licea
Sigma-Alimentos, Torre Sigma, San Pedro Garza Garcia 66269, Nuevo Leon, Mexico

Ena Deyla Bolaina-Lorenzo
Food Research Department, School of Chemistry, Universidad Autonoma de Coahuila, Saltillo City 25250, Coahuila State, Mexico

Charu Lata Mahanta
Department of Food Engineering and Technology, School of Engineering, Tezpur University, Assam, India

Manabendra Mandal
Department of Molecular Biology and Biotechnology, School of Sciences, Tezpur University, Assam, India

Guillermo Niño-Medina
Laboratorio de Química y Bioquímica, Facultad de Agronomía, Universidad Autónoma de Nuevo León, Francisco Villa S/N, Col. Ex-Hacienda El Canadá, C.P. 66050, General Escobedo, Nuevo León, México

Emilio Mendez-Merino
Sigma-Alimentos, Torre Sigma, San Pedro Garza Garcia 66269, Nuevo Leon, Mexico

Genrietta E. Merzlaya
Pryanishnikov All-Russian Scientific Research Institute of Agrochemistry, d. 31A, Pryanishnikova Street, Moscow 127550, Russia

Elizabeth Carvajal-Millan
Laboratory of Biopolymers, Research Center for Food and Development (CIAD, AC), Carretera Gustavo Enrique Astiazarán Rosas No. 46, Col. La Victoria, Hermosillo 83304, Sonora, Mexico

Arockiasamy Arun Prince Milton
Division of Animal Health, ICAR Research Complex for NEH Region, Umiam, Meghalaya 793103, India

Abhinav Mishra
Department of Food Science and Technology, University of Georgia, Athens, GA 30602, USA

Kasanchi M. Momin
Division of Animal Health, ICAR Research Complex for NEH Region, Umiam, Meghalaya 793103, India

Kona Mondal
Department of Food Engineering and Technology, Tezpur University, Napaam, Sonitpur, Assam 784028, India

Hilda Aguayo Morales
Research Group of Chemist-Pharmacist-Biologist, School of Chemistry, Autonomous University of Coahuila, Blvd. Venustiano Carranza, Col. República Oriente, Saltillo, Coahuila 25280, Mexico

Juan Pablo Bracho-Oliveros
Research and Development Center for Industrial Fermentations, CINDEFI (CONICET, La Plata, UNLP), Calle 47 y 115 B1900ASH, La Plata, Argentina

Gaston Ezequiel-Ortiz
Research and Development Center for Industrial Fermentations, CINDEFI (CONICET, La Plata, UNLP), Calle 47 y 115 B1900ASH, La Plata, Argentina

Ana Irene Ledesma-Osuna
Departamento de Investigación y Posgrado en Alimentos, Universidad de Sonora, Hermosillo, Sonora 83000, Mexico

Jayanta Pal
Department of Food Engineering and Technology, School of Engineering, Tezpur University, Napaam, Assam 784028, India

Ana Mayela Ramos-de-la-Peña
Tecnológico de Monterrey, School of Engineering and Sciences, Av. Eugenio Garza Sada 2501 Sur, Monterrey 64849, Nuevo Leon, Mexico

Govindarajan Bhuvana Priya
College of Agriculture, Central Agricultural University (Imphal), Kyrdemkulai, Meghalaya 793104, India

Luis Enrique Cobos Puc
Research Group of Chemist-Pharmacist-Biologist, School of Chemistry, Autonomous University of Coahuila, Blvd. Venustiano Carranza, Col. República Oriente, Saltillo, Coahuila 25280, Mexico

Crystel Aleyvick Sierra Rivera
Research Group of Chemist-Pharmacist-Biologist, School of Chemistry, Autonomous University of Coahuila, Blvd. Venustiano Carranza, Col. República Oriente, Saltillo, Coahuila 25280, Mexico

Concepción Lorenia Medina-Rodríguez
Departamento de Investigación y Posgrado en Alimentos, Universidad de Sonora, Hermosillo, Sonora 83000, Mexico

Jesús Enrique Gerardo-Rodríguez
Departamento de Investigación y Posgrado en Alimentos, Universidad de Sonora, Hermosillo, Sonora 83000, Mexico

Maria del Carmen Rodríguez Salazar
Research Group of Chemist-Pharmacist-Biologist, School of Chemistry, Autonomous University of Coahuila, Blvd. Venustiano Carranza, Col. República Oriente, Saltillo, Coahuila 25280, Mexico

Michail O. Smirnov
Pryanishnikov All-Russian Scientific Research Institute of Agrochemistry, d. 31A, Pryanishnikova Street, Moscow 127550, Russia

Preetisagar Talukdar
Department of Chemical Engineering, IIT Guwahati, Assam, India

José Ezequiel Viveros Valdez
Department of Chemistry, School of Biological Sciences, Autonomous University of Nuevo León, Pedro de Alba S/N, between Av. Alfonso Reyes and Av. Fidel Velázquez, University City. San Nicolás de los Garza, Nuevo León 66455, Mexico

Benjamín Ramírez-Wong
Departamento de Investigación y Posgrado en Alimentos, Universidad de Sonora, Hermosillo, Sonora 83000, Mexico

Abbreviations

DPPH	2,2-Diphenyl-1-picrylhydrazyl
AA	amino acid
ACE	angiotensin-l-converting enzyme
AX	arabinoxylans
ANN	artificial neural network
AAD	associated diarrhoea
CDI	*C. difficile* infection
CPA	*C. perfringens* alpha toxin
CPB	*C. perfringens* beta toxin
CPE	*C. perfringens* enterotoxin
COS	chito-oligosaccharides
CA	community associated
DW	dry weight
EN	enteritis necroticans
EIA	enzyme immune assay
EMC	equilibrium moisture content
ER	equivalent radius
EFSA	European Food Safety Authority
FRAP	ferric reducing antioxidant property
FAX	ferulate esterase treated AX
FA	ferulic acid
FDA	Food and Drug Administration
FTIR	Fourier transform infrared spectroscopy
GA	gallic acid
GIT	gastrointestinal tract
GRAS	generally recognized as safe
GlcN	glucosamine
GH	glycoside hydrolases
GMP	good manufacturing practices
HACCP	hazard analysis critical control point
HPLC	high-performance liquid chromatography
HTST	high-temperature short-time
IMC	initial moisture content
IL-1	interleukin-1

ISO	International Organization for Standardization
LAMP	loop-mediated isothermal amplification
LDL	low-density lipoproteins
MDG	maize distillers grains
MSE	mean square error
MFC	microfibrillated cellulose
MAE	microwave-assisted extraction
MC	moisture content
GlcNAc	*N*-acetylglucosamine
NMR	nuclear magnetic resonance
PSD	particle size distribution
PAN	polyacrylonitrile nanofibrous
PEF	pulsed electric fields processing
ROS	reactive oxygen species
RH	relative humidity
RSM	response surface methodology
RP-HPLC	reverse-phase high-performance liquid chromatography
AgNPs	silver nanoparticles
SHIME	simulator of the Human Intestinal Microbial Ecosystem
SNLs	solid lipid nanoparticles
SD	sporadic diarrhea
SIDS	sudden infant death syndrome
TLC	thin layer chromatography
TA	titrable acidity
TFC	total flavonoid content
TPC	total phenolic content
TSS	total soluble sugars
TNF-α	tumor necrosis factor alpha
UAE	ultrasound-assisted extraction
USFDA	United States Food and Drug Administration
VBNC	viable but nonculturable cells

Preface

This book discusses food quality and safety standards that are critically important not only for developed but also for developing economies, where the consumers' safety is among the primary issues to be considered in food supply chain management. After the rapid development of many economies, quality standards have focused on consumers' demand for safe food and beverage.

Food safety is a multifaceted subject, using microbiology, chemistry, standards, and regulations, and risk management to address issues involving bacterial pathogens, chemical contaminants, natural toxicants, additive safety, allergens, and more. It allows the reader to take what knowledge is required for understanding food safety at a wide range of levels.

Food chemistry and safety is the study of the underlying properties of foods and food ingredients. It seeks to understand how chemical systems behave in order to better control them to improve the nutritional value, safety, and culinary presentation of food. As a core subject in food science, food chemistry, and safety is the study of the chemical composition, processes, and interactions of all biological and nonbiological components of foods.

This book provides an overview of the risk and hazard analysis of different foods and the important advances in technology that have become indispensable in controlling hazards in the modern food industry. It covers critically important topics and organizes them in a manner to facilitate learning for those who are, or who may become, food safety professionals.

This is an outstanding research-oriented book for students and a valuable reference for professionals in food processing, as well as for those working in fields that service, regulate, or otherwise interface with the food industry.

Food quality concerns the "fitness for purpose" of our food in terms of appearance (e.g., color and surface qualities, texture, flavor, and odor) and how these can be improved. Food safety considers the physical, microbiological, and chemical aspects of our food, which may be harmful to human health and how these can be minimized. This volume emphasizes the inter-relationship between these areas and their equal importance in food production.

Key features of this volume are

- strengthening food control regulatory frameworks;
- providing scientific advice to support the standards;
- enhancing food safety management along food chains to prevent diseases and trade disruptions;
- promoting food safety emergency preparedness to build resilient agri-food chains;
- developing a platform for information sharing and tools to support food safety management.

Chapter 1 reviewed innovative methods in soil phosphorus research.

In Chapter 2, the content of exchange potassium in soil with a prolonged application of fertilizer with a particular application on food safety is discussed in detail.

Foodborne pathogenic anaerobes are the group of bacteria that are not capable to grow in the presence of oxygen and causes foodborne diseases in human beings. The genus *Clostridium* represents the major food pathogenic anaerobes and the *Clostridia* of food safety relevance include *Clostridium botulinum, Clostridium perfringens,* and *Clostridium difficile.* The ubiquitous presence, ability to produce heat resistant endospores, wide growth temperature range, ability to adapt in diverse food environments, and the different diseases caused make these bacteria the significant health hazard causing potentially lethal foodborne illnesses in humans. *C. botulinum* causes neurotoxin-mediated foodborne botulism, infant botulism, and wound botulism. *C. perfringens* causes food poisoning, necrotic enteritis, antibiotic-associated diarrhea, and gas gangrene in human beings. *C. difficile* causes severe pseudomembranous colitis and *C. difficile* antibiotic-associated diarrhea. Chapter 3 gives a detailed outlook of these three potentially important pathogenic anaerobes as well as describes strategies to control them in food.

In Chapter 4, microwave-assisted extraction of Ceylon olive was analyzed with the four independent variables, the independent variables were microwave power, irradiation time, solid-to-solvent ratio, and solvent concentration. Solvent-to-solid ratio and the irradiation time have the maximum influence on the microwave-assisted extraction process. The optimized conditions for the process were microwave power 300 W, irradiation time 58 s, solid-to-solvent ratio 1:26.19, solvent concentration 58.05% ethanol, and the experimental response values were total phenolic content (TPC) 300.125 mg GAE/100g DW, 2,2-diphenyl-1-picrylhydrazyl (DPPH) 80.325% RSA. Among conventional heat reflux and microwave-assisted extraction process microwave-assisted extraction methods had the maximum efficiency to extract the phenolic compound.

In Chapter 5, phenolic compounds (anthocyanin) were extracted from the bracts of *Musa balbisiana* inflorescence using a surfactant-mediated ultrasound-assisted extraction process. The parameters of the extraction process were optimized using response surface methodology (RSM). The extraction kinetics of the surfactant-mediated ultrasound-assisted extraction process was investigated and compared with other extraction processes. The phytochemical and antioxidant properties of the extract were also investigated.

Chapter 6 mainly focuses on the brief overview of bio-based origin of the nanostructured material and their development processes, property analysis, and application into the food sector. Along with this, the various safety concerns, available rules, and regulations relating to the nanomaterials will be discussed as they are susceptible to cause health issues. However, the possibility of showing toxicity is caused by these nanostructured materials either in a chronic or acute way. Although the application of newly developed nanomaterials in food is likely to be consumed at a low level over an extended period of time, where the occurrence of the adverse effect is less. Furthermore, nanostructured material can also alter human health by damaging the microbial cell, which generally affects the human gastrointestinal tract. Despite these limitations, the benefits of nanotechnology are promising for the development of novel functional materials, and product processing at micro and nanoscale level, where the design of materials, methods, and instrumentation are very crucial for food safety and biosecurity.

Campylobacteriosis: Emerging Foodborne Zoonosis is discussed in detail in Chapter 7.

Enzymatic modification of ferulic acid content in arabinoxylans from maize distillers grains is investigated in Chapter 8.

In Chapter 9, the probiotic characteristics of the three different strains of *Lactobacilli* and their survivability in freshly prepared juices from four different fruits were studied. The potential isolates of the *Lactobacilli* and a suitable carrier juice were identified based on the viability of the bacteria, pH, total soluble sugars, titrable acidity, and color of the probiotic fruit juices. The viability of the bacterial cells and quality parameters of the probiotic juices were also studied in refrigerated storage conditions. The effects of the addition of probiotics on the phytochemical and antioxidant properties of the selected fruit juices were also studied. The probiotic juices were analyzed for TPC, total flavonoid content (TFC), and antioxidant properties like DPPH radical scavenging activity, ferric reducing antioxidant property (FRAP).

Reverse-phase high-performance liquid chromatography of the probiotic juices were also carried out to study the changes in the profile of polyphenols, organic acids, and sugars during storage. Mineral constituents of the juice were also determined by atomic absorption spectroscopy.

In Chapter 10, experiments were carried out in the Department of Food Engineering and Technology of Tezpur University to estimate the equilibrium moisture content of Bengal gram, black-eyed peas, chickpeas, and mung beans at known relative humidity by the chemical method, at temperatures from 30 °C to 50 °C. The equilibrium moisture data is related to storage relative humidity by the GAB equation as well as by artificial neural network (ANN) model. The seven input neurons corresponded to the seven input variables, a_w, temperature, ash content, dietary fiber, crude protein, fat content, and carbohydrates while the output neuron represented the equilibrium moisture content. The highest R-value and the least mean square error value corresponding to the hidden neurons with the neural network are obtained.

At given moisture content, water activity value is estimated by the fitted GAB model in the moisture range from 5% to 20 %, at all the temperatures. Based on these predicted data of water activity, enthalpy–entropy compensation analysis of the four samples were carried out to estimate isokinetic temperature, and develop the generalized isotherm model involving isokinetic temperature and two other parameters, namely, K_1 and K_2. These two parameters are later on used for hydration kinetics study of the same product. Also, an overall K_1 and K_2 containing all the data of the four samples together was calculated and obtained.

Finally, for studying the kinetics of hydration of Bengal gram, black-eyed peas, chickpeas, and mung beans soaking experiments at different temperatures were carried out with measurement of moisture content at known intervals. For explaining the moisture migration during the soaking process, the principle of one-dimensional mass transfer to the spherical body was considered with the use of equivalent diameter. For expressing the dependence of the diffusivity value on temperature and moisture, the model involved an expression containing the parameters K_1 and K_2 and heat of sorption. Simulations were carried out to estimate a value of the pre-exponent of diffusivity expression (D_o).

This work validated the application of the approach of enthalpy–entropy compensation analysis of sorption behavior to analyze hydration behavior of the product under consideration. The developed model is applicable to predict the endpoint of soaking to obtain the desired final moisture content.

A comprehensive review on enzymatic production of chito-oligosaccharides and D-glucosamine by fungal chitosanases from *Aspergillus* spp is presented in Chapter 11.

Chapter 12 aims to make a quick review to understand and start the biotechnological path for the design and development of future foods based on isomaltulose.

The aim of Chapter 13 is to provide a general perspective for the use of pulsed electric field processing (PEF) on pasteurization, sterilization, and enzyme inactivation. In addition, other purposes of PEF in food are described, such as extraction, drying, freezing, etc. Available equipment, consumer's point of view, environmental impact, and legislation frameworks are considered to increase the knowledge and interest about the use of this novel technology and the impact of the final product on consumers, industry, and environment.

Security and bioavailability of derivatives from medicinal plants in food consumption are discussed deeply in Chapter 14.

Viscoelastic measurement using oscillatory dynamic strain test has been shown to be a useful tool to know the integrity and changes of chemical components caused on each step of the bread-making process discussed in Chapter 15. In bread making it is useful to be aware of the effect of added ingredients and make decisions about their use. Measurement of viscoelastic behavior is useful to know in order to obtain parameters of the bread-making process such as the optimal fermentation, and mixing time and the damages caused by freezing and storage steps to proteins and starch. Dough is the material that suffers most of the physicochemical changes; this allows knowing in what step and temperature would occur these changes and how it affects baking speed and time.

In Chapter 16, physicochemical characteristics and gelling properties of arabinoxylans recovered from maize wastewater are discussed in detail.

CHAPTER 1

Transformation of Phosphorus in Soils of Agroecosystems in Long-Term Experiments: Sustainability Challenges of Phosphorus and Food

RAFAIL A. AFANAS'EV, GENRIETTA E. MERZLAYA, and MICHAIL O. SMIRNOV*

Pryanishnikov All-Russian Scientific Research Institute of Agrochemistry, d. 31A, Pryanishnikova Street, Moscow 127550, Russia

**Corresponding author. E-mail: user53530@yandex.ru*

ABSTRACT

In field experiments with the long-term systematic application of fertilizers (prolonged trial), new regularities of the dynamics of the mobile phosphorus content in soils of various agroecosystems have been established. It is shown that the increase in the content of mobile phosphorus in different soils occurred only in the first rotation of crop rotations. In the future, despite the positive balance of phosphorus, the content of its mobile forms did not increase or even had a tendency to decrease due to the transition of phosphorus to an inactive state. In the case of a negative balance, the content of mobile phosphorus in soils was compensated by the reserve of inactive phosphates. Thanks to these processes of phosphate transformation in agricultural ecosystems supported the ecological balance, protecting against loss of phosphorus with surface and subsurface runoff of the element in the external environment and, thus, reduces the risk of water body eutrophication. The influence of phosphorus fertilizers on the biodiversity of soil microflora is also established. The dynamics of mobile phosphorus in various soils revealed in long-term field experiments can serve as a model of phosphorus transformation, which entered the soil from fertilizers in production conditions.

1.1 INTRODUCTION

The content of mobile phosphorus in soils in the years of intensive chemicalization was directly related to the level of fertilizer application. If in Russia in 1965, before the beginning of intensive chemicalization, the application of mineral fertilizers (NPK) per 1 ha of arable land per year averaged 20 kg of active substance, then for five years their use increased: in 1966–1970 up to 28 kg/ha, in 1976–1980 up to 65 kg/ha, and in 1986–1990 up to 99 kg/ha. With the increase in the use of minerals, including phosphorus, fertilizers in the country increased soil fertility, in particular the availability of their mobile phosphates. Thus, during the period from 1971 to 1999, the share of arable land with a low content of mobile phosphorus decreased from 52% to 22% with a corresponding increase in the share of medium and well-endowed soils (Sychev and Mineev, 2011).

However, increasing the content of mobile phosphorus in soils is not the only effect of the intensive use of phosphate fertilizers. A significant part of the phosphorus applied in excess of phosphorus removal by crops passed into inactive forms, creating a reserve of phosphorus nutrition of plants (Cook, 1970; Black, 1973; Ginsburg, 1981). According to Sychev and Shafran (2013), for 25 years it was applied over the removal of about 300 kg/ha of phosphorus, which remained in the soil. This amount of phosphorus is sufficient to produce 2 t/ha of grain crops for 25–30 years if you do not apply phosphorus-containing fertilizers.

Since 1995, when the balance of phosphorus in the country's agriculture began to develop with the excess of removal over the application, there has been a certain tendency to reduce mobile phosphates in arable soils, although not as obvious as previously expected. At the same time, the regularities of the transformation of phosphate fertilizers have not been sufficiently studied. This, in particular, is evidenced by the statement of one of the leading experts in this field of knowledge K. E. Ginzburg: "The absorption capacity of soils with respect to phosphates confuses our calculations about increasing the content of mobile phosphates in the soil, since when soluble phosphates are applied into the soil, an unknown, but a significant part of them passes into poorly soluble and less accessible forms for plants" (Ginsburg, 1981, p 124). Long-term dynamics, the nature of the transition of mobile phosphorus in soils in inactive forms can be traced by the example of long-term field experiments with fertilizers, which is the subject of this work.

1.2 METHODS

The research method is based on the generalization of the materials of long-term field experiments, as well as the results of their own research. The authors analyzed the impact of the systematic use of high doses of organic and mineral fertilizers, including phosphorus, on the change in the content of mobile phosphorus in various soils: sod-gleyic clay loam (Lithuania), sod-podzolic clay loam (Moscow region), sod-podzolic sandy loam (Smolensk region), sod-podzolic loamy sand (Belarus), and ordinary chernozem (Stavropol region).

The economic balance of phosphorus on the main variants of field experiments is calculated taking into account the applied amount of phosphorus on rotations of field crop rotations and the influence of the long-term systematic application of phosphorus fertilizers on the content of mobile phosphorus in soils is shown.

Determination of the content of phosphorus mobile forms in various soils was carried out according to the methods adopted in agrochemistry (Agrochemical Methods of Investigation of Soils, 1975). The main methods for determining the content of mobile phosphorus were: Egner–Rima method—extraction from soil with 0.04 n solution of calcium lactic acid, $CH_3CHOHCOO)_2Ca{\bullet}5H_2O$ at pH 3.5–3.7; Kirsanov method—extraction from soil with 0.2 normal solution; HCl and Machigin method—extraction with 1% solution $(NH4)_2CO_3$. Methods for determining the content of mobile phosphorus are specified in the description of the research results.

1.3 DISCUSSION

In a field experiment in the Lithuanian Scientific Research Institute of Agriculture (Plesyavichius, 1982), the research was conducted to study the efficacy of continuous application of mineral fertilizers on drained sod-gleyic clay loamy soil. In the variant with the application of the rotation N225P324K350 with an annual application of phosphorus fertilizer on average in seven deployed to the nature of the fields of crop rotation at the economic balance of phosphorus—68 kg/ha in the first rotation and 73 kg/ha in the second; the amount of mobile P2O5 (by Egner–Rome) at the end of the rotation was 42 and 43 mg/kg soil, respectively (Figure 1.1).

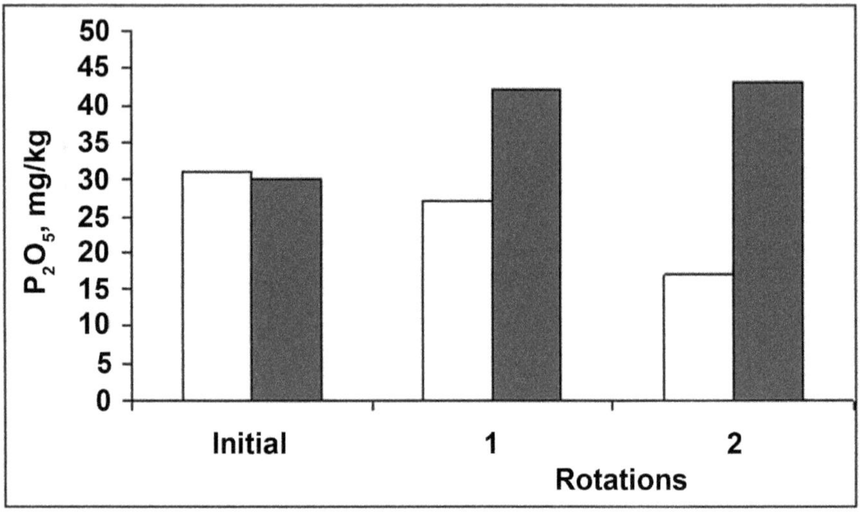

FIGURE 1.1 Dynamics of phosphorus in sod-gleyic clay loamy dried soil according to the rotations of crop rotation (Lithuania).
Light columns—control, dark columns—N225P324K350 variant.

The change in this indicator at the end of rotations compared with their beginning was equal to 12 and 1 mg/kg, respectively, with the productivity of the first rotation (1961–1968) on average, 44.8 C/ha of grain units and for the second (1969–1974), 46.5 C/ha. Thus, almost about 140 kg/ha of P_2O_5, applied in excess of removal for two rotations of the crop rotation, at the end of the second rotation passed into the soil in forms not extracted by the method of Egner–Rome.

According to the calculation, the cost of phosphorus fertilizers to increase the content of mobile phosphorus in the arable soil layer by 10 mg/kg in the first rotation was 57 kg/ha P_2O_5, in the second—by a mathematical order higher.

Quite pronounced was also the change in the content of mobile phosphorus in sod-podzolic clay loam soil in the long-term experiment conducted at the Central Experimental Station of Pryanishnikov All-Russian Scientific Research Institute of Agrochemistry (Moscow region) (Efremov, 2011).

Of the seven rotations of the four-field crop rotation, or for 28 years of systematic fertilizer application, the highest content of mobile phosphorus (according to Kirsanov) in the version of the organomineral system was observed in the fourth rotation, after which it decreased (Figure 1.2), despite the fact that the positive phosphorus balance for all 28 years of the application was 2.7 t/ha.

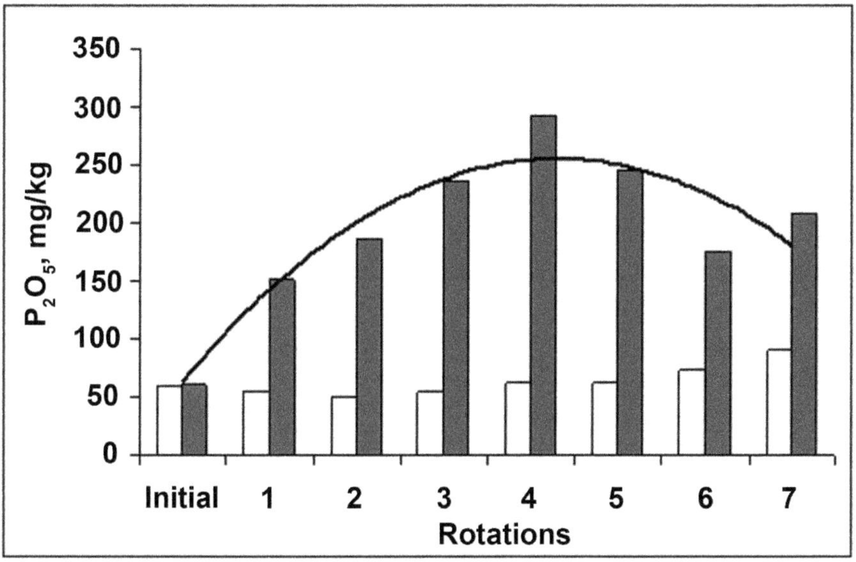

FIGURE 1.2 Dynamics of mobile phosphorus in sod-podzolic clay loam soil by rotation of crop rotation (Moscow region).
Light columns—control, dark columns—variant—NPK+manure.

The consumption of phosphorus of mineral fertilizers, applied in excess of removal, to increase the content of mobile phosphorus in the arable soil layer by 10 mg/kg per 1 ha was on average 119 kg P_2O_5.

The study of the 12-year aftereffect of fertilizers applied in this variant showed that the yield of winter wheat in an average of three rotations of the crop rotation was 17.2 C/ha compared to 12.0 C/ha at the control, that is, due to previously applied fertilizers only from winter wheat more than 5 C/ha of grain products were obtained.

In the field experiments conducted for 30 years by Pryanishnikov All-Russian Scientific Research Institute of Agrochemistry on sod-podzolic sandy loam soil (Smolensk region), with the systematic use of mineral fertilizers for all crops, except perennial grasses, in the first rotation of the crop rotation (1979–1989) in the variant with the mineral fertilizer system was applied N990P990K990, in the second (1990–1995) and the third rotation

(1996–2001) on N450P450K450, and in the fourth rotation (2002–2008), N405P405K405.

The content of mobile phosphorus in the arable soil layer (according to Kirsanov) for the first two rotations at the economic balance of phosphorus 943 kg/ha P_2O_5 increased from 149 to 210 mg/kg, or by 61 mg/kg. At the end of the fourth rotation at the balance of phosphorus in the amount of the third and fourth rotation 523 kg/ha P_2O5 observed even a decrease in the content of mobile phosphorus in the soil from 210 to 174 mg/kg, that is, by 36 mg/kg (Figure 1.3, Table 1.1).

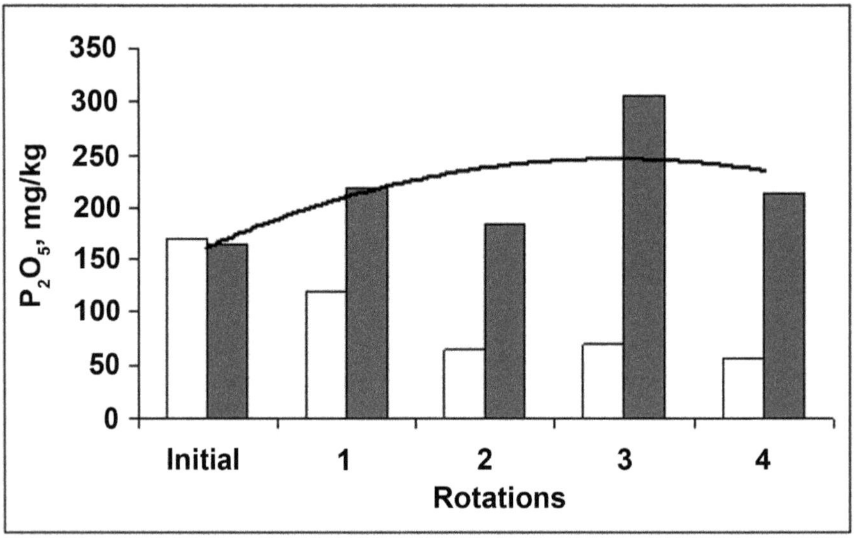

FIGURE 1.3 Content of mobile phosphorus in sod-podzolic sandy loam soil in the rotations of crop rotation (Smolensk region).
Light columns—control, dark columns—option—NPK + manure.

TABLE 1.1 Influence of Fertilizers on Crop Rotation Productivity and Mobile Phosphorus Content in Sod-podzolic Sandy Loam Soil (Smolensk Region)

Indicators	Control Without Fertilizers	NPK	Manure	NPK + Manure
	1–2 Rotations (17 years)			
Productivity in an average year (C g.u./ha)	24.0	34.4	28.6	34.6
Applied P_2O_5 (kg/ha)	–	1440	340	1780

TABLE 1.1 *(Continued)*

Indicators	Control Without Fertilizers	NPK	Manure	NPK + Manure
The removal of P_2O_5 (kg/ha)	369	497	422	500
Balance P_2O_5 (kg/ha)	−369	+943	−82	+1280
The content of P_2O_5 in the soil at the beginning of the rotation (mg/kg)	170	149	143	166
The content of P_2O_5 in the soil at the end of the rotation (mg/kg)	65	210	85	185
A change in the content of P_2O_5 in soil (mg/kg)	−105	+61	−58	+19
3–4 Rotations (30 years)				
Productivity in an average year (C g.u./ha)	20.1	29.6	26.0	26.0
Applied P_2O_5 (kg/ha)	–	855	252	107
The removal of P_2O_5 (kg/ha)	222	332	287	300
Balance P_2O_5 (kg/ha)	−222	+523	−35	+807
The content of P_2O_5 in the soil at the beginning of the rotation (mg/kg)	65	210	85	85
The content of P_2O_5 in the soil at the end of the rotation (mg/kg)	56	174	160	213
A change in the content of P_2O_5 in soil, mg/kg	−9	−36	+75	+28

Note: g.u., grain unit equivalent to 1 kg of wheat.

In the subsurface soil layer, the content of mobile phosphorus (P_2O_5) in the control variant of the experiment at the end of the fourth rotation (2008) compared to the middle of the first rotation (1983) decreased from 1.8 to 1.2 t/ha due to the removal of the element with the crop (Figure 1.4).

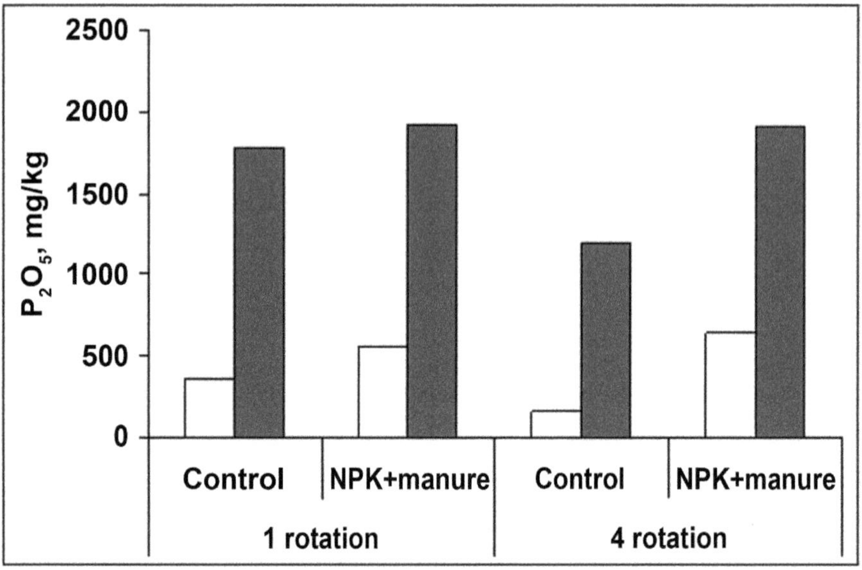

FIGURE 1.4 Dynamics of the content (supply) of mobile phosphorus in the profile layers sod-podzolic sandy loam soil (Smolensk region).
Light columns—0–20 cm, dark columns—20–100 cm.

In the variant with the joint application of mineral fertilizers and manure during the same period, the supply of mobile phosphates in the subsurface soil layers (20–100 cm) remained at the level of 1.9 t/ha, that is, there was no noticeable change in the content of mobile phosphates in the subsurface soil layers.

This demonstrates the transformation of migratory species of phosphorus in the fully inactive form unless you consider its migration outside of the controlled layer of soil.

Studies conducted by the Stavropol Research Institute of Agriculture (Shustikova and Shapovalova, 2011) in conditions of ordinary loamy Chernozem, showed that the annual application of mineral fertilizers for crops of six-field crop rotation in doses of N120P90K120 provided a positive balance of phosphorus in the first rotation (according to the method Machigin) 270 kg/ha P_2O_5 (Table 1.2, Figure 1.5).

The content of mobile phosphorus in the arable soil layer at the end of rotation increased to 26 mg/kg, or by 13 mg/kg compared to the beginning of the rotation.

TABLE 1.2 Influence of Mineral Fertilizers on Crop Rotation Productivity and Content of Mobile Phosphorus in Ordinary Chernozem (Stavropol Region)

Indicators	Control Without Fertilizers	N120P30K120	N120P30K120	N120P30K120
		1 Rotation (6 years)		
Productivity in an average year (C g.u./ha)	27.7	34.3	35.2	35.7
Applied P_2O_5 (kg/ha)	–	150	450	750
The removal of P_2O_5 (kg/ha)	139	177	180	182
Balance P_2O_5 (kg/ha)	−139	−27	+270	+568
The content of P_2O_5 in the soil at the beginning of the rotation (mg/kg)	13	13	13	13
The content of P_2O_5 in the soil at the end of the rotation (mg/kg)	13	21	26	58
A change in the content of P_2O_5 in soil (mg/kg)	0	+8	+13	+45
		2–3 Rotations (18 years)		
Productivity in an average year (C g.u./ha)	26.5	32.7	34.5	33.3
Applied P_2O_5 (kg/ha)	–	300	900	1500
The removal of P_2O_5 (kg/ha)	277	342	358	350
Balance P_2O_5 (kg/ha)	−277	−42	+542	+1150
The content of P_2O_5 in the soil at the beginning of the rotation (mg/kg)	13	21	26	58
The content of P_2O_5 in the soil at the end of the rotation (mg/kg)	12	28	52	76
A change in the content of P_2O_5 in soil (mg/kg)	−1	+7	+26	+18
		4–5 Rotations (30 years)		
Productivity in an average year (C g.u./ha)	19.0	23.2	24.5	24.6
Applied P_2O_5 (kg/ha)	–	180	540	900
The removal of P_2O_5 (kg/ha)	226	281	290	291

TABLE 1.2 *(Continued)*

Indicators	Control Without Fertilizers	N120P30K120	N120P30K120	N120P30K120
Balance P_2O_5 (kg/ha)	−226	−101	+250	+609
The content of P_2O_5 in the soil at the beginning of the rotation (mg/kg)	12	28	52	76
The content of P_2O_5 in the soil at the end of the rotation (mg/kg)	20	30	54	70
A change in the content of P_2O_5 in soil (mg/kg)	+8	+2	+2	−6

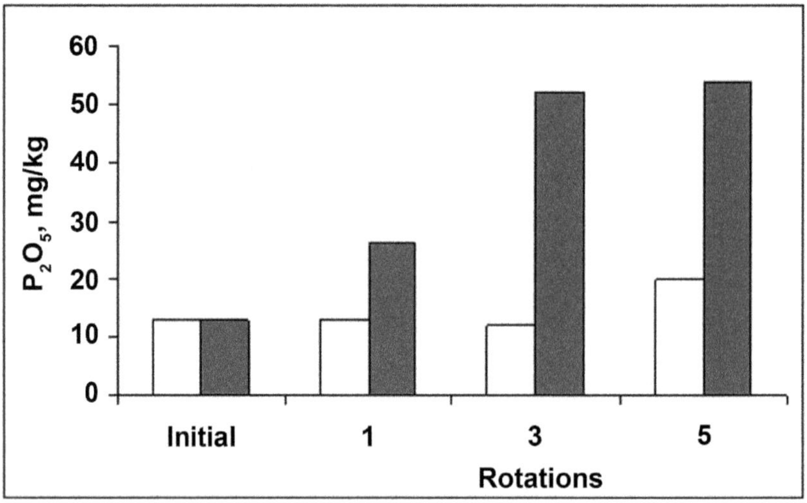

FIGURE 1.5 Content of mobile phosphorus in ordinary chernozem (Stavropol region). Light columns—control, dark columns—N120P90K120.

During the second and third rotations at the same annual doses of mineral fertilizers and the total phosphorus balance of 542 kg/ha, the content of mobile phosphorus in the soil of this variant increased to 52 mg/kg, increasing compared to the beginning of the second rotation by 26 mg/kg. However, by the end of the fifth rotation, that is, 12 years after the completion of the third rotation, despite the systematic application of mineral fertilizers at the same doses with the phosphorus balance over the years of 250 kg/ha, the content of mobile phosphorus in the arable layer increased by only 2 mg/kg.

At a higher dose of phosphorus in the N120P150K120 variant and phosphorus balance for the first rotation of 568 kg/ha, the total balance for the first three rotations of 1718 kg/ha and the fourth and fifth rotations of 609 kg/ha, the content of mobile phosphorus in the arable layer was, respectively, 58, 76, and 70 mg/kg. Therefore, over the last two rotations of the field rotation compared to the end of the third rotation, the application of 609 kg/ha of phosphorus fertilizers in excess of removal did not only increase the content of mobile phosphorus in the arable soil layer but even led to its reduction by 6 mg/kg.

Analysis of the content of mobile phosphorus, not only in arable but also in the subsurface layers of the soil profile of ordinary chernozem, showed that the intensity of its transition to inactive forms was in close dependence ($r = 0.99$) on the value of the positive balance in the agroecosystem. From Figure 1.6, it is seen that with the increase in the balance of P_2O_5 in the variant N120P90K120 increased the amount of mobile phosphorus, transformed into an inactive state. In general, for five rotations of field crop rotation more than 1000 kg/ha of P_2O_5 were transformed into inactive forms.

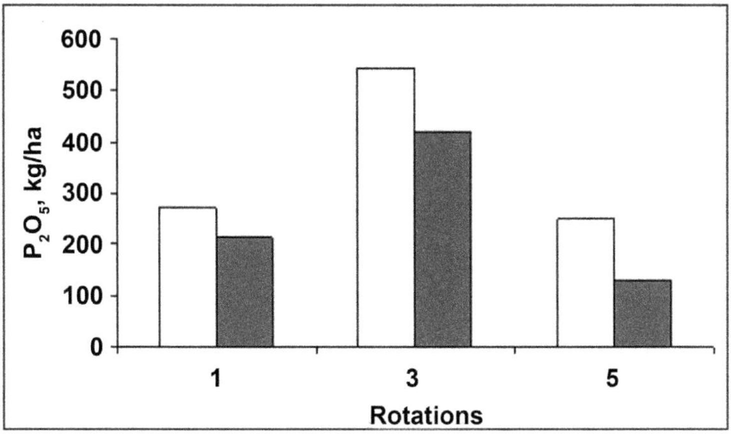

FIGURE 1.6 The dependence of the transformation of mobile phosphorus in inactive forms (loss) in 1-m layer of ordinary chernozem in variant N120P90K120 as a function of the magnitude of the phosphorus positive balance in the rotations of the rotation (Stavropol region) crop.
Light columns—balance columns of the dark—the decline of mobile phosphorus.

According to microbiological studies conducted by Pryanishnikov All-Russian Scientific Research Institute of Agrochemistry together with Lomonosov Moscow State University, the unilateral use of mineral fertilizers reduced the total number of microorganisms in the soil and the Shannon biodiversity index (Table 1.3).

TABLE 1.3 Influence of Fertilizers on the Content of Micro-organisms in Sod-podzolic Sandy Loam Soil (Smolensk Region)

Indicators	Control Without Fertilizers	NPK	Manure	NPK + Manure
Proteobacteria (cells/g $\times 10^6$)	13.6	17.1	22.0	20.3
Actinobacteria (cells/g $\times 10^6$)	19.3	13.4	18.1	13.3
Firmicutes (cells/g $\times 10^6$)	11.9	5.1	20.4	22.5
Bacteroidetes (cells/g $\times 10^6$)	2.0	2.5	5.9	3.7
The total number of microorganisms (cells/g $\times 10^6$)	46.8	38.1	66.4	59.8
Biodiversity index	5.0	4.7	4.6	4.9

The application of manure increased the total number of microorganisms but reduced the biodiversity index. When using the organomineral fertilizer system, an increase in the total number of microorganisms was observed with virtually no decrease in their biodiversity. The reduction in the number of microorganisms noted in Table 1.3 was mainly due to representatives of *Actinobacteria*, and the increase was due to *Proteobacteria* and *Bacteroidetes*.

Under the conditions of fertilizer aftereffect, the number of proteolytic bacteria actively transforming organic phosphates, such as *Pseudomonas fluorescens, Pseudomonas putida, Brevundimonas vesicularis*, increased 1.5–2 times (Merzlaya et al., 2012). In general, the microbial community of the soil was represented by more than 40 species belonging to 34 genera. The high number and microbial diversity indicate a sufficient degree of culture of the studied sod-podzolic soil.

In soils of light granulometric composition, the dynamics of mobile phosphorus in accordance with the increase in the positive balance of phosphorus in agrocenoses, in contrast to clay and clay loamy soils, can go on increasing. This can be seen from the data on the dynamics of mobile phosphorus in loamy sand soil of the long-term field experiment of the Grodno Agricultural Experimental Station (Belarus) (Shuglya, 1982).

From Figure 1.7, it follows that the content of mobile phosphorus (according to Kirsanov) due to the systematic application of mineral fertilizers increased from 44 mg/kg at the beginning of the first rotation to 216 mg/kg at the end of the fourth rotation of the four full-field crop rotation. This trend is explained by the fact that the phosphate capacity of loamy sand soils is several times lower than the capacity of the heavier granulometric composition of soil differences (Ginsburg, 1981), which inhibits the transition of mobile forms of phosphorus in less mobile.

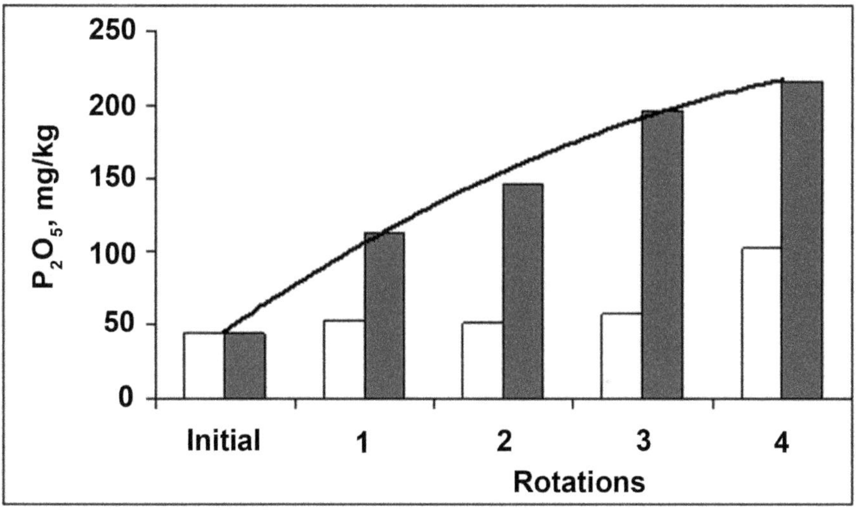

FIGURE 1.7 Dynamics of mobile phosphorus in sod-podzolic loamy sand soil in the rotations of crop rotation (Belarus).
Light columns—control, dark columns—NPK.

At the same time, the consumption of phosphorus fertilizers to increase the content of mobile phosphorus in the arable soil layer by 10 mg/kg increased even with an increase in the content of mobile forms. Thus, in the third rotation of the crop rotation, these costs per 1 ha amounted to 22 kg of P_2O_5, but in the fourth rotation of the crop rotation they increased to 56 kg.

Analysis of the dynamics of the content of mobile phosphorus in loamy and clay soils revealed a general pattern of its transition over time in inactive forms with a decrease in the concentration of mobile forms. Depending on the agrochemical properties of soils, the terms of this transition occurred in different periods: on the sod-gleyic clay loam (Lithuania) after 7 years (in the second rotation of the crop rotation), on the sod-podzolic sandy loam soil (Smolensk region) after 17 years (in the third and fourth rotations), on ordinary chernozem after 18 years (in the fourthand fifth rotations).

Since the soil is a bioinert system, largely formed under the influence of biological factors, it has a function of preservation and transformation of substances (Williams, 1951; Shein and Milanovsky, 2001), and it reacts to soluble phosphorus fertilizers according to the well-known Le Chatelier principle: if the equilibrium system is affected in any way, as a result of the processes occurring in it, the equilibrium will shift in the direction of reducing this impact (Glinka, 1976). In different types of soils, these functions are manifested in accordance with the original natural factors.

On sod-podzolic soils, characterized by a high content of iron and aluminum compounds, applied phosphorus fertilizers are converted into phosphates of sesquioxides oxides, while in chernozems and chestnut soils enriched with carbonates, the phosphorus preservation function is manifested in the formation of multibasic phosphates, including the insoluble type of apatite. The specific reasons causing a slight increase or even decrease in the content of mobile phosphorus in the last rotations of these crop rotations are obviously associated with an increase in the intensity of the processes of converting phosphorus fertilizers into slightly soluble forms from the first rotations of crop rotations to the subsequent due to the increased reaction of the soil as a system to the increasing total rotation of the excess application of water-soluble phosphorus fertilizers.

From all the above, it follows that the traditional agrochemical soil survey, that is, the determination of mobile forms of phosphorus in them, which are an indicator of effective fertility, does not give a sufficient idea of their potential fertility.

Thus, there are reports (Fertilizers, 1982) that the presence of 600–800 kg of residual phosphates per 1 ha on almost all soils guarantees their optimal phosphate state. Therefore, for an objective assessment of the phosphate level of soils, it is advisable to take into account data not only on the content of mobile forms of phosphorus in them but also the created reserves of less mobile forms that can be used by crops in the future. It is also important for the economic assessment of arable soils to take into account their potential fertility.

1.4 CONCLUSIONS

1. Long-term systematic application of fertilizers with phosphorus doses exceeding the element removal by crops increases the content

of mobile phosphates in soils of agroecosystems only at the beginning of intensive application of fertilizers. In the future, due to the transition of phosphorus to the deposited forms, the content of mobile phosphorus in soils increases very slightly or even tends to decrease. The intensity of the transition of mobile phosphates to an inactive state depends largely on the granulometric composition of the soil: the lowest intensity of such a transformation is characteristic of light soils, in particular loamy sand, as well as the duration of the interaction of mobile phosphates with the soil and the magnitude of the positive phosphorus balance.

2. The transition of the phosphorus applied with fertilizers into the precipitated soil phosphates as a result of their immobilization provides phosphorus nutrition of plants for a number of years, even when the application of phosphorus fertilizers is stopped. These processes largely determine the level of crop production, including grain, observed in the country in the post-perestroika years.
3. The use of fertilizers generally increases soil fertility, enriches them with soil microflora. Unilateral fertilization reduces the biodiversity of microbiocenosis, while the use of complex organomineral fertilizer contributes to its preservation. At the same time, the processes of phosphorus transformation under the conditions of fertilizer aftereffect in agroecosystems are influenced by many types of proteolytic bacteria.

KEYWORDS

- **agroecosystems**
- **soils**
- **fertilizers**
- **mobile phosphorus**

REFERENCES

Agrochemical Methods of Investigation of Soils. Science: Moscow, 1975; p 656 (in Russian).
Black, K. A. *Plant and Soil*. Ear: Moscow, 1973; p 501 (in Russian).
Cook, J. W. *Soil Fertility Regulation*. Ear: Moscow, 1970; p 520 (in Russian).

Efremov, V. F. Study of the Role of Organic Manure Organic Matter in Raising Sod-podzol Soil Fertility. The Results of Long-termed Researchers in the System of Geographical Network of Experiments with Fertilizers in Russian Federation. Pryanishnikov All-Russian Scientific Research Institute of Agrochemistry: Moscow, 2011; pp 47–71 (in Russian).
Fertilizers, their Properties and How to Use Them; Koren'kov, Ed.; Ear: Moscow, 1982; p 415 (in Russian).
Ginsburg, K. E. *Phosphorus of General Soil Types in U.S.S.R.* Ear: Moscow, 1981; p 542 (in Russian).
Glinka, N. L. *General Chemistry*. Chemistry: Leningrad; 18th ed.; 1976; p 728 (in Russian).
Merzlaya, G. Ye.; Verkhovtseva, N. V.; Seliverstova, O. M.; Makshakova, O. V.; Voloshin, S. P. Interconnection of the Microbiological Indices of Sod-podzolic Soil on Application of Fertilizer Over Long Period of Time. *Problems Agrochem. Ecol.* 2012, *2*, pp 18–25 (in Russian).
Plesyavichius, K. I. Fertilizer Application Systems Comparison in Heavy Soils. The Results of Long-termed Researchers with Fertilizers in Regions of the Country. Proceedings of Pryanishnikov All-Russian Scientific Research Institute of Agrochemistry. 1982, Issue 12, pp 4–82 (in Russian).
Shein, E. V.; Milanovsky, E. Yu. Spatial Heterogeneity of Properties on Different Hierarchic Levels as a Basis of Soils Structure and Functions. In *Scale Effects in the Study of Soils*. Publishing house of Moscow state University: Moscow; 2001; pp 47–61 (in Russian).
Shuglya, Z. M. Fertilizer System in Crop Rotation. The Results of Long-termed Researchers with Fertilizers in Regions of the Country. Proceedings of Pryanishnikov All-Russian Scientific Research Institute of Agrochemistry. 1982, Issue II. pp 94–118 (in Russian).
Shustikova, E. P.; Shapovalova, N. N. Productivity of Ordinary Chernozems while Long-termed Systematic Mineral Fertilizer Applying. The Results of Long-termed Researchers in the System of Geographical Network of Experiments with Fertilizers in Russian Federation. Pryanishnikov All-Russian Scientific Research Institute of Agrochemistry: Moscow, 2011, Issue 1, pp 331–351 (in Russian).
Sychev, V. G.; Mineev, V. G. Role of Pryanishnikov All-Russian Scientific Research Institute of Agrochemistry in Solving Complex Problems of Agriculture Chemicalization. *Fertility* **2011,** *3*, pp 2–4 (in Russian).
Sychev, V. G.; Shafran, S. A. *Agrochemical Properties of Soils and Mineral Fertilizer Efficiency*. Pryanishnikov All-Russian Scientific Research Institute of Agrochemistry: Moscow, 2013; p 296 (in Russian).
Williams, V. R. *The Collected Works*. Agricultural Edition: Moscow, 1951, Vol. 6, p 576 (in Russian).

CHAPTER 2

The Content of Exchange Potassium in Soil with Prolonged Application of Fertilizers: Impact of Soil on Food Safety

RAFAIL A. AFANAS'EV[1*] and GENRIETTA E. MERZLAYA[2]

Pryanishnikov All-Russian Scientific Research Institute of Agrochemistry, d. 31A, Pryanishnikova St., Moscow, 127550, Russia

[*]Corresponding author. E-mail: rafail-afanasev@mail.ru

ABSTRACT

The influence of potash fertilizers on the dynamics of potassium exchange in arable and subsurface layers was established in long-term field experiments with the systematic application of fertilizers on different soils. It is shown that with a positive potassium balance, the content of exchange potassium in the arable layer changes due to the transition of exchange forms to nonexchange ones. With a negative balance of potassium, the reverse process occurs, that is the transition of nonexchange forms to exchange ones.

2.1 INTRODUCTION

Potassium belongs to the first group of the periodic table of elements. Along with sodium, also included in this group, potassium is of great biological importance for plants and animals. The evolution shared priorities in the consumption of sodium and potassium between plants and animals. For animals the first priority is sodium and for plants first priority is potassium. Widespread in nature, potassium is among other elements the seventh after sodium, which belongs to the sixth. The earth's crust contains 2.1% potassium, mainly in the form of potassium feldspar. In natural soil, its content is determined by both primary minerals (feldspar and mica) and products of their weathering, that is secondary minerals: illite, vermiculite, smectite, kaolinite, etc. Depending on the type and granulometric composition of mineral soils in the arable layer, the gross content of K_2O can vary from 0.6 to 2.5%. The poorest gross forms of potassium peat (0.1%–0.4%) and sod-podzolic sandy (0.6%–1.4%) soils are the most rich-loamy varieties (1.8%–2.5%). As weathering occurs in the process of soil formation, potassium leaching occurs and its mobility increases, which determines the availability of the element to living organisms (Emsley, 1993). The fate of potassium and sodium entering the soil from rocks is not the same. It was found that out of 100 potassium ions only two reach the sea basins, and 998 are absorbed by the soil cover. Seawater itself contains 379×10^{-4}% potassium and 1.5% sodium (Prokoshev and Deryugin, 2000). Sodium compounds are easily washed out of the soil, concentrated in the seas and oceans. Potassium ions are mostly absorbed by the soil, and in varying degrees of mobility. At dynamic equilibrium between different forms of potassium ions fixation they can be easily consumed by plants (Tsitovich, 1970). In agronomic practice, there are four states of potassium, including water-soluble, exchange, nonexchange, and potassium of mineral skeleton, which are inextricably linked to each other in the soil and are constantly changing to achieve the equilibrium typical of the soil. The property of potassium to move in the soil from one form to another determines the security of plants with this element. As it is known, the most accessible are water soluble and exchange potassium, which are determined by the agrochemical service by various chemical methods: in sod-podzolic and gray forest soils by Kirsanov, in leached and typical chernozems—by Chirikov, in carbonate soils (typical and southern chernozems, chestnut soils)—by Machigin. The most powerful influence on the availability of potassium to plants is the use of fertilizers, primarily potassium and organic (Pchelkin, 1966; Sobachkin et al., 1978). However, potassium applied with fertilizers is included in the processes peculiar to different soils.

According to the agrochemical service, in the Russian Federation from 1966 to the present, the potassium balance in agriculture, with the exception of the five years 1981–1985, was characterized by negative values, especially in recent decades (Romanenko, 2010). The demand of agricultural crops for potash nutrition earlier, and even more so now is provided mainly by soil resources, including nonexchange forms of the element. The soils with very low content of exchange potassium include 21% of arable land, 36% of arable land is characterized by increased and high content of exchange potassium in soils, 35%-average. However, the agrochemical service almost universally noted the depletion of soil exchange potassium (Sychev and Mineev, 2011; Sychev, 2005; Sychev et al., 2005). In this regard, it is of interest to study the dynamics of mobile forms of potassium in both fertilized and nonfertilized soils, as the content of exchange potassium is the main factor determining the need for fertilizers for crops.

2.2 *METHODS*

To determine the dynamics of exchange potassium in soils, the analysis of the results of five long-term field experiments were conducted by scientific institutions in different soils, including sod-gleyic clay loam (Lithuanian Research Institute of Grain), sod-podzolic clay loam (Central Experimental Station of the Research Institute of Agricultural Chemistry), sod-podzolic sandy loam (Research Institute of Agricultural Chemistry, Smolensk Research Institute of Agriculture), sod-podzolic sandy loam (Grodno Experimental Station, Belarus), and ordinary clay loam chernozem (Stavropol Research Institute of Agriculture). Dynamics of potassium exchange in soils was considered by rotations of field rotations in typical variants of field experiments. Analysis of the exchange potassium content in soils was carried out according to the methods adopted in each of the regions. Results on potassium content in soils and plants are given in potassium oxide (K_2O).

2.3 DISCUSSION

In the sod-podzolic clay loam drained loam soil (Lithuania) (Plesyavicyus, 1982), the fully expanded seven-field crop rotation in the variant N225P324K350 for each of the two rotations was made in the soil at 350 kg/ha of K_2O, or in the amount of 700 kg/ha. The removal of potassium by crops for the first rotation and the second was 869 and 893 kg/ha, respectively, or in the amount of 1762 kg/ha. The result of this variant on average seven fields was a negative balance of potassium, formed during the first rotation of 519 kg/ha, for the second rotation of 544 kg/ha, and in total, over two rotations of 1063 kg/ha K_2O. At the same time, the content of exchange potassium (according to Egner, Rome) in the arable soil layer in the first rotation for the seven-year period decreased from 93 to 91 mg/kg, that is, by only 2 mg/kg. In the second rotation, despite the negative balance of the element, the potassium content in the soil even increased—from 91 to 99 mg/kg (Table 2.1, Figure 2.1). It follows that the provision of crop rotation with potassium in each rotation was mainly due to the transition of the element from nonexchange forms to exchange ones. With an annual negative potassium balance in the long-term experience on sod-podzolic loamy sand soil (Belarus) (Shuglya, 1982) in the first three rotations of the 4-full field crop rotation, the content of exchange potassium (according to Kirsanov) also increased slightly due to nonexchange forms (Figure 2.2). At the end of the fourth rotation, with depletion of their supply, there was a decrease in the content of exchange potassium in the arable layer, due to the lower saturation of loamy sand soils with nonexchange forms of the element.

TABLE 2.1 Influence of Mineral Fertilizers on Crop Rotation Productivity and the Content of Potassium Exchange in Sod-Gleyic Clay Loam Soil (According to the Lithuanian Research Institute of Agriculture)

Indicator	Crop Rotation Fields							
	1	2	3	4	5	6	7	Average of 7 fields
				1st rotation				
Productivity per year on average, C g.u/ha	48.7	54.0	37.6	38.9	45.9	45.1	43.1	44.8
Applied K_2O, kg/ha	350	350	350	350	350	350	350	350
The removal of K_2O, kg/ha	920.8	831.5	752.2	875.0	879.9	855.3	967.8	868.9
The balance of K_2O, kg/ha	–570.8	–484.5	–402.2	–525.0	–529.9	–505.3	–617.8	–518.9
K_2O content in soil at the beginning of rotation, mg/kg	52	55	79	103	117	155	90	93
K_2O content in soil at the end of rotation, mg/kg	84	64	66	103	121	70	129	91
The change in the content of K_2O in soil, mg/kg	+32	+9	–13	0	+4	–85	+39	–2
				2nd rotation				
Productivity per year on average, C g.u/ha	46.9	46.4	40.7	44.0	39.5	46.1	52.1	46.5

TABLE 2.1 *(Continued)*

Indicator	Crop Rotation Fields 1	2	3	4	5	6	7	Average of 7 fields
Applied K_2O, kg/ha	350	350	350	350	350	350	350	350
The removal of K_2O, kg/ha	940.6	939.7	781.5	979.7	906.6	844.3	857.7	892.9
The balance of K_2O, kg/ha	−590.6	−589.7	−431.5	−629.7	−566.6	−494.3	−507.7	−544.3
K_2O content in soil at the beginning of rotation, mg/kg	84	64	66	103	121	70	129	91
K_2O content in soil at the end of rotation, mg/kg	95	75	112	145	99	96	68	99
The change in the content of K_2O in soil, mg/kg	+11	+11	+46	+42	−22	+26	−61	+8

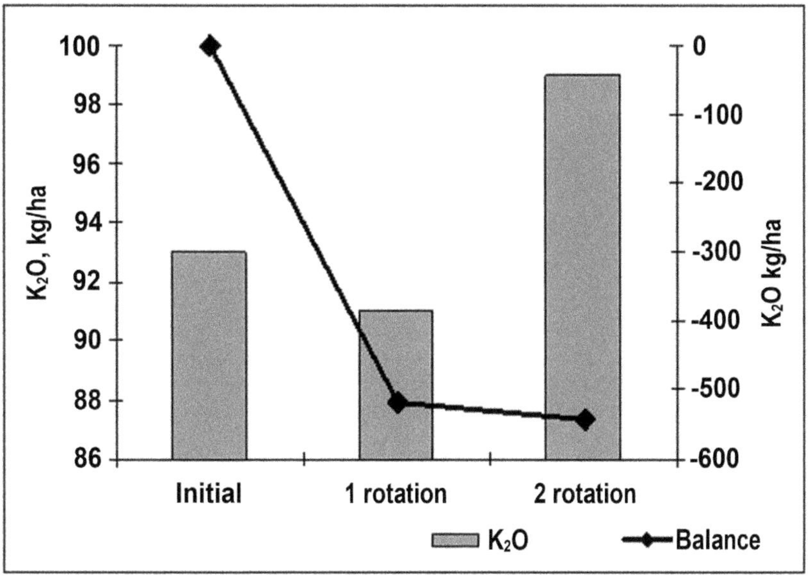

FIGURE 2.1 The balance and content of exchange potassium (Egner–Rome) in sod-gleyic clay loam drained soil in the NPK variant (Lithuania).

In the study of the dynamics of potassium exchange (by Machigin) in terms of ordinary Chernozem (Stavropol region) (Shustikova and Shapovalova, 2011) in the variant N1205P120K150 in the first three rotations of 6-full field crop rotation with a positive potassium balance of 896 kg/ha, this Figure 2.2 in the arable layer increased from 201 to 288 mg/kg, or 87 mg/kg. In the fourth and fifth rotations, this used only nitrogen and phosphorus fertilizers, the negative balance of potassium was 789 kg/ha, and the content of potassium exchange decreased by only 3 mg/kg—from 288 to 285 mg/kg (Table 2.2, Figure 2.3). Thus, with a positive potassium balance for the first three rotations of the crop rotation, an increase in the exchange potassium by 10 mg/kg required 103 kg of potassium fertilizers, and to maintain the potassium content in the fourth and fifth rotations, in fact, at the initial level, about 790 kg/ha of potassium was transferred from the soil fund to the exchange form. In a meter layer of soil for the fourth and fifth rotation with a negative balance of potassium, its content (reserve) not only did not change, but even increased by 253 kg/ha.

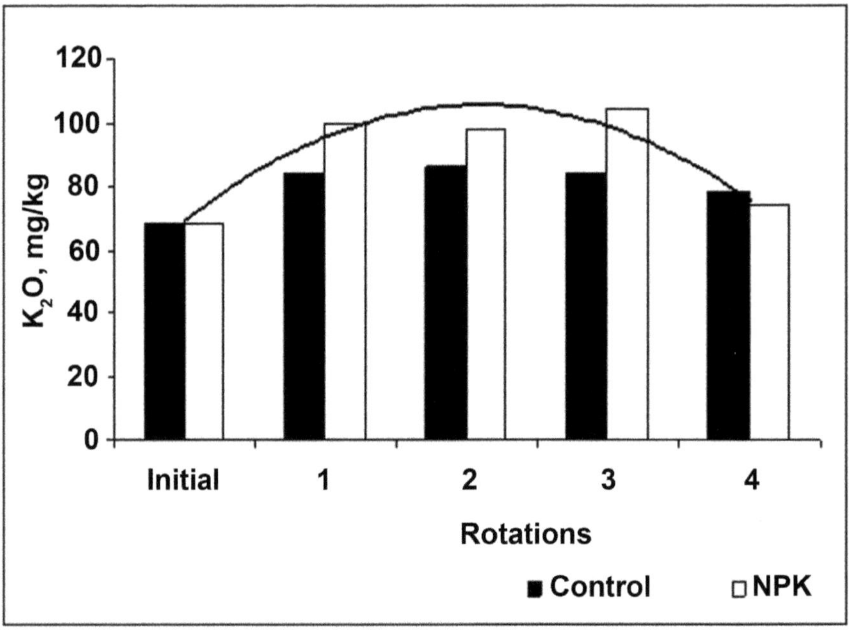

FIGURE 2.2 Dynamics of potassium exchange (by Kirsanov) in sod-podzolic sandy loam soil (Belarus).

TABLE 2.2 The Influence of Mineral Fertilizers on the Productivity of Crop Rotation and the Maintenance of Potassium Exchange in Common Chernozem (According to the Stavropol Research Institute of Agriculture)

Indicator	1 rotation (6 years)				2–3 rotations (18 years)				4–5 rotations (30 years)			
	Control	N120P120K30	N120P120K90	N120P120K150	Control	N120P120K30	N120P120K90	N120P120K150	Control	N120P120K30	N120P120K90	N120P120K150
Productivity per year on average, C g.u/ha	27.7	33.0	33.5	33.7	26.5	32.8	33.8	33.6	19.0	24.3	25.4	25.0
Applied K_2O, kg/ha	–	150	450	750	–	300	900	1500	–	–	–	–
The removal of K_2O, kg/ha	350	436	446	447	717	895	916	907	621	768	797	789
The balance of K_2O, kg/ha	–350	–286	4	303	–717	–595	–16	593	621	–768	–797	–789
K_2O content in soil at the beginning of rotation, mg/kg	201	201	201	201	201	222	225	276	203	240	256	288
K_2O content in soil at the end of rotation, mg/kg	201	222	225	276	203	240	256	288	216	211	235	285
The change in the content of K_2O in soil, mg/kg	–	21	24	75	2	18	31	12	13	–29	–21	–3

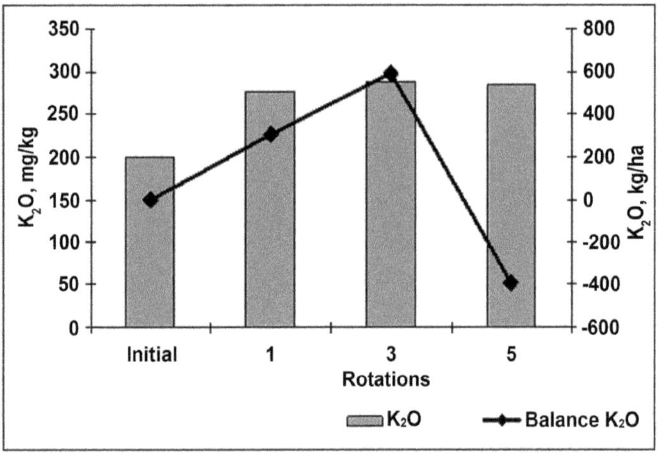

FIGURE 2.3 The balance and content of exchange potassium (by Machigin) in ordinary clay loam chernozem soil in the NPK variant (Stavropol region).

In the conditions of sod-podzolic clay loam soil (Moscow region) (Efremov, 2011) with a constant positive balance of potassium in the rotations of the rotation in the variant of the organic-mineral system of fertilizers (NPK + manure) of the seven rotations of the four-field crop rotation the content of exchange potassium (by Maslova) increased only in the first three rotations, and then began to decline (Figure 2.4). As a result, for 28 years with positive potassium balance in the amount of 3600 kg/ha, the content

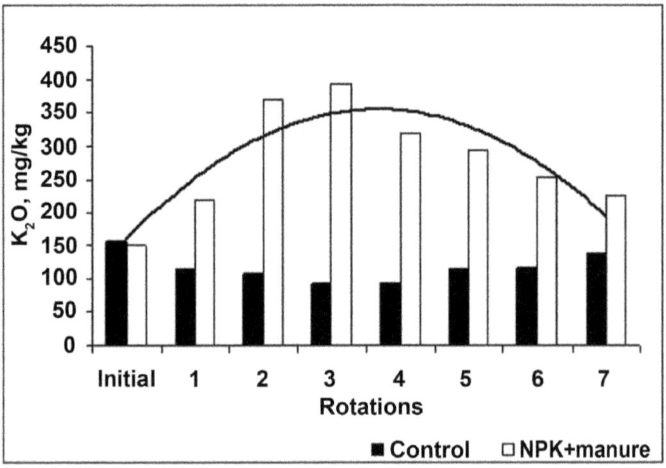

FIGURE 2.4 Dynamics of the content of potassium exchange (by Maslova) in sod-podzolic clay loam soil (Moscow region).

of potassium exchange increased only by 76 mg/kg. The cost of potassium fertilizer made in excess of the removal, to improve the content of K_2O in soil at 10 mg/kg to 470 kg/ha, whereas a simple calculation that requires only 30 kg/ha. Unaccounted amount of potassium—more than 3 t/ha—moved to inactive forms and may partially migrated in the subsurface horizons.

On sod-podzolic sandy loam soil (Smolensk region) (Merzlaya, Polunin, and Gavrilova, 1991) the content of exchangeable potassium (according to Kirsanov) in the arable layer in the NPK + manure variant was associated with the nature of the element balance (Figure 2.5, Table 2.3), which, in turn, is due to different levels of crop yields by rotations of crop rotation (with stable doses of fertilizers and in the control version). In contrast to clay loam in this case there was no growth in the content of potassium exchange. For 30 years of experience with a positive balance of potassium in the amount of 1280 kg/ha there was even a decrease in its content from 146 to 115 mg/kg. In the meter layer of soil in the first rotation of the crop rotation with a positive balance of potassium 563 kg/ha its supply increased to 1350 kg/ha, and at the end of the fourth rotation, that is 30 years of fertilizer application, with a total balance of 1270 kg/ha its content has not increased. In other words, the processes of transition of metabolic forms of the element in the soil in nonexchange form affected not only arable, but also deep subsurface layers.

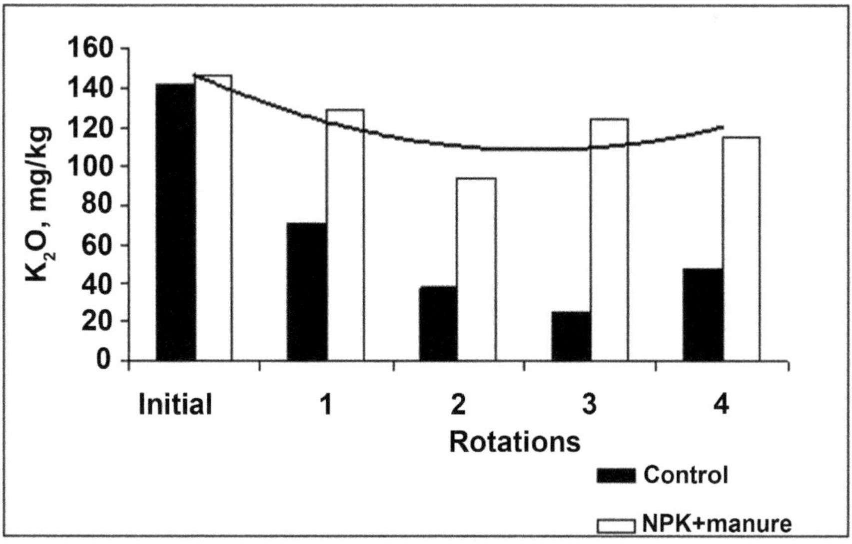

FIGURE 2.5 Dynamics of the content of potassium exchange (according to Kirsanov) in sod-podzolic sandy loamy soil (Smolensk region).

TABLE 2.3 Influence of Fertilizers on Productivity of Crop Rotation and the Maintenance of Exchange Potassium in Sod-Podzolic Light Loamy Soil (According to the Research Institute of Agricultural Chemistry and the Smolensk Research Institute of Agriculture)

Indicator	1–2 rotations (17 years)				3–4 rotations (30 years)			
	Control	NPK	Manure	NPK + manure	Control	NPK	Manure	NPK + manure
Productivity per year on average, C g.u/ha	24.0	34.4	28.6	34.6	20.1	29.6	26.0	26.0
Applied K_2O, kg/ha	–	1440	1071	2511	–	855	792	1647
The removal of K_2O, kg/ha	1410	1869	1619	1865	829	1167	1061	1014
The balance of K_2O, kg/ha	−1410	−429	−548	+646	−829	−312	−269	+633
K_2O content in soil at the beginning of rotation, mg/kg	14.2	13.3	13.8	14.6	3.8	9.4	3.5	9.4
K_2O content in soil at the end of rotation, mg/kg	3.8	9.4	3.5	9.4	4.7	15.8	7.7	11.5
The change in the content of K_2O in soil, mg/kg	−10.4	−3.9	−10.3	−5.2	+0.9	+6.4	+4.2	+2.1

2.4 CONCLUSIONS

1. The study of the dynamics of the exchangeable potassium content in different soils of field experiments showed that it depends on the intensity and duration of fertilizer use, and on the availability of soil reserves of nonexchange potassium.
2. With a positive potassium balance, the content of metabolic forms of the element in the arable soil layer in the first rotations of crop rotations, as a rule, increases, then either stabilizes or even decreases. With a negative balance on clay loam soils, the content of exchangeable potassium in the arable layer can be maintained or even increased for a certain time due to the intensive transformation of potassium from nonexchange forms.
3. To a large extent, the direction of the transformation of exchange and nonexchange potassium depends on the intensity of its balance and granulometric composition of soils. At the same time, the transformation processes affect not only the arable, but also the subsurface layers. If on loamy soils the decrease in the content of exchange potassium in the last rotations of field experiments with a positive balance occurred due to the transition to nonexchange forms, on light soils with a negative balance, the content of exchange potassium after several rotations significantly decreased due to the depletion of nonexchange forms.
4. In general, the dynamics of exchange potassium in soils of long-term experiments corresponds to the trends observed in successive rounds of agrochemical soil survey of agricultural land in different regions. The results of the studies can serve as a basis for regulating the potassium regime of soils through the use of fertilizers, taking into account the reserves of nonexchange potassium.

KEYWORDS

- **dynamics**
- **balance**
- **exchange potassium**
- **fertilizers**
- **crop rotation**

- **soil**
- **long-term field experiments**
- **transition**

REFERENCES

Efremov V.F. Study of the role of organic manure organic matter in improvement of sod-podzol soil fertility. The results of long-termed researches in the system of geographical series of experiments with fertilizers in Russian Federation. Moscow: Pryanishnikov All-Russian Scientific Research Institute of Agrochemistry, 2011. pp. 47–71 (in Russian).

Emsley J. Elements. Translation from English. Moscow: World, 1993. p. 256 (in Russian).

Merzlaya G.E., Polunin S.F., Gavrilova V.A. Influence of various combinations and doses of organic and mineral fertilizers on the fertility of sod-podzolic soil.Agrochemistry, 1991. Issue 9. pp. 43–48.

Pchelkin V.U. Soil potassium and potassic fertilizers. Moscow: Ear, 1966. p. 336 (in Russian).

Plesyavicyus K.I. Comparison of the system of fertilizers used on heavy soils of different mechanical composition. The results of long-termed researches with fertilizers in regions of the country. Moscow: Proceedings of Pryanishnikov All-Russian Scientific Research Institute of Agrochemistry, 1982. Issue 12. pp.4–82 (in Russian).

Prokoshev, V., Deryugin, I. P. Potassium and potash fertilizers. M.: Ledum, 2000. 185 p. (in Russian).

Romanenko G.A. The concept of development of agrochemistry and agrochemical service of agriculture of the Russian Federation for the period till 2010. [Ed. Romanenko G.A.]. Moscow: Pryanishnikov All-Russian Scientific Research Institute of Agrochemistry, 2005. pp. 80. (in Russian).

Shuglya Z.M. Fertilizer system in crop rotation. The results of long-termed researches with fertilizers in regions of the country. Moscow: Proceedings of Pryanishnikov All-Russian Scientific Research Institute of Agrochemistry, 1982. Issue II. pp. 94–118 (in Russian).

Shustikova E.P., Shapovalova N.N. Productivity of ordinary chernozems after prolonged mineral fertilization. The results of long-termed researches in the system of geographical series of experiments with fertilizers in Russian Federation. Moscow: Pryanishnikov All-Russian Scientific Research Institute of Agrochemistry, 2011, Issue 1. pp. 331–351 (in Russian).

Sobachkin A.A., Zabavskaya K.M., Cheban V.M. et al. Research on the efficiency of potassic fertilizers at crop rotation saturation with intensive cultures (field crop rotation). he results of long-termed researches with fertilizers in regions of the country. Moscow: Proceedings of Pryanishnikov All-Russian Scientific Research Institute of Agrochemistry. 1978, Issue V. pp.100–120 (in Russian).

Sychev V. G. Advisory agrochemical service in the Russian Federation. Results and prospects (40 years of Agrochemical service). [Ed. Sychev V.G.]. Moscow: Pryanishnikov All-Russian Scientific Research Institute of Agrochemistry, 2005. p. 569 (in Russian).

Sychev V.G., Kuznetsov A.V., Pavlikhina A.V., Lobas N. I., Kruchinina L. K., Demina N. A., Vasilyeva N. M. Agrochemical characteristics of soils of agricultural lands of the Russian

Federation (as of January 1, 2004) Moscow: Pryanishnikov All-Russian Scientific Research Institute of Agrochemistry, 2005. p. 183. (in Russian).
Sychev V.G., Mineev V.G. Role of Pryanishnikov All-Russian Scientific Research Institute of Agrochemistry in solving complex problems of agriculture chemicalization. Fertility, 2011. Issue 3. pp. 2–4 (in Russian).
Tsitovich I.K., The chemistry and agricultural analysis. M.: Ear, 1970. p. 535 (in Russian).

CHAPTER 3

Foodborne Pathogenic Anaerobes

AROCKIASAMY ARUN PRINCE MILTON[1*],
GOVINDARAJAN BHUVANA PRIYA[2], KASANCHI M. MOMIN[1],
MADESH ANGAPPAN[1], SANDEEP GHATAK[1], and PORTEEN KANNAN[3]

[1]*Division of Animal Health, ICAR Research Complex for NEH Region, Umiam, Meghalaya 793103, India*

[2]*College of Agriculture, Central Agricultural University (Imphal), Kyrdemkulai, Meghalaya 793104, India*

[3]*Department of Veterinary Public Health and Epidemiology, Madras Veterinary College, Chennai, Tamil Nadu 600007, India*

[*]*Corresponding author. E-mail: vetmilton@gmail.com*

ABSTRACT

Foodborne pathogenic anaerobes are the group of bacteria that are not capable to grow in the presence of oxygen and causes foodborne diseases in human beings. The genus *Clostridium* represents the major food pathogenic anaerobes and the *Clostridia* of food safety relevance include *Clostridium botulinum, Clostridium perfringens,* and *Clostridium difficile.* The ubiquitous presence, ability to produce heat resistant endospores, wide growth temperature range, ability to adapt in diverse food environments and the different diseases caused make these bacteria the significant health hazard causing potentially lethal foodborne illnesses in humans. *C. botulinum* causes neurotoxin-mediated foodborne botulism, infant botulism, and wound botulism. *C. perfringens* causes food poisoning, necrotic enteritis, antibiotic associated diarrhea, and gas gangrene in human beings. *C. difficile* causes severe pseudomembranous colitis and *C. difficile* antibiotic-associated diarrhoea. This chapter gives a detailed outlook of these three potentially important pathogenic anaerobes as well as describes strategies to control them in food.

3.1 INTRODUCTION

Foodborne pathogenic anaerobes are the group of bacteria that are unable to grow in the aerobic atmosphere and causes foodborne diseases in human beings. The genus *Clostridium* represents the major food pathogenic anaerobes *viz., Clostridium botulinum, Clostridium perfringens* and *Clostridium difficile* (emerging). They are ubiquitously found in the entire environment and gastrointestinal tract of humans and animals, thus consequently contaminates the raw food materials. The persistence of *Clostridia* in food chain is due to its stress resistant spores. The ubiquitous presence, ability to produce heat resistant endospores, wide growth temperature range, ability to adapt in diverse food environments, and the different diseases caused make these bacteria the significant health hazard causing potentially lethal foodborne illnesses in humans. *C. botulinum* causes neurotoxin-mediated foodborne botulism, infant botulism, and wound botulism. All forms of botulism cause neurological damage, paralysis, respiratory failure, and death. *C. perfringens* causes food poisoning, necrotic enteritis, antibiotic associated diarrhea, and gas gangrene in human beings. *C. difficile* causes severe pseudomembranous colitis and *C. difficile* antibiotic-associated diarrhoea. Foodborne transmission of *C. difficile* has been assumed as a probable source for community associated infections; nevertheless, the proof to confirm or contest this hypothesis is deficient.

3.2 CLOSTRIDIUM BOTULINUM

3.2.1 HISTORY

During the late 1700s, many human deaths due to botulinum intoxication were reported in Europe. One of the major reasons for these deaths was of food sanitary measures due to poverty triggered by Napoleonic war. Smoked blood sausage was the primary source of botulinum toxin, hence called as "sausage poisoning." In 1811, the Kingdom of Wurttemberg attributed prussic acid as a cause of sausage poisoning (Erbguth and Naumann, 2000). By 1871, the disease was named as botulism (Latin word "botulus" meaning sausage). Botulism outbreaks continued to occur in the United States till 1900 (Shapiro et al., 1998). This frequent outbreak has led to the development of innovative food canning methods during the 1920s by Dole Food Company, Inc, which is one of the world's largest suppliers of canned foods today. Justinus Kerner, a German poet and physician published case studies and monograph on botulinum intoxication (Erbguth and Naumann, 1999). He has made quite a few significant observations even valid today about the botulinum toxin upon conducting daring experiments upon himself. They were (1). toxin develops in sausages under anaerobic conditions, (2) toxin is lethal even in smaller doses, (3) it affects the transmission of motor signals in the peripheral and autonomic nervous system. In 1895, Emile Pierre van Ermengem correlated botulism poisoning with a bacterium present. It was simultaneously detected in salted raw pork and tissue of died victims. After isolation, he had named it as bacillus botulinus. In later years, it was further renamed as *Clostridium botulinum* (van Ermengem, 1979).

3.2.2 THE BACTERIUM

Clostridium botulinum is a Gram positive rod-shaped motile obligate anaerobe. It is a spore-forming organism with its spore located at subterminal part of the bacterium. The usual habitats of the spores are soil, intestine of mammals, and marine environment (Hielm et al. 1998; Dahlenborg et al. 2001; Myllykoski et al. 2006). The genome size varies from 3.76 to 4.26 Mb, and the G + C content is 27.4%–28.5%. Based on phenotypic and genetic traits, these bacterial strains are grouped into four distinct groups, named I to IV (Keto-Timonen et al., 2006). Human botulism is attributed to strains of group I and II, which produces toxin types A, B, E, and F. Group III strains, producing toxin types C and D are involved with botulism in animals. Group IV, which produces type G toxin has not been linked with any illness and *Clostridium argentinense* is the new name proposed to this group (Suen et al., 1988). Besides *C. botulinum*, few strains of *Clostridium baratii* and *Clostridium butyricum* also produce type F and E neurotoxins, respectively, and are responsible for few foodborne botulism outbreaks. Groups I and II are proteolytic and saccharolytic, respectively. By producing several extracellular enzymes, Group I strains degrade proteins and carbohydrates and thus results in spoilage of food. On the contrary, Group II ferments sugars and does not have any enzymatic activity (Lindström and Korkeala, 2006). However, both the groups are capable of producing the neurotoxin and are accountable for the most foodborne botulisms. Group I strains carry a single or two neurotoxin genes for A, B, F, or H neurotoxin types, whereas the majority of Group II saccharolytic strains produce types B, E, or F. Group I proteolytic strains are mesophilic and can grow and produce neurotoxin at 35–37 °C; nevertheless, they can also produce toxins at 12 °C. Group II strains are psychrotrophic with the optimum temperature for toxin production and growth of 25 °C and can also produce toxin at 3.0–3.3 °C. Proteolytic and nonproteolytic strains do not grow below pH 4.6 and 5, respectively. Neurotoxin produced by group I strains is activated by protease produced on its own, while host proteases activate Group II strains since they do not produce protease (Lindström and Korkeala, 2006).

3.2.3 BOTULINUM NEUROTOXIN (BONT) AND MECHANISM OF ACTION

BoNT is produced by *C. botulinum* during its vegetative growth and the peak toxin production occurs at late logarithmic phase (Chen et al., 2008). *C. botulinum* secretes eight antigenically distinct BoNT types, designated as A, B, C, D, E, F, G, and H (H recently discovered by Barash and Arnon, 2004)). Toxin gene cluster along with numerous nontoxin accessory genes encodes each toxin. Hemagglutinin (*ha*) and *orfX* are the two known toxin gene clusters. Toxin types A, B, C, D, and G are encoded in *ha*, while different subtypes of A, E, F, and H are encoded in *orfX* gene cluster. These neurotoxin genes are positioned either on the chromosome or on plasmid or the bacteriophage. The genes for toxin types A, B, E, F, and H are located on the chromosome, type G on a plasmid (81 MDa) and BoNT types C and D on bacteriophages. BoNT types A, B, E, H, and, in the sporadic cases, F attributed to human botulism, while types C and D are responsible for animal botulism. BoNT is secreted when the vegetative cells undergoes bacteriophage induced lysogeny or cell autolysis during late logarithmic growth phase. Important factors affecting the toxin production are availability of nutrients, cultural conditions including the presence of certain amino acids, peptides, metal ions, temperature, pH, and the cell density. BoNT is derived from a progenitor inactive form and it is a 150 kDa polypeptide. The active form of the toxin is an A–B type toxin with two subunits that are joined by a disulfide bond. The two subunits are, A (50 kDa light chain) and B (100 kDa heavy chain). The principal function of BoNT is to block acetylcholine (neurotransmitter) release at the myoneural junction, thus affecting the propagation of nerve impulse leading to flaccid paralysis (Rossetto et al., 2014; Bhunia, 2018).

3.2.4 VARIOUS FORMS OF DISEASES

Based on the route of entry of the toxin, five forms of the disease are established. Foodborne botulism is typically caused by ingesting preformed BoNT contaminated food (mostly canned). The toxins usually survive in the proteolytic environment of the GI tract and get absorbed from intestines (Johnson and Montecucco, 2008). Infant botulism occurs after the consumption of clostridial spores contaminated food and the spores germinate in the intestine (Koepke et al., 2008). Intestinal colonization in infants is accelerated by less competitive resident microflora of infants compared with robust bowel microbiota of adults (Johnson and Montecucco, 2008; Koepke et al., 2008). Upon colonization, BoNTs are secreted for prolonged time unless treated with antibiotics. A rare form of adult type infant botulism has also been recognized in persons who have structural or functional bowel abnormalities or consequent to antibiotic treatment, in both the conditions, the *Clostridia* is protected from being outcompeted by the resident microbiota of intestine (Johnson and Montecucco, 2008). In both, foodborne and infant botulism, the potent BoNTs pass the mucus layer of the intestine and reaches the general blood circulation (Johnson and Montecucco, 2008). Further, it reaches the cholinergic nerve endings and paralyses them, which eventually causes typical botulism. Another form is the wound botulism, which is the outcome of clostridial spores contaminating the injured tissues. In majority cases, it is associated with injecting drug use (Wenham, 2008). Iatrogenic botulism results from excessive exposure to BoNTs used for therapeutic or cosmetic procedures (Chertow et al., 2006). BoNTs bypass the GI tract in both the cases and enters the systemic circulation. Botulinum neurotoxins are 100 to 1000 times more toxic in injection route when compared to oral administration as the LD_{50} in injection route ranges from 0.1 to 1 ng/kg in laboratory mice (Smith and Sugiyama, 1988). Lastly, in inhalational botulism, the BoNT enters through the inhalation route; yet, delivery via aerosols is ineffective (Johnson and Montecucco, 2008). Foodborne and infant botulism are the major forms of the botulism in humans, and the other forms are seldom encountered.

3.2.5 CLINICAL PRESENTATION

The incubation period of the botulism is toxin dose dependent. It varies from 12 to 72 h, and at times it lasts even for some weeks (Gao et al., 1990).

Botulism has got distinctive clinical symptoms characterized by symmetrical cranial nerve palsies with ophthalmoparesis and ptosis, followed by symmetrical descending flaccid paralysis of voluntary muscles affecting the following part in order: neck muscles, shoulders, proximal followed by distal upper extremities, proximal and then distal lower extremities. This may even advance to respiratory arrest and death due to paralysis of accessory breathing muscles and diaphragm (Shapiro et al., 1998). Ultimate constipation is almost a universal symptom; Paralysis of cranial nerves III, IV, and VI causes extraocular muscle paralysis which present as frank diplopia or blurry vision and problem in near vision. A noticeable symptom is ptosis. Cranial nerve VII paralysis induces loss of facial expression and paralysis of cranial nerve IX produces dysphagia, which may manifest as regurgitation of food. Prominent symptoms are dysarthria, dry mouth, and throat with anhydrosis and postural hypotension in some cases, otherwise normal blood pressure is maintained. Occasionally, pharyngeal collapse may warrant intubation. GI symptoms like nausea and vomiting may precede nervous signs in foodborne botulism especially when caused by toxin types B and E. The triggering factor for such vomiting is unknown. It may be due to non-neurotoxin products of *C. botulinum* or other contaminants of the spoiled food. These GI symptoms have never been recorded in wound botulism patients (Werner et al., 2000) nor have been observed in primate experiments, wherein pure toxin was administered intravenously or intragastrically (Sobel, 2005).

3.2.6 INCIDENCE AND VEHICLE FOODS

In the European Union, during 2008, 150–300 botulism cases were reported with 0.02 per 1 lakh population notification rate. While in the USA, 17–43 (median number, 23) foodborne botulism cases were reported during 1990–2000. Most reported cases are not part of any outbreak, but outbreaks in restaurants do occur (Sobel et al., 2004). Globally, type B toxin contributes to 52% cases; 34% and 12% cases were attributed to toxin types A and E, respectively (Hatheway, 1995). Cases due to type F are very rare (1%; Harvey et al., 2002). Although foodborne botulism outbreaks are frequently sporadic and limited to a small number of people, outbreaks involving more than one country and affecting more people have also been reported. In 2006, in Thailand, a typical foodborne botulism outbreak affecting 209 people was witnessed. Out of 209, 42 people developed respiratory failure and the incriminated food was bamboo shoots contaminated with preformed type A toxin (Kongsaengdao et al., 2006). Another widespread outbreak in the USA and Canada in 2006 affected six people after consuming a commercial carrot juice contaminated with type A toxin (Sheth et al., 2008). High incidence of foodborne botulism was reported in Poland, Russia, China, Kyrgyzstan, and the Republic of Georgia. The normal habitat of *C. botulinum* is soil and intestinal tract of animals. High risk bearing foods of foodborne botulism are home canned or under processed canned food items, spices, mushrooms, herbs, bamboo shoots, blood sausage, fish, cream cheese, yoghurt, jarred peanuts, baked potatoes, home canned vegetables like carrots, asparagus, corn, beans and condiments like garlic in oil, sautéed onions, cheese sauce, and chilli sauce. Generally, these foods are contaminated by clostridial spores, as they are heat resistant (survive below 120 °C), can survive, germinate and produce toxin in low oxygen, low acidity (pH > 4.6) and high water content environment. Honey bees can also carry spores and in honey, the acceptable limit of clostridial spores is <7 per 25 g (Bhunia, 2018).

3.2.7 DETECTION METHODS

Rapid diagnosis is crucial in botulism as it is a life-threatening food poisoning condition. Although diagnosis is being done on the basis of unique clinical manifestation, laboratory diagnosis is required for confirmatory diagnosis. The standard method of foodborne botulism diagnosis is the detection of neurotoxin in the patient's feces or serum. In case of infant botulism, simple detection of *C. botulinum* bacteria in the patient's sample is sufficient for the diagnosis (Lindström and Korkeala, 2006). The classical method for botulism neurotoxin detection is the mouse lethality assay (Solomon and Lilly, 2001). This test involves the injection of test sample into mice through intraperitoneal route and interpretation based on the development of peculiar symptoms like fuzzy hair and wasp-waist appearance due to respiratory failure (CDC, 1998). Toxin typing is done by toxin neutralization with specific antitoxins. Mice injected with specific antitoxin survive, whereas the others develop symptoms of botulism. Serological diagnostic assays were also deployed in the form of enzyme linked immune sorbent assay (ELISA) (Dezfulian and Bartlett, 1984; Rodriguez and Dezfulian, 1997), passive hemagglutination assay (Johnson et al., 1966) and gel diffusion assay (Ferreira et al., 1981). Main drawbacks of immunological tests are unavailability of good quality antibodies and false positive results due to inactivated or biologically inactive toxins. ELISAs have been widely used with toxic *C. botulinum* cultures, purified botulinum toxin, and foods associated with outbreaks or spiked with *C. botulinum* or BoNT (Lindström and Korkeala, 2006).

The conventional method of diagnosis involves isolation of bacteria employing cultural methods followed by toxin typing based on mouse lethality assay. Enrichment liquid broth media generally used are cooked-meat medium, chopped-meat–glucose–starch medium, fastidious anaerobe broth, and reinforced clostridial medium. After enrichment, the positive samples are streaked on routine solid media like blood agar and egg yolk agar (Lindström and Korkeala, 2006). Samples like feces and serum can be cultured directly or after preheating or pretreated with ethanol to recover only spores after eliminating vegetative bacteria and nonspore formers (Nordic Committee on Food Analysis, 1991). Heating temperatures recommended for recovering group I and group II spores are 80 °C for 10 min and 60 °C for 10–20 min, respectively (Smith and Sugiyama, 1988). The addition of lysozyme may boost the germination of spores which are heat stressed (Peck et al., 1993). For rapid toxin typing of isolated cultures, multiplex PCR methods are often being deployed in recent days. This multiplex PCR (Lindström et al., 2001) simultaneously detect genes (*botA, botB, botE, and botF*) encoding the botulism toxins in a single reaction. The main advantages, which make them a very suitable assay are rapidity, less labor intensive and nonrequirement of laboratory animals. Real-time PCR has also been applied in the simple qualitative detection of *C. botulinum* toxin types in foodborne, infant, and wound botulism cases (Akbulut et al., 2004; 2005).

3.2.8 TREATMENT, PREVENTION, AND CONTROL

In the case of foodborne botulism, early detection and injection of antitoxin neutralize the neurotoxin and helps in the prevention of much damage. At the same time, successful treatment depends on the amount of toxin ingested and the intervention becomes unsuccessful if the toxin has already bound to the neuronal receptors. The antitoxin can neutralize the toxin which is circulating in the blood. Apart from the antitoxin, mechanical ventilation support is required to facilitate easy breathing and to combat neurological damage. The patients who are recovering can endure permanent neuronal damage or can suffer from general fatigue and dry mouth for weeks or months. Edrophonium chloride (anticholinesterase drug) and guanidine and 4-aminopyridine (4-AP) may be helpful to some patients to improve limb and ocular muscle strength. Antibiotic coverage is needed in the case of wound, infant and hidden botulism to eliminate the bacteria from the system (Bhunia et al., 2018).

Some of the crucial prevention strategies to circumvent botulism include early diagnosis, proper boiling, and canning of canned food products prepared in home as they present serious threat and restraining from feeding children with honey. BoNT is heat labile, thus proper heating of food products before consumption for minimum 85 °C for 15 min can prevent botulism. BoNTs do not cause any food spoilage signs as they are tasteless and odorless. However, Group I strains, as they are proteolytic in nature produce off-flavors and off-odors and consequent spoilage of the food product and therefore rejection by the consumer. Group II nonproteolytic strains are not associated with any sensory spoilage and do not produce any warning signs to the consumer. To destroy *C. botulinum* Group I and II spores, sterilization at >100 °C and pasteurization at 70 °C – 90 °C are required, respectively. In the canning industry, to control the spores of Group I and II strains, 12D (to eliminate the risk by a factor of 10^{12}) and 6D (to eliminate the risk by a factor of 10^{6}) processes are recommended (Food Standards Agency, 2008). However, they are often ignored to favor the sensory properties (Lindström et al. 2003). Other steps to prevent BoNT formation in foods are 1. Store food at <3.0 °C, 2. Store food at a pH $\leq$ 5.0, 3. Store food at a NaCl concentration $\geq$ 3.5%, 4. Store food at an aw $\leq$ 0.97 (Bhunia et al., 2018).

3.3 CLOSTRIDIUM PERFRINGENS

3.3.1 HISTORY

In 1891, at John Hopkins Hospital, USA, W.H. Welch isolated *Bacillus aerogenes capsulatus* during a postmortem examination of a man. Later, in 1892, Welch and G. Nuttall reported the bacterium as *Bacillus welchii,* which was later renamed as *Clostridium welchi*aka *Clostridium perfringens.* Although the foodborne illness caused by *C. perfringens* was proposed 100 years ago, only in the 1960s and 1970s the sufficient evidence were accumulated regarding the association between enterotoxin production and sporulation of *C. perfringens* in the intestine and subsequent infection in the affected persons (McDonel, 1986).

3.3.2 THE BACTERIUM

C. perfringens is a Gram-positive rod-shaped nonmotile anaerobic bacterium. It forms large, round, regular, shiny, and slightly opaque colonies on the surface of agar plates. The typical *C. perfringens* colonies present double zone hemolysis on blood agar plates with an outer hazy zone due to alpha toxin and inner clear zone due to theta toxin. Despite its anaerobic nature, it has the ability to grow at Eh of +350 mV and reducing the environment to less than −400 mV (Brynestad and Granum, 2002). *C. perfringens* is capable of growing between the temperature of 15 °C and 50 °C with a 37 °C to 45 °C optimum zone for most of the strains. It has got a faster growth rate with the generation time below eight minutes at 33–49 °C temperatures. The bacterium usually does not grow below 15 °C, except some atypical strains which can grow at 6 °C. It can grow well in the pH range of 5–9. *C. perfringens* can tolerate and grow in the 4%–6% salt concentration and 300 ppm sodium nitrate (Juneja et al., 2001). However, it is a fastidious organism requiring more than twelve amino acids and vitamins for their growth; hence it grows well in meat and its products. The size of the genome is 3.6 Mb and all the virulent genes positioned either on a plasmid or the chromosome (Canard et al., 1992). Based on the major toxin production, *C. perfringens* is divided into five toxin types (A–E). The major toxins comprise alpha, beta, iota, and epsilon toxins. In addition to the major toxins, most strains produce a variety of other toxins or virulence factors, such as the *C. perfringens* enterotoxin (CPE). Types A and C have been reported to cause diseases in human, types A, B, and D have been associated with animal diseases especially in small ruminants and poultry (Heikinheimo et al., 2006; Lindstrom et al., 2011).

3.3.3 *CLOSTRIDIUM PERFRINGENS* ENTEROTOXIN

CPE is encoded by the gene *cpe* which is located either on the bacterial chromosome or plasmids. CPE is synthesized during spore formation in the intestine. CPE is a polypeptide of size 35 kDa. This toxin is sensitive to acidic pH but resistant to proteolytic digestive enzymes like chymotrypsin and trypsin, which takes off 37 and 25 amino acids, respectively, from the N terminal end; thus rendering the toxin highly active. Only 5% of *C. perfringens* isolate produces CPE. Among different toxin types, type A strains are the majority producer of CPE. However, C and D types also produce CPE. Type B strains have not been shown to produce CPE, whereas type E carries a variant or silent *cpe* gene. CPE producing type A are accountable for the widely reported human food poisoning. In the ileum of the human small intestine, CPE acts as a pore forming toxin and causes morphological damage and transport loss (Fernández Miyakawa et al., 2005). CPE gets inserted into the plasma membrane of the enterocytes and forms a complex (90 kDa) with claudin 3 and claudin 4 receptors in the tight junctions (Fujita et al., 2000; Smedley and McClane, 2004). Residue N149, essential in the binding of CPE to claudin-4, is conserved in mammalian and nonmammalian species. It implies that CPE may act on a wide range of animals (Robertson et al., 2010). After binding, the CPE localizes to a bigger complex of 155 kDa. It affects the membrane permeability leading to an influx of calcium into the cell. Depending on the amount of CPE, calcium levels inside the cell may increase rapidly and the cells die eventually via apoptosis (Smedley and McClane, 2004). The infection manifests with severe diarrhoea and abdominal cramps (Lindström et al., 2011). Until recently, food poisoning incidences were only associated with chromosomal *cpe*-positive (Sparks et al., 2001). The major reason for this is that the resistance of such strains to low and high temperatures, salt, nitrites which are usually employed to preserve foods (Sarker et al., 2000; Grant et al., 2008; Lindström et al., 2011). However, the recent investigation studies of German and Finnish outbreaks revealed the involvement of the same plasmid borne *cpe* positive strain in human food poisoning cases and the suspected food sample, establishing the initial epidemiological link between the patient and vehicle food for the first time (Lahti et al., 2008). Further detection of plasmid-borne *cpe* positive *C. perfringens* strains in food poisoning patients and retail foods in Japan (Tanaka et al., 2003) and Europe (Grant et al., 2008) supports its role in food poisoning.

3.3.4 HUMAN DISEASES CAUSED BY CPE, CPA, AND CPB

3.3.4.1 CPE FOOD POISONING

Food poisoning caused by *C. perfringens* is classically a toxicoinfection rather than intoxication and it is self-limiting disease which resolves within 12–24 h (McClane, 2007). Preformed CPE is not responsible for the infection, as the toxin may be degraded during its passage through the alimentary tract. CPE is a heat labile toxin which gets destroyed at 60 °C for 10 min. *C. perfringens* spores are resistant to heat and it survives in cooking. Heating of food removes the oxygen and an anaerobic environment is created. Slow cooling of food after cooking favors the germination of spores. In addition, storing food at room temperature permits the vegetative cells to multiply swiftly to reach an infective dose of 10^7–10^9cells. Following the consumption of contaminated food, several vegetative cells die due to stomach acid, but some spores survive and form spores. Upon sporulation, the mother vegetative cells endure lysis and release CPE and spores in the lumen of the small intestine. Spores with high resistance incline to produce augmented levels of enterotoxin nd damages the epithelial cells. CPE damages the epithelial cell damage. Diarrhoea and abdominal cramps appear within 8–12 h of ingestion. These symptoms were followed by spontaneous recovery within 24 h. Death usually occurs in elderly and debilitated individuals (Smith, 1998; Bos et al., 2005; García and Heredia, 2011; Lindström et al., 2011).

3.3.4.2 ANTIBIOTIC-ASSOCIATED DIARRHOEA (AAD) AND SPORADIC DIARRHOEA (SD)

Besides acute food poisoning, *C. perfringens* type A has also been described to cause AAD and SD in 15% human cases (Modi and Wilcox, 2001; Asha and Wilcox, 2002). Most of the times, AAD and SD are considered to be independent of contaminated food (McClane and Chakrabarti, 2004; Raju and Sarker, 2005). AAD usually develops following antibiotic treatment with cephalosporins, penicillin, cotrimoxazole or trimethoprim, whereas SD has been reported to be nonantibiotic related. Generally, large amount of *cpe* carrying *C. perfringens* is obligatory to cause food poisoning, whereas small number of plasmid-borne *cpe* harbouring *C. perfringens* followed by conjugative transfer of *cpe* to other normal gut *C. perfringens* (*cpe* negative) is sufficient to cause AAD and SD (Sparks et al., 2001; Heikinheimo et al., 2006).

3.3.4.3 SUDDEN INFANT DEATH SYNDROME (SIDS)

SIDS accounts for sudden or unexpected deaths in new-born babies under the age of 1 year (Mage et al., 2009). CPE positive *C. perfringens* type A is attributed to the SIDS (Murrell et al., 1993; Lindsay et al., 1993). In contrast to the local damage of the intestinal epithelium caused by CPE food poisoning, in SIDS the CPE gets absorbed into the systemic blood circulation followed by rapid death of infants. Symptoms after administration of

CPE through intravenous route comprised prolonged and slowed thorax expansion, and reduced blood pressure, heart rate, and respiration. Higher CPE concentrations caused immediate death without any noticeable trauma or stress (Siarakas et al., 1995).

3.3.4.4 ENTERITIS NECROTICANS (EN) OR PIGBEL

The CPE-positive *C. perfringens* type C strains are attributed to causing enteritis necroticans or pigbel, an endemic disease of Southeast Asia. The 35 kDa *C. perfringens* beta toxin (CPB) produced by type C strains is vital for developing the disease. It is encoded by the gene *cpb* located on 65–110 kb sized large virulence plasmids, which also harbours other toxin genes. The characteristic symptoms of EN are vomiting, bloody diarrhoea, abdominal pain and in severe cases rapid death due to toxaemia. EN is characterized by hemorrhagic diarrhoea in animals (Bhunia, 2018).

3.3.4.5 GAS GANGRENE

C. perfringens alpha toxin (CPA) aka lecithinase or phospholipase C is produced by all types of *C. perfringens* (A–E). CPA is a 43 kDa toxin accountable for gas gangrene in humans and animals. Gas gangrene is characterized by high fever, oedema, pain, and myonecrosis. CPA and perfringolysin O are necessary for causing gas gangrene (Simonds, 1917; Bhunia, 2018).

3.3.5 INCIDENCE AND VEHICLE FOODS

CPE food poisoning is not a notifiable disease; but, in the USA, the CDC has estimated that 250,000 cases of CPE food poisoning occur annually in the United States (Lynch et al., 2006). In the 1990s, *C. perfringens* was recorded as the most common foodborne pathogen in Norway. Likewise, the prevalence in other nations like Japan and the UK is also high. This organism is found in soil, water, dust, sewage, sludge, spices, contaminated food equipment, raw, and processed foods. In human/animal feces, it is present in low numbers (10^3–10^6 spores/g). Its ubiquitous distribution, low generation time, enterotoxin production and heat resistance of spores have made *C. perfringens* as a popular foodborne pathogen. It grows and multiplies well in ground meat especially in beef. Most of the times, the CPE food poisoning involves a large number of victims and are often associated with temperature abused muscle foods. Therefore, the meat animals have been indicated as the main reservoir of *cpe* carrying *C. perfringens*. Nevertheless, a study conducted in Finland reported a large number of *cpe*-harboring *C. perfringens* in the alimentary tract of healthy humans, suggesting that workers of food establishments as a major reservoir for *cpe*-positive *C. perfringens* and consequently play a key role in contamination of foods (Heikinheimo et al., 2006). So it seems that any foods allowing the survival and growth of *C. perfringens* can act as vehicles for food poisoning outbreaks.

3.3.6 DETECTION METHODS

Cultural, serological, and molecular methods are routinely employed in detection and toxin typing of *C. perfringens*. To isolate, *C. perfringens* from stool samples, the samples are usually suspended in thioglycolate or Robertson cooked meat broth for selective enrichment. Heat shock at 70–75 °C for 15–20 min is usually given to kill vegetative cells. Enriched inoculum is streaked onto brain heart infusion agar with sheep blood (10%) or tryptose-sulfite-cycloserine agar containing 10% egg yolk and incubated for 18–24 h under an anaerobic condition at 37 °C. Multiplex PCR targeting *cpa, cpb, etx*, and *iota* are usually employed for toxin typing of *C. perfringens* isolates based on the presence of specific genes encoding particular toxins. Two immunoassays are commercially available for the detection of CPE in sporulating culture (PET-RPLA, Oxoid) and stool samples (ELISA-TECHLAB) (Bhunia, 2018).

3.3.7 TREATMENT, PREVENTION, AND CONTROL

Generally, in food poisoning cases, bed rest and fluid therapy are indicated as the common treatment option. Specifically in CPE food poisoning, antibiotic therapy is usually discouraged as it destroys the normal microbiota of the gut, stimulates Clostridium bloom and enterotoxin production in the colon. However, antibiotic coverage is advised for other forms of diseases like gas gangrene and enteritis necroticans. In order to prevent *C. perfringens* food poisoning, the food should be cooked adequately, chilled rapidly and it should not be stored for a longer time at room temperatures. Addition of chemical preservatives like organic acid salts (sodium acetate, sodium lactate, and sodium citrate) can extend the shelf life of food by inhibiting the outgrowth and germination of *C. perfringens* (Bhunia, 2018). Sabah et al. (2003, 2004) observed that sodium citrate (0.5%–4.8%) suppressed the growth of *C. perfringens* in cooked vacuum-packaged beef that was cooled slowly from 54.4 °C to 7.2 °C and oregano oil in combination with organic acids was found to inhibit the growth of the pathogen during cooling of sous-vide cooked beef products. In another study, Juneja and Thippareddi (2004) found that organic acid salts such as 1% sodium acetate, 1% sodium lactate, or 1%buffered sodium citrate (with or without sodium diacetate) have efficiently inhibited *C. perfringens* outgrowth and germination of spores during the chilling of marinated turkey breast. Incorporation of 0.1% carvacrol, thymol, oregano oil, and cinnamaldehyde into the beef has totally inhibited spore germination and outgrowth of *C. perfringens* during exponential cooling of cooked beef within 12 h (Juneja et al., 2006). *C. perfringens* spores can also be killed by hypochlorite solution (at a pH < 8.5) or by the use of ultraviolet light (Brynestad and Granum, 2002).

3.4 *CLOSTRIDIUM DIFFICILE*

Clostridium difficile, once considered as only a nosocomial pathogen has assumed a significant status in both hospitals associated and community associated (CA) infections in human due to its increasing mortality and severity in infection (Rupnik et al., 2009). Data for this dramatic rise are available in European and North American Countries (Kuijper et al., 2006; Loo et al., 2005). Findings of similar strains of *C. difficile* in food animals, foods of animal origin, and human put forward the probable transmission through foods. It is a Gram positive, spore-forming rod-shaped obligate anaerobic bacterium. The size of *C. difficile* genome is about 4.3 Mb. It produces two important toxins, TcdA (*C. difficile* toxin A) and TcdB (*C. difficile* toxin A) encoded by *tcdA* and *tcdB*genes, respectively, positioned on a pathogenicity island of 19.6 kb. Apart from these two major toxins, *C. difficile* also produces a binary toxin with sequence similarity to *C. perfringens* iota toxin called *C. difficile* toxin (CDT). And only the hypervirulent strain expresses the CDT (Gould and Limbago, 2010; Gerding et al., 2014).

C. difficile causes asymptomatic colonization to life-threatening infection like severe pseudomembranous colitis and *C. difficile* antibiotic-associated diarrhoea in humans and animals. Although *C. difficile* infection (CDI) was very common among patients under antibiotic therapy or admitted in hospitals, now CDI has also been alarmingly recognized among outpatients and people without any hospital exposure. Such infections are called as community associated CDI. CA–CDIs have also been reported in low risk populations like young children and those without any antibiotic coverage. The population at high risk are patients admitted in hospitals with prolonged antibiotic treatment, immune compromised and elderly individuals (Rupnik et al., 2009). *C. difficile* has been isolated from the range of domestic animals and wildlife including food animals like pigs, poultry, cattle, rabbits, etc. The prevalence rate of *C. difficile* from pork, poultry and beef is around 3%–12% and this organism has also been isolated from milk and ready to eat salads. Since 2003, CDI cases have risen vividly with abundant fatalities in Europe and North America (Kuijper et al., 2006; Loo et al., 2005). In the USA, 3875 deaths were documented due to CDI in 2007. According to the CDC, in 2011 alone, *C. difficile* has infected half a million patients of America with 29,000 deaths approximately. Out of these, 15,000 deaths were incriminated directly to CDI (Abt et al., 2016). Another worry is the emergence of *C. difficile* strains demonstrating PCR ribotype027, which produce a drastically high amount of toxins and are resistant to fluoroquinolones (Kuijper et al., 2006).

Diarrhoea caused by *C. difficile* is usually nonbloody and self-limiting. Symptoms generally begin after the commencement of antibiotic treatment and come to an end along with the treatment. However pseudomembranous colitis is associated with the most severe form of CDI causing fever, bloody diarrhea, and abdominal pain. Fifty percent of CDI patients suffer from pseudomembranous colitis. This occurs mainly due to the disturbance of cytoskeletal structures in colon, ulceration, leakage of serum proteins, mucus forming plaques along with inflammatory cell accumulates. About 3% of the CDI patients show the most severe and life threatening form of CDI, called fulminant colitis. It is associated with toxic megacolon and perforation of the intestine. Fulminant colitis patients manifest symptoms like high fever, diffuse abdominal pain, chills, tachypnea, hypertension, with/without diarrhoea along with marked leukocytosis (Hookman and Barkin, 2009).

Diagnosis of CDI involves an endoscopic examination of the colon for the formation of the pseudomembrane. However, anaerobic fecal culture for isolation of *C. difficile* and toxin detection (TcdA/TcdB) is the standard method of diagnosis (Cohen et al., 2010). PCR-based methods and enzyme immune assays are also available and routinely used as diagnostic tools (Sinh et al., 2011). For the selective isolation of *C. difficile* from foods, suitable enrichment broths and agar media supplemented with antibiotics, namely, norfloxacin and cycloserine or moxalactam are used.

For treatment of mild cases of CDI, Metronidazole (500 mg TID, PO for 10–14 days) is recommended. For treating more severe cases and for those who are not able to tolerate metronidazole, vancomycin (125 mg, QID, PO) is indicated for treatment. In treating recurrent CDI, vancomycin followed by rifampicin (400 mg, BID for 10–14 days) was found successful. Use of tigecycline (100 mg followed by 50 mg BID, IV) in refractory CDI has revealed promising results (Sinh et al., 2011). Surgical removal of the part of the infected intestine is indicated in fulminant colitis patients or those refractory to antibiotic therapy (Kariv et al., 2011). Fecal bacteriotherapy has also been proven to be an efficacious treatment option for recurrent CDI (O'Horo et al., 2014; Abt et al., 2016). It involves repopulating the patient gut with healthy microbiota through rectal or nasogastric route. Prevention strategies include barrier nursing, antibiotic stewardship and disinfection of hospitals with sporicidal agents like sodium hypochlorite, hydrogen peroxide (in high concentration). Alcohol and ammonium-based cleaning agents are not encouraged as they encourage sporulation instead of acting against them (Mayfield et al., 2000; Sinh et al., 2011).

KEYWORDS

- **anaerobe**
- **foodborne**
- ***Clostridium***
- ***C. botulinum***
- ***C. perfringens***
- ***C. difficile***

REFERENCES

Abt, M. C.; McKenney, P.T.; Pamer, E.G. *Clostridium difficile* Colitis: Pathogenesis and Host Defence. Nat. Rev. Microbiol. 2016, 14, 609–620.

Akbulut, D.; Grant, K. A.; McLauchlin, J. Development and Application of Real-Time PCR Assays to Detect Fragments of the *Clostridium botulinum* types A, B, and E Neurotoxin Genes for Investigation of Human Foodborne and Infant Botulism. Foodborne. Path. Dis. 2004, 1, 247–257.

Akbulut, D.; Grant, K. A.; McLauchlin, J. Improvement in Laboratory Diagnosis of Wound Botulism and Tetanus Among Injecting Illicit Drug Users by Use of Real-Time PCR Assays for Neurotoxin Gene Fragments. J. Clin. Microbiol. 2005, 43, 4342–4348.

Asha N. J.; Wilcox M. H.; Laboratory Diagnosis of *Clostridium perfringens* Antibiotic-associated Diarrhoea. J. Med. Microbiol. 2002, 51, 891–894

Barash, J.R.; Arnon, S.S. Dual Toxin-producing Strain of *Clostridium botulinum* type Bf Isolated from a California Patient with Infant Botulism. J. Clin. Microbiol. 2004, 42, 1713–1715.

Bhunia, A.K. In Foodborne Microbial Pathogens: Mechanisms and Pathogenesis. Second Edition. Springer Publisher. 2018. https://doi.org/10.1007/978-1-4939-7349-1

Brynestad, S.; Granum, P. E. *Clostridium perfringens* Foodborne Infections. Int. J. Food Microbiol. 2002, 74, 195–202.

Canard, B.; Saint-Joanis, B.; Cole, S. T. Genomic Diversity and Organization of Virulence Genes in the Pathogenic Anaerobe *Clostridium perfringens*. Mol. Microbiol. 1992, 6, 1421–1429.

Centers for Disease Control and Prevention (CDC). Botulism in the United States (1899–1996). Handbook for Epidemiologists, Clinicians, and Laboratory Workers. U.S. Department of Health, Education, and Welfare, CDC, Atlanta, GA. 1998.

Chen, Y.; Korkeala, H.; Lindén, J.; Lindström, M. Quantitative Real-Time Reverse Transcription-PCR Analysis Reveals Stable and Prolonged Neurotoxin Cluster Gene Activity in a *Clostridium botulinum* type E Strain at Refrigeration Temperature. Appl. Environ. Microbiol. 2008, 74, 6132–6137.

Chertow, D. S.; Tan, E. T.; Maslanka, S. E.; Schulte, J.; Bresnitz, E. A.; Weisman, R. S.; Bernstein, J.; Marcus, S. M.; Kumar, S.; Malecki, J.; Sobel, J. Botulism in 4 Adults Following Cosmetic Injections with an Unlicensed, Highly Concentrated Botulinum Preparation. J. Am. Med. Ass. 2006, 296, 2476–2479.

Cohen, S. H.; Gerding, D. N.; Johnson, S.; Kelly, C. P.; Loo, V. G.; McDonald, L. C.; Pepin, J.; Wilcox, M. H. Clinical Practice Guidelines for *Clostridium difficile* Infection in Adults: 2010 Update by the Society for Healthcare Epidemiology of America (SHEA) and the Infectious Diseases Society of America (IDSA) Infection Control and Hospital Epidemiology. 2010, 31, 431–455.

Dahlenborg, M.; Borch, E.; Rådström, P. Development of a Combined Selection and Enrichment PCR Procedure for *Clostridium botulinum* types B, E, and F and its Use to Determine Prevalence in Faecal Samples from Slaughtered Pigs. Appl. Environ. Microbiol. 2001, 67, 4781–4788.

Dezfulian, M.; Bartlett, J. G. Detection of *Clostridium botulinum* type A toxin by Enzyme-Linked Immuno Sorbent Assay with Antibodies Produced in Immunologically Tolerant Animals. J. Clin. Microbiol. 1984, 19, 645–648.

Erbguth, F.J.; Naumann, M. On the First Systematic Descriptions of Botulism and Botulinum Toxin by Justinus Kerner (1786–1862). J. Hist. Neurosci. 2000, 9, 218–220.

Fernández Miyakawa, M.E.; Pistone Creydt, V.; Uzal, F.A.; McClane, B.A.; Ibarra, C. *Clostridium perfringens* Enterotoxin Damages the Human Intestine *in vitro*. Infect. Immun. 2005, 73, 8407–8410.

Ferreira, J. L.; Hamdy, M. K.; Zapatka, F. A.; Hebert, W. O. Immuno diffusion Method for Detection of type A *Clostridium botulinum*. Appl. Environ. Microbiol. 1981, 42, 1057–1061.

Food Standards Agency. Food Standards Agency guidance on the safety and shelf-life of vacuum and modified atmosphere packed chilled foods with respect to non-proteolytic *Clostridium botulinum*. Food Standards Agency, United Kingdom. 2008. Available at: http://food.gov.uk/multimedia/pdfs/publication/vacpacguide.pdf

Fujita, K.; Katahira, J.; Horiguchi, Y.; Sonoda, N.; Furuse, M.; Tsukita, S. *Clostridium perfringens* Enterotoxin Binds to the Second Extracellular Loop of Claudin3, a Tight Junction Integral Membrane Protein. FEBS. Lett. 2000, 476, 258–261.

Gao, Q. Y.; Huang, Y.F.; Wu, J.G.; Liu H.D.; Xia. H.Q. A Review of Botulism in China. Biomed. Environ. Sci. 1990, 3, 326–336.

García, S.; Heredia, N. *Clostridium perfringens*: A Dynamic Foodborne Pathogen. Food. Bioprocess. Technol. 2011, 4, 624–630.

Gould, L. H, Limbago, B. *Clostridium difficile* in Food and Domestic Animals: A New Foodborne Pathogen? Clin. Infect. Dis. 2010, 51: 577–582

Grant, K.A.; Kenyon, S.; Nwafor, I.; Plowman, J.; Ohai, C.; Halford-Maw, R.; Peck, M.W.; McLauchlin, J. The Identification and Characterization of *Clostridium perfringens* by Real-Time PCR, Location of Enterotoxin Gene, and Heat Resistance. Foodborne Pathog. Dis. 2008, 5, 629–639.

Harvey, S. M.; Sturgeon, J.; Dassey, D. E. Botulism due to *Clostridium baratii* type F toxin. J. Clin. Microbiol. 2002, 40, 2260–2262.

Hatheway, C. L. Botulism: The Present Status of the Disease. Curr. Top. Microbiol. Immunol. 1995, 195, 55–75.

Heikinheimo, A.; Lindström, M.; Granum, P. E.; Korkeala, H. Humans as Reservoir for Enterotoxin Gene-carrying *Clostridium perfringens* type A. Emerg. Infect. Dis. 2006, 12, 1724–1729.

Hielm, S.; Hyytiä, E.; Andersin, A.B.; Korkeala, H. A high Prevalence of *Clostridium botulinum* type E in Finnish Freshwater and Baltic Sea Sediment Samples. J. Appl. Microbiol. 1998, 84, 133–137.

Hookman, P.; Barkin, J. S.; *Clostridium difficile* Associated Infection, Diarrhea and Colitis. World. J. Gastroenterol. 2009; 15, 1554–1580.

Johnson, H. M.; Brenner, K.; Angelotti, R.; Hall, H. E. Serological Studies of types A, B, and E Botulinum Toxins by Passive Hemagglutination and Bentonite Fluctuation. J. Bacteriol. 1966, 91, 964–974.

Johnson, E. A.; Montecucco, C. In Handbook of Clinical Neurology(ed. Engel, A. G.). Elsevier, 2008, 91, 333–368

Juneja, V. K.; Novak, J. S.; Marks, H. M.Gombas, D. E. Growth of *Clostridium perfringens* from Spore Inocula in Cooked Cured Beef: Development of a Predictive Model. Innov. Food. Sci. Emerg. Technol, 2001, 2, 289–301.

Juneja, V. K.; Thippareddi, H. Control of *Clostridium perfringens* in a Model Roast Beef by Salts of Organic Acids During Chilling. J. Food Saf. 2004, 93, 155–163.

Juneja, V. K.; Thippareddi, H.; Friedman, M. Control of *Clostridium perfringens* in Cooked Ground Beef by Carvacrol, Cinnamaldehyde, Thymol, or Oregano oil During Chilling. J. Food. Prot. 2006, 69, 1546–1155.

Kariv, R.; Navaneethan, U.; Venkatesh, P. G. K.; Lopez. R.; Shen, B. Impact of *Clostridium difficile* Infection in Patients with Ulcerative Colitis. J. Crohns. Colitis. 2010, 1, 34–40.

Keto-Timonen, R.; Heikinheimo, A.; Eerola, E.; Korkeala, H. Identification of *Clostridium* species and DNA fingerprinting of *Clostridium perfringens* by amplified Fragment Length Polymorphism Analysis. J. Clin. Microbiol. 2006, 44, 4057–4065.

Koepke, R.; Sobel, J.; Arnon, S. S. Global Occurrence of Infant Botulism, 1976–2006. Pediatrics. 2008, 122, e73–e82.

Kongsaengdao, S.; Samintarapanya, K.; Rusmeechan, S.; Wongsa, A.; Pothirat, C.; Permpikul, C.; Pongpakdee, S.; Puavilai, W.; Kateruttanakul, P.; Phengtham, U.; Panjapornpon, K.

An Outbreak of Botulism in Thailand: Clinical Manifestations and Management of Severe Respiratory Failure. Clin. Infect. Dis. 2006, 43, 1247–1256.
Kuijper, E. J.; Coignard, B.; Tull, P. Emergence of *Clostridium difficile*-associated Disease in North America and Europe. Clin. Microbiol. Infect. 2006, 12, 2–18.
Lahti, P.; Heikinheimo, A.; Johansson, T.; Korkeala, H. *Clostridium perfringens* Type A Isolates Carrying Plasmid-borne Enterotoxin Gene (genotypes IS 1151-cpe or IS 1470-like-cpe) are a Common Cause of Food Poisonings. J. Clin. Microbiol. 2008, 46, 371–373.
Lindsay, J. A.; Mach, A. S.; Wilkinson, M. A.; Martin, L. M.; Wallace, F. M.; Keller, A. M.; Wojciechowski, L. M. *Clostridium perfringens* type A Cytotoxic Enterotoxin (s) as Triggers for Death in the Sudden Infant Death Syndrome: Development of a Toxico-infection Hypothesis. Curr. Microbiol, 1993, 27, 51–59.
Lindström, M.; Heikinheimo, A.; Lahti, P.; Korkeala, H. Novel Insights into the Epidemiology of *Clostridium perfringens* type A Food Poisoning. Food. Microbiol 2011, 28, 192–198.
Lindström, M.; Keto, R.; Markkula, A.; Nevas, M.; Hielm, S.; Korkeala, H. Multiplex PCR Assay for Detection and Identification of *Clostridium botulinum* types A, B, E, and F in Food and Fecal Material. Appl. Environ. Microbiol. 2001, 67, 5694–5699.
Lindström, M.; Korkeala, H. Laboratory Diagnostics of Botulism. Clin. Microbiol. Rev. 2006, 19, 298–314.
Lindström, M.; Nevas, M.; Hielm, S.; Lähteenmäki, L.; Peck, M.W.; Korkeala, H. Thermal Inactivation of Nonproteolytic *Clostridium botulinum* type E Spores in Model Fish Media and in Vacuum-packaged Hot-smoked Vacuum-packaged Fish Products. Appl. Environ. Microbiol. 2003, 69, 4029–4035.
Loo, V. G.; Poirier, L.; Miller, M. A.; Oughton, M.; Libman, M. D.; Michaud, S.; Bourgault, A. M.; Nguyen, T.; Frenette, C.; Kelly, M.; Vibien, A. A Predominantly Clonal Multi-institutional Outbreak of *Clostridium difficile*-associated Diarrhea with High Morbidity and Mortality. N. Engl. J. Med. 2005, 353, 2442–2449.
Lynch, M.; Painter, J.; Woodruff, R.; Braden, C. Surveillance for Foodborne-disease Outbreaks—United States, 1998–2002. MMWR Surveill. Summ. 2006, 55, 1–42.
Mage, D. T.; Cohen, M.; Donner, M. Sudden Infant Death Syndrome. N. Engl. J. Med. 2009, 24, 2581.
Mayfield, J. L.; Leet, T.; Miller, J.; Mundy, L. M. Environmental Control to Reduce Transmission of *Clostridium difficile*. Clin. Infect. Dis. 2000, 31, 995–1000.
McClane, B. A. *Clostridium perfringens*. In M. Doyle, & L. Beuchat (Eds.), Food Microbiology, Fundamentals and Frontiers (3rd edition). 423–445. Washington, DC, USA: American Society for Microbiology, 2007.
McClane, B.A.; Chakrabarti, G. New Insights into the Cytotoxic Mechanisms of *Clostridium perfringens* Enterotoxin. Anaerobe. 2004, 10, 107–114.
McDonel, J. L. (1986). Toxins of *Clostridium perfringens* types A, B, C, D, and E. Pharmacol. Bacter. Toxins. 1986, 477–517.
Modi, N.; Wilcox, M.H. Evidence for Antibiotic Induced *Clostridium perfringens* diarrhoea. J. Clin. Pathol. 2001, 54, 748–751.
Murrell, W. G.; Stewart, B. J.; O'Neill, C.; Siarakas, S.; Kariks, S. Enterotoxigenic Bacteria in the Sudden Infant Death Syndrome. J. Med. Microbiol. 1993, 39, 114–127.
Myllykoski, J.; Nevas, M.; Lindström, M.; Korkeala, H. The Detection and Prevalence of *Clostridium botulinum*in Pig Intestinal Samples. Int. J. Food Microbiol. 2006, 110, 172–177.
Nordic Committee on Food Analysis. *Clostridium botulinum.* Detection in foods and other test materials. Method no. 80, 2nd ed. Nordic Committee on Food Analysis, Espoo, Finland. 1991.

Peck, M. W.; Fairbairn, D. A.; Lund, B. M. Heat-resistance of Spores of non-proteolytic *Clostridium botulinum* Estimated on Medium Containing Lysozyme. Lett. Appl. Microbiol. 1993, 16, 126–131.

Raju, D.; Sarker. M.R. Comparison of the Levels of Heat Resistance of Wild-type, *cpe* Knockout, and *cpe* plasmid-cured *Clostridium perfringens* type A Strains. Appl. Environ. Microbiol. 2005, 71, 7618–7620.

Rodriguez, A.; Dezfulian, M. Rapid Identification of *Clostridium botulinum* and Botulinal toxin in Food. Folia. Microbiol. 1997, 42, 149–151.

Rossetto, O.; Pirazzini, M.; Montecucco, *C. botulinum* Neurotoxins: Genetic, Structural and Mechanistic Insights. Nat. Rev. Microbiol. 2014, 12, 535–549.

Rupnik, M.; Wilcox, M. H.; Gerding, D. N. *Clostridium difficile* infection: New Developments in Epidemiology and Pathogenesis. Nat. Rev. Microbiol. 2009, 7, 526–536.

Sabah, J. R.; Juneja, V. K.; Fung, D. Y. Effect of Spices and Organic Acids on the Growth of *Clostridium perfringens* During Cooling of Cooked Ground Beef. J. Food. Prot. 2004, 67, 1840–1847.

Sabah, J. R.; Thippareddi, H.; Marsden, J. L.; Fung, D. Y. Use of Organic Acids for the Control of *Clostridium perfringens* in Cooked Vacuum-packaged Restructured Roast Beef During an Alternative Cooling Procedure. J. Food. Prot. 2003, 66, 1408–1412.

Sarker, M. R.; Shivers, R. P.; Sparks, S.G.; Juneja, V.K.; McClane, B. A. Comparative Experiments to Examine the Effects of Heating on Vegetative Cells and Spores of *Clostridium perfringens* Isolates Carrying Plasmid Genes Versus Chromosomal Enterotoxin Genes. Appl. Environ. Microbiol. 2000, 66, 3234–3240.

Shapiro, R. L.; Hatheway, C.; Swerdlow, D. L. Botulism in the United States: AClinical and Epidemiologic Review. Ann. Intern. Med. 1998, 129, 221–228.

Sheth, A.N.; Wiersma, P.; Atrubin, D.; Dubey, V.; Zink, D.; Skinner, G.; Doerr, F.; Juliao, P.; Gonzalez, G.; Burnett, C.; Drenzek, C.; Shuler, C.; Austin, J.; Ellis, A.; Maslanka, S.; Sobel, J. International Outbreak of Severe Botulism with Prolonged Toxemia caused by Commercial Carrot Juice. Clin. Infect. Dis. 2008, 47, 1245–1251.

Siarakas, S.; Damas, E.; Murrell, W. G. Is Cardiorespiratory Failure Induced by Bacterial Toxins the Cause of Sudden Infant Death Syndrome? Studies with an Animal Model (the rabbit). Toxicon. 1995, 33, 635–649.

Simonds, J.P. Studies on *Bacillus welchii* with Special Reference to Gas Gangrene. J. Exp. Med. 1917, 25, 819–836.

Sinh, P.; Barrett, T.A.; Yun, L. *Clostridium difficile* Infection and Inflammatory Bowel Disease: A Review. Gastroenterol. Res. Pract. 2011, 2011, 136064.

Smedley, J. G.; McClane, B. A. Fine Mapping of the N-terminal CytotoxicityRegion of *Clostridium perfringens* Enterotoxin by Site-directed Mutagenesis. Infect. Immun. 2004, 72, 6914–6923.

Smith, L. D. S.; Sugiyama, H. Botulism: The Organism, its Toxins, the Disease. Charles C. Thomas Publisher, 1988.

Sobel, J. Botulism. Clin. Infect. Dis. 2005, 41, 1167–1173.

Sobel, J.; Tucker, N.; McLaughlin, J.; Maslanka, S. Foodborne Botulism in the United States, 1990–2000. Emerg. Infect. Dis. 2004, 10, 1606–1611.

Solomon, H. M.; Lilly, T. Jr. *Clostridium botulinum. In* R. I. Merker (ed.), Bacteriological Analytical Manual Online, Revision A. Center for Food Safety and Applied Nutrition, Food and Drug Administration, MD. USA, 2001.

Sparks, S. G.; Carman, R.J.; Sarker, M. R., McClane, B. A. Genotyping of Enterotoxigenic *Clostridium perfringens* FecalIsolates Associated with Antibiotic Associated Diarrhea and Food Poisoning in North America. J. Clin. Microbiol. 2001, 39, 883–888.

Suen, J.C.; Hatheway, C.L.; Steigerwalt, A.G.; Brenner, D.J. *Clostridium argentinense*, sp. nov: A Genetically Homogenous group Composed of All Strains of *Clostridium botulinum* type G and Some Nontoxigenic Strains Previously Identified as *Clostridium subterminale*or *Clostridium hastiforme*. Int. J. Syst. Bacteriol. 1988, 38, 375–381.

Tanaka, D.; Isobe, J.; Hosorogi, S.; Kimata, K.; Shimizu, M.; Katori, K.; Gyobu, Y.; Nagai, Y.; Yamagishi, T.; Karasawa, T.; Nakamura, S. An Outbreak of Foodborne Gastroenteritis Caused by *Clostridium perfringens* Carrying the *cpe* gene on a Plasmid. Jpn. J. Infect. Dis. 2003, 56, 137–139.

van Ermengem, E. Classics in Infectious Diseases. A New AnaerobicBacillus and its Relation to Botulism. Rev. Infect. Dis. 1979, 1, 701–19.

Wenham, T. N.; Botulism: A Rare Complication of Injecting Drug Use. Emerg. Med. J. 2008, 25, 55–56.

Werner, S. B.; Passaro, D. J.; McGee J.; Schechter R.; Vugia D. J. Wound Botulism in California, 1951–1998: A Recent Epidemic in Heroin Injectors. Clin. Infect. Dis. 2000, 31, 1018–1024.

CHAPTER 4

Microwave-Assisted Extraction of Phenolic Compounds from Ceylon Olive (*Elaeocarpus serratus*)

JAYANTA PAL and LAXMIKANT S. BADWAIK*

Department of Food Engineering and Technology, School of Engineering, Tezpur University, Napaam, Assam 784028, India

**Corresponding author. E-mail: laxmikantbadwaik@gmail.com*

ABSTRACT

Phenols are secondary plant metabolites, widely spread throughout the plant kingdom. Substantial developments in research focused on the extraction, identification, and quantification of phenolic compounds as medicinal and dietary molecules. Organic solvent extraction is the main method used to extract phenolics. Microwave-assisted extraction (MAE) is one of the efficient extraction process among all of the modern techniques. The process of extraction of phenolic compounds from Ceylon olive was optimized using conventional heat reflux method and MAE process. The independent variables like temperature (50–100 °C), time (60–240 min), solid-to-solvent ratio (1:10 to 1:50), and solvent concentration (20%–100%) were taken for conventional heat reflux method. Whereas, for MAE process the independent variables were taken as microwave power (300–600 W), irradiation time (30–90 s), solid-to-solvent ratio (1:10 to 1:30), and solvent concentration (50%–70%). Optimized value of the total phenolic content for conventional heat reflux method and MAE were 239.256 and 315.247 mg GAE/g dry weight of the sample, respectively. Time and solid-to-solvent ratio require for the conventional heat reflux method were 72.42 min and 33.82 mL/g, respectively; while in MAE the time was only 90 s and solid-to-solvent ratio was 19.53 mL/g. Also, the high-performance liquid chromatography result shows a reasonable increase in phenolic compound extract by microwave-assisted extraction.

4.1 INTRODUCTION

Phenolic compounds consist of a hydroxyl group directly bonded with an aromatic hydrocarbon group. Most simple class of phenolic compounds is phenol, this phenol is also known as carbolic acid C_6H_5OH (Amorati et al., 2012; Khoddami et al., 2013). Dietary plant and medicinal herbs are rich in phenolic compounds. Some of these phenolic compounds are flavonoids, phenolic acids, stilbenes, tannins, coumarins, curcuminoids, quinones, lignans, etc. (Huang et al., 2009). Phenolic compounds are generally extracted by using both the inorganic and organic solvent. Extraction yield of phenolics depends on various parameters; these are temperature, extraction time, solvent type, sample from which phenolics to be extract, number of repetitions of sample extraction, active compound, and solvent-to-sample ratio. Various solvents useful for extraction such as ethanol, methanol, propanol, acetate, water, acetone, and mixture of the solvent also influence the extraction yield (Garcia Salas et al., 2010).

Conventional processes of phenolic compound extraction are heat reflux, Soxhlet, and maceration. These processes are widely used to extract phenolics from solid samples (Khoddami et al., 2013). Nowadays there are some modern techniques that are used to extract phenolic compounds which provide high yield of extraction as they need less time, less temperature, and to some extent reduces the solvent requirement. The modern techniques used are microwave-assisted extraction (MAE), ultrasound-assisted extraction, subcritical water extraction, supercritical fluid extraction (SFE), and high hydrostatic pressure processing (Kalia et al., 2008; Biesaga, 2011; Bimakr et al., 2011). Among all of the modern extraction processes, MAE is a widely used process.

Microwave-assisted extraction is a process of heating sample or solvents with microwave energy to collect the compounds which need to be extracted from the sample. In MAE, synergistic combination of two basic transport phenomena that is heat and mass transfer gradient works in a same direction which helps to get rapid and high yield of extraction. (Chemat et al., 2009). Microwave-assisted extraction process is highly depending upon the solid matrix and the solvent using for the extraction. Generally, solvents have a wide range of polarities. Some solvents which have high dielectric constant strongly absorb the microwave energy (Kaufmann and Christen,

2002). Microwave heat up the moisture content available inside the cell, as a result the moisture evaporates and produces a pressure on the cell wall. This pressure causes rupture of cell wall and leaching out of the component into the solvent (Mandal et al., 2007). Many researchers used MAE process for extraction of phenolic compound from tomato (Li et al., 2012), extraction of phenolics from plant *Rosmarinus officinalis, Origanum dictamnus, Origanum majorana*, and *Teucrium polium* (Proestos and Komaitis, 2008), and extraction of phenolic alkaloid from *Nelumbo nucifera* Gaertn (Lu et al., 2008).

There are numerous plants in India which have medicinal benefits. The plant *Elaeocarpus serratus* (Ceylon olive) belonging to the family *Elaeocarpaceae*, is used as a cardiovascular stimulant. The leaves are used in the treatment of rheumatism and as antidote to poison, while the fruits are locally prescribed for the treatment of diarrhea and dysentery. The plant is a small, evergreen tree species commonly found in tropical countries. Its fruit, which is rich in vitamin C, is a popular fruit among the well-known minor fruits of Bangladesh. When green, the fruit is edible fresh and can also be used in preserves, mainly chutney.

Indian medicinal plants are used as a local medicine and commercially can offer a low-cost medicine. Which is rich in phenolic compound, extraction of phenolic compound from this plant by a modern and more effective method of extraction like MAE is necessary. Therefore, the current study is planned for the extraction of phenolic compounds from Ceylon olive using conventional extraction process and MAE process and their comparison.

4.2 MATERIALS AND METHODS

4.2.1 MATERIALS

4.2.1.1 COLLECTION AND PRETREATMENT OF SAMPLES

Ceylon olive fruits were collected from local market and village and brought to the laboratory. The fresh fruits were washed with the tap water and all the rotten samples were removed. Then the samples were dried at 50 °C for 2 days to get moisture content not more than 15%. This dried sample was then grinded by using a mixer grinder and sieve with a 40 mesh size sieve to get the powder of less than 400 μm. This powdered sample was then kept in sealed polyethylene packet and stored in a cool and dry place as it must not gain moisture from the environment.

4.2.2 EXTRACTION OF PHENOLIC COMPOUND BY CONVENTIONAL HEAT REFLUX METHOD

Extraction of dry powder sample by using conventional heat reflux method was done as follows: 40 mL of solvent was added to 1 g of dried sample in a round bottom flask. The mixture was stirred carefully. In each sample nitrogen was bubbled for ca. 40–60 s. The extraction mixture was then reflux in a water bath at 90 °C for 2 h. Each extraction was repeated three times. All this work was carried out in the dark (flasks was covered with aluminum foil) and under nitrogen atmosphere (the headspace above the plant extract was under inert atmosphere created by a stream of nitrogen) to prevent oxidation. Before the estimation of total phenolic content (TPC), the samples were kept at cool and dark environment.

4.2.3 EXPERIMENTAL PROCESS FOR OPTIMIZATION OF CONVENTIONAL HEAT REFLUX PROCESS

A Box Behnken Design was used to optimize the extraction of the sample to optimize by the response surface methodology (RSM) in respect of the TPC and antioxidant activity (DPPH radical scavenging activity). Four independent variables were considered, temperature 50–100 °C, time 60–240 min, solid-to-solvent ratio 1:10–1:50 w/v, and solvent concentration 20%–100% ethanol. The levels of the independent variable with coded and uncoded form are shown in Table 4.1.

TABLE 4.1 Level and Values of the Independent Variable for Conventional Heat Reflux Method in Coded and Uncoded Form

Independent Variable	**Range and Levels**		
	−1	**0**	**+1**
Temperature (°C)	50	75	100
Time (min)	60	150	240
Solid solvent ratio (w/v)	1:10	1:30	1:50
Solvent concentration (% ethanol)	20	60	100

The Box–Behnken design (BBD) design is prepared using the Design expert 7.0 Software. These four independent variables of three levels give 29 total experimental combinations including five center point combination.

The full experimental design with coded (xi) and uncoded (Xi) is shown in Table 4.2. Optimization of the combination of these four variables for maximum extraction was obtained by analyzing the responses such as TPC and DPPH of the extract.

According to the general model of polynomial equation is used to predict the data to the experimental data:

$$Y = B_0 + \sum_{i=1k}^{k} BiXi + \sum_{i=1k}^{k} BiiX^2 + \sum_{i>1k}^{k} BijXiXj + E$$

where Y represents the response function (TPC yield); B_0 is a constant coefficient; *Bi*, *Bii*, and *Bij* are the coefficients of the linear, quadratic, and interactive terms, respectively, and *Xi* and *Xj* represent the coded independent variables.

According to the analysis of variance, the regression coefficients of individual linear, quadratic, and interaction terms were determined. In order to visualize the relationship between the response and experimental levels of each factor and to deduce the optimum conditions, the regression coefficients were used to generate 3D surface plots and contour plots from the fitted polynomial equation. The factor levels were coded as −1 (low), 0 (central point or middle), and 1 (high), respectively. The variables were coded according to the following equation:

$$x_i = \frac{X_i - X_0}{\Delta X}$$

where x_i is the (dimensionless) coded value of the variables X_i; X_0 is the value of ΔX at the center point and, X is the step change.

4.2.4 MICROWAVE-ASSISTED EXTRACTION FOR PHENOLIC COMPOUNDS

A domestic microwave of 20 L capacity 2450 kHz working frequency was used. Radiation time and microwave power were controlled manually as per the domestic oven. The oven was modified in order to condensate into the sample the vapors generated during extraction. For extraction, 1 g of dried sample powder was placed in around bottom flask of 250 mL containing water–ethanol mixture at different ethanol concentrations (as per the experimental design (v/v)). The suspension was irradiated at regular intervals

TABLE 4.2 BBD Experimental Design in Coded and Uncoded Values for Conventional Heat Reflux Method

Experiment No.	Coded Values of the Independent Variable				Real Values of the Independent Variable			
	Temperature	Time	Solid-to-Solvent Ratio	Solvent Concentration	Temperature	Time	Solid-to-Solvent Ratio	Solvent Concentration
	(°C)	(min)	(mL/g)	(%)	(°C)	(min)	(mL/g)	(%)
	x_1	x_2	x_3	x_4	X_1	X_2	X_3	X_4
1	0	0	−1	+1	75	150	10	100
2	+1	0	+1	0	100	150	50	60
3	0	−1	0	+1	75	60	30	100
4	−1	0	0	−1	50	150	30	20
5	0	0	0	0	75	150	30	60
6	−1	0	−1	0	50	150	10	60
7	0	−1	0	−1	75	60	30	20
8	0	+1	0	+1	75	240	30	100
9	0	0	0	0	75	150	30	60
10	−1	−1	0	0	50	60	30	60
11	+1	−1	0	0	100	60	30	60
12	0	0	0	0	75	150	30	60
13	0	0	+1	+1	75	150	50	100
14	0	+1	+1	0	75	240	50	60
15	+1	0	0	+1	100	150	30	100
16	0	+1	−1	0	75	240	10	60
17	−1	0	0	+1	50	150	30	100

TABLE 4.2 *(Continued)*

Experiment No.	Coded Values of the Independent Variable				Real Values of the Independent Variable			
	Temperature	Time	Solid-to-Solvent Ratio	Solvent Concentration	Temperature	Time	Solid-to-Solvent Ratio	Solvent Concentration
	(°C)	(min)	(mL/g)	(%)	(°C)	(min)	(mL/g)	(%)
	x_1	x_2	x_3	x_4	X_1	X_2	X_3	X_4
18	−1	0	+1	0	50	150	50	60
19	0	−1	+1	0	75	60	50	60
20	+1	0	−1	0	100	150	10	60
21	+1	0	0	−1	100	150	30	20
22	−1	+1	0	0	50	240	30	60
23	0	0	+1	−1	75	150	50	20
24	0	+1	0	−1	75	240	30	20
25	0	0	0	0	75	150	30	60
26	0	0	0	0	75	150	30	60
27	0	−1	−1	0	75	60	10	60
28	+1	+1	0	0	100	240	30	60
29	0	0	−1	−1	75	150	10	20

according to oven operation. Different solvent, irradiation time, microwave power, and solvent-to-solid ratio were used. At the end of microwave irradiation, the volumetric flask was allowed to cool to room temperature (Dahmoune et al., 2014). After extraction, the extract was recovered by filtration on a Büchner funnel through No. 1 Whatman filter paper and collected in a volumetric flask. The extract was stored at 4 °C until use and analyzed for the TPC. The extract obtained under the optimum conditions by RSM was analyzed also for the antioxidant capacity.

4.2.5 EXPERIMENTAL PROCESS FOR OPTIMIZATION OF MAE PROCESS

Optimization of MAE processes the factor that influences the extraction. Four independent variables were considered which have the influence on MAE are: microwave power (W) 300–600 W, Irradiation time (s) 30–90 s, solid-to-solvent ratio (SS ratio) 10–30 mL/g, and solvent concentration (SC) 50%–70% ethanol. To get a successive optimization RSM based on BBD was conducted. RSM was prepared by considering the TPC and DPPH radical scavenging activity of the sample. Design expert software was used to prepare the BBD and RSM optimization process.

The levels of the independent variable which have the influence on MAE with coded and uncoded values are given in Table 4.3. BBD design for the conventional heat reflux extraction method is given in Table 4.3.

TABLE 4.3 Level and Values of the Independent Variable for MAE in Coded and Uncoded Form

Independent Variables	Range and Levels		
	–1	0	+1
Microwave power (W)	300	450	600
Irradiation time (s)	30	60	90
Solid solvent ratio (w/v)	1:10	1:20	1:30
Solvent concentration (% ethanol)	50	60	70

The BBD design was produced by using the Design Expert 7.0 Software and which gives the total number of 29 experimental setups for four variables and has five center point experimental setup. The actual full design of the experimental combination for the four independent variables is given in Table 4.4, where both the coded (xi) and uncoded (Xi) are shown. Also optimization of the extraction is obtained by the analysis of the TPC and DPPH of the extract obtained according to the experimental combination of the four independent variables. To describe the process mathematically a general second-order polynomial equation was used which is given as

$$Y = B_0 + \sum_{i=1k}^{k} BiXi + \sum_{i=1k}^{k} BiiX^2 + \sum_{i>1k}^{k} BijXiXj + E$$

where Y represents the response function (TPC yield); B_0 is a constant coefficient; Bi, Bii, and Bij are the coefficients of the linear, quadratic and interactive terms, respectively, and Xi and Xj represent the coded independent variables.

4.2.6 ANALYSIS OF EXTRACTS

4.2.6.1 TOTAL PHENOLIC CONTENT ESTIMATION

The TPC was determined using the Folin–Ciocalteu method which as described by Wei et al. (2010). Briefly, 0.2 mL of extract and 0.3 mL distilled water were mixed with 0.5 mL of Folin–Ciocalteu reagent; 2.5 mL of sodium carbonate (20%, w/w) solution was added to the mixture. After 30 min at room temperature, the absorbance was measured at 725 nm using a UV/visible spectrophotometer and a calibration curve was drawn using data from standard solutions of gallic acid (GA). The TPC of the extract was calculated and expressed as milligram GA equivalents per gram of dry weight (DW) (mg GAE/100 g DW) based on the GA standard curve. The extraction abilities of phenolics were calculated based on the value of the phenolic content.

4.2.6.2 DPPH RADICAL SCAVENGING ACTIVITY

The antioxidant activity of the extract was determined using DPPH radical-scavenging assay. DPPH is a stable free radical and has a dark violet color. It has maximum absorption at 517 nm. Absorption of the DPPH is decreased in presence of an antioxidant. In a test tube containing 50 μL of extract with concentration of 1 mg/mL, 3.95 mL of methanol and 1 mL 0.2 mM of the DPPH methanol solution were added. After 30 min of incubation in the dark at room temperature, the absorbance was measured against a blank

TABLE 4.4 BBD Experimental Design in Coded and Uncoded Values MAE Method

Experiment No.	Coded Values of the Independent Variable				Real Values of the Independent Variable			
	Microwave Power	Irradiation Time	Solid-to-solvent Ratio	Solvent Concentration	Microwave Power	Irradiation Time	Solid-to-solvent Ratio	Solvent Concentration
	(W)	(s)	(mL/g)	(%)	(W)	(s)	(mL/g)	(%)
	x_1	x_2	x_3	x_4	X_1	X_2	X_3	X_4
1	0	0	+1	−1	450	60	30	50
2	0	−1	0	−1	450	30	20	50
3	0	−1	+1	0	450	30	30	60
4	+1	+1	0	0	600	90	20	60
5	0	0	+1	+1	450	60	30	70
6	−1	0	−1	0	300	60	10	60
7	+1	0	+1	0	600	60	30	60
8	0	−1	0	+1	450	30	20	70
9	−1	+1	0	0	300	90	20	60
10	0	0	−1	+1	450	60	10	70
11	0	+1	−1	0	450	90	10	60
12	0	0	0	0	450	60	20	60
13	−1	0	0	+1	300	60	20	70
14	0	0	0	0	450	60	20	60
15	0	0	0	0	450	60	20	60
16	0	0	0	0	450	60	20	60

TABLE 4.4 *(Continued)*

Experiment No.	Coded Values of the Independent Variable				Real Values of the Independent Variable			
	Microwave Power	Irradiation Time	Solid-to-solvent Ratio	Solvent Concentration	Microwave Power	Irradiation Time	Solid-to-solvent Ratio	Solvent Concentration
	(W)	(s)	(mL/g)	(%)	(W)	(s)	(mL/g)	(%)
	x_1	x_2	x_3	x_4	X_1	X_2	X_3	X_4
17	+1	0	0	−1	600	60	20	50
18	0	0	−1	−1	450	60	10	50
19	0	−1	−1	0	450	30	10	60
20	+1	0	0	+1	600	60	20	70
21	0	+1	0	−1	450	90	20	50
22	−1	0	0	−1	300	60	20	50
23	−1	−1	0	0	300	30	20	60
24	+1	−1	0	0	600	30	20	60
25	−1	0	+1	0	300	60	30	60
26	0	+1	0	+1	450	90	20	70
27	0	0	0	0	450	60	20	60
28	0	+1	+1	0	450	90	30	60
29	+1	0	−1	0	600	60	10	60

(methanol) at 517 nm using the UV/visible spectrophotometer. Inhibition of DPPH radical was calculated as a percentage (%) using the following equation:

$$\text{Inhibition } (\%) = \frac{A_{\text{control}} - A_{\text{sample}}}{A_{\text{control}}} \times 100$$

where A_{sample} is the absorbance of the test sample and A_{control} is the absorbance of the control. A control was prepared using the same procedure as applied to the samples, where 50 μL of the extract was replaced by the pure solvent (the solvent varied between batches of extracts).

4.2.6.3 HIGH-PERFORMANCE LIQUID CHROMATOGRAPHY

The HPLC analysis was carried out over the extract sample to determine the constituents which are responsible for their antioxidant activity. The HPLC analysis was performed by the method (Pfundstein et al., 2010) of characterization of polyphenolic compound with little modification in run time. Analytical HPLC was conducted on fitted with a reverse-phase C18 column. Samples of dried *Terminalia chebula* extracts were dissolved in methanol (10 mL) and, when necessary, further diluted prior to injection (20 μL) into the HPLC. The mobile phase consisted of 0.1% orthophosphoric acid in water (solvent A) and methanol (solvent B) with the following gradient profile: 95% A to 20% A for 30 min; reduced to 20% A for 30–33 min; and from 20% A to 95% A from 33 to 35 min; and 0% A over 5 min; continuing at 0% A until completion of the run. The flow rate of the mobile phase was 0.8 mL/min. Phenolic compounds in the eluant were detected at 280 and 360 nm with a diode-array UV detector. Phenolic compounds were detected in the raw extracts and fractions using standard curves (optical absorbance vs. concentration for the range 0.05–1.0 mM) prepared using either authentic commercial samples (gallic acid, methyl gallate, and ellagic acid) or by the following previously purified and characterized compounds: 3-*O*-methyl ellagic acid; 3,30-di-*O*-methyl ellagic acid; 30-*O*-methyl-4-*O*-b-D-xylopyranosylellagic acid; 3,30-di-*O*-methyl-4-*O*-b-D-xylopyranosyl ellagic acid; 1,6-di-*O*-galloyl-b-D-glucopyranose; 3,4,6-tri-*O*-galloyl-D-glucopyranose; 1,3,4,6-tetra-*O*-galloyl-b-D-glucopyranose; and 1,2,3,4,6-penta-*O*-galloyl-b-D-glucopyranose (Pfundstein et al., 2010).

4.3 RESULTS AND DISCUSSION

4.3.1 OPTIMIZATION OF EXTRACTION FROM CEYLON OLIVE BY CONVENTIONAL HEAT REFLUX METHOD

4.3.1.1 MODEL FITTING

Optimization of the process was done according to the responses (TPC and DPPH) which were analyzed for each of 29 experimental setups as shown in Table 4.5. Each model equation is observed for all of the independent variables to their linear, quadratic, and interaction effect on the responses (Tables 4.6 and 4.7) (Zhang et al., 2013). Model F-value of the experiments were 65.42 and 9.14 for the responses TPC and DPPH, respectively. It shows that all the models are significant and there are only 0.01% chances that this large "Model F-value" could occur due to noise.

"Prob > *F*" values are less than 0.0001 for all the responses is an indication of very high significance of the model terms. According to the Fisher *F*-test, such low probability value demonstrates that the experiment to be a very high significant regression model. The goodness of the fitting of the model is generally shown by the correlation coefficients being significantly high (R^2>0.9). In this experiment, R^2 for each response is 0.9849 and 0.9330, respectively, which indicates that the model is to be fitted with the experiment with a very high significance. The adjusted R^2 values are also very high enough to fit the model significantly. Adjusted R^2 are to correct the R^2 values for the number of terms and sample size of the model. Adjusted R^2 values are 0.9699 and 0.8660, respectively. The predicted R^2 values (0.9175 and 0.6383) are also in reasonable agreement with the "adjusted R^2 values" (Karazhiyan et al. 2011). CV% indicated the deviation between the predicted and experimental value. In this experiment low CV% values (8.39 and 13.08) indicates that the deviation between experimental and predicted values are low enough (Chen et al., 2012).

The polynomial equations developed through the design to correlate the relationship between the independent variable and the responses for Conventional Heat Reflux Method for Ceylon olive are given in Eqs. (4.1) and (4.2).

$$\text{TPC} = +225.00 + 30.21\,X_1 - 2.25\,X_2 + 41.11\,X_3 - 24.82\,X_4 - 14.43\,X_1X_2 + 14.62\,X_1X_3 + 2.00X_1X_4 + 1.31\,X_2X_3 + 57.34\,X_2X_4 - 43.49\,X_3X_4 - 58.08X_1^2 - 21.82X_2^2 - 63.47\,X_3^2 - 70.75 \times X_4^2 \tag{4.1}$$

TABLE 4.5 Values of Experimental Data for the Conventional Heat Reflux Extraction of Ceylon Olive

Experiment No.	Real Process Variable				Experimental Response	
	Temperature (X_1) (°C)	Time (X_2) (min)	Solid-to-Solvent Ratio (X_3) (mL/g)	Solvent Concentration (X_4) (%)	TPC (mg GAE/100 g DW)	DPPH (% RSA)
1	75	150	10	100	63.256	32.68
2	100	150	50	60	182.758	56.3
3	75	60	30	100	61.145	40.36
4	50	150	30	20	82.268	52.78
5	75	150	30	60	218.365	78.78
6	50	150	10	60	57.154	26.45
7	75	60	30	20	232.856	82.45
8	75	240	30	100	150.457	67.69
9	75	150	30	60	220.356	81.91
10	50	60	30	60	95.751	67.78
11	100	60	30	60	180.861	71.61
12	75	150	30	60	231.445	84.08
13	75	150	50	100	73.268	35.55
14	75	240	50	60	175.336	68.23
15	100	150	30	100	106.26	52.11
16	75	240	10	60	90.485	42.08

TABLE 4.5 *(Continued)*

Experiment No.	Real Process Variable				Experimental Response	
	Temperature (X_1) (°C)	Time (X_2) (min)	Solid-to-Solvent Ratio (X_3) (mL/g)	Solvent Concentration (X_4) (%)	TPC (mg GAE/100 g DW)	DPPH (% RSA)
17	50	150	30	100	35.41	26.94
18	50	150	50	60	95.335	35.87
19	75	60	50	60	178.489	46.35
20	100	150	10	60	86.115	25.81
21	100	150	30	20	145.109	69.36
22	50	240	30	60	142.235	62.33
23	75	150	50	20	209.305	81.68
24	75	240	30	20	92.821	53.74
25	75	150	30	60	226.305	76.2
26	75	150	30	60	228.522	86.29
27	75	60	10	60	98.885	42.77
28	100	240	30	60	169.621	59.62
29	75	150	10	20	25.32	23.93

$$DPPH = +81.45 + 5.22\ X_1 + 0.20\ X_2 + 10.86\ X_3 - 9.05\ X_4 - 1.63\ X_1 X_2 + 5.27\ X_1X_3 + 2.15\ X_1X_4 + 5.64\ X_2X_3 + 14.01\ X_2X_4 - 13.72\ X_3X_4 - 15.88\ X_1^2 - 3.62\ X_2^2 - 27.03\ X_3^2 - 14.34\ X_4^2 \quad (4.2)$$

TABLE 4.6 Estimated Regression Coefficient and ANOVA for Response Surface Quadratic Model of TPC of Ceylon Olive Extract by Conventional Heat Reflux Extraction Method

Source	Estimated Coefficients	Sum of Squares	DF	*F* Value	*P*-value
Model		119917.3	14	65.41517	<0.0001
Intercept	225	–	–	–	–
X_1-temp	30.21	10954.81	1	83.66224	<0.0001
X_2-time	−2.25	60.89409	1	0.46505	0.5064
X_3-solid-to-solvent ratio	41.11	20276.77	1	154.8543	<0.0001
X_4-solvent Concentration	−24.82	7394.523	1	56.47221	<0.0001
X_1X_2	−14.43	833.015	1	6.361763	0.0244
X_1X_3	14.62	854.4514	1	6.525473	0.0229
X_1X_4	2.00	16.03602	1	0.122468	0.7316
X_2X_3	1.13	6.882752	1	0.052564	0.8220
X_2X_4	57.34	13150.01	1	100.4271	<0.0001
X_3X_4	−43.49	7566.651	1	57.78676	<0.0001
X_1^2	−58.08	21878.65	1	167.0879	<0.0001
X_2^2	−21.82	3088.028	1	23.58337	0.0003
X_3^2	−63.47	26133.35	1	199.5812	<0.0001
X_4^2	−70.75	32470.74	1	247.9801	<0.0001
Lack of Fit		1711.937	10	5.648305	0.0549
Pure Error		121.2355	4		
R^2	0.9849				
Adjusted R^2	0.9699				
Predicted R^2	0.9175				
CV%	8.39				
Adequate Precision	23.573				

TABLE 4.7 Estimated Regression Coefficient and ANOVA for Response Surface Quadratic Model of DPPH of Ceylon Olive Extract by Conventional Heat Reflux Extraction Method

Source	Estimated Coefficients	Sum of Squares	DF	*F*-Value	P-value
Model		6457.119	14	9.142686	<0.0001
Intercept	81.45	–	–	–	–
X_1-temp	5.22	1.140833	1	0.022614	0.8826
X_2-time	0.20	240.397	1	4.76532	0.0466
X_3-solid-to-solvent ratio	10.86	471.128	1	9.339034	0.0085
X_4-solvent concentration	−9.05	70.47053	1	1.396917	0.2569
X_1X_2	−1.63	27.19623	1	0.539103	0.4749
X_1X_3	5.27	69.30563	1	1.373825	0.2607
X_1X_4	2.15	9.61	1	0.190496	0.6692
X_2X_3	5.64	670.2921	1	13.28701	0.0027
X_2X_4	14.01	49.4209	1	0.979656	0.3391
X_3X_4	−13.72	737.6656	1	14.62253	0.0019
X_1^2	−15.88	886.3421	1	17.5697	0.0009
X_2^2	−3.62	373.2558	1	7.398941	0.0166
X_3^2	−27.03	3207.825	1	63.58779	< 0.0001
X_4^2	−14.34	1284.005	1	25.45247	0.0002
Lack of fit		606.7701	10	2.439509	0.2023
Pure error		99.49052	4		
R^2	0.9330				
Adjusted R^2	0.8660				
Predicted R^2	0.6383				
CV%	13.08				
Adequate Precision	11.674				

4.3.1.2 INFLUENCES OF INDEPENDENT VARIABLES ON THE RESPONSES DURING EXTRACTION FROM CEYLON OLIVE BY CONVENTIONAL HEAR REFLUX METHOD ON

4.3.1.2.1 Total Phenolic Content

The results in Table 4.6 clearly define the linear, quadratic, and interaction effect of the four independent variables on the TPC. It was seen that the independent variables (temperature and solid-to-solvent ratio) and interaction term of time and solvent concentration have positive effect on the TPC of the fruit extract. However, the maximum effect is of interaction term of time and solvent concentration. Also, the linear term temperature and solid-to-solvent ratio have the maximum positive effect on the TPC. The other terms that have positive effect are interaction term of temperature and solid-to-solvent ratio, temperature and solvent concentration, and time and solid-to-solvent ratio. Negative effect contributing terms are interaction of temperature and time, solid-to-solvent ratio and solvent concentration, and quadratic term of all four variables (temperature, time, solid-to-solvent ratio, and solvent concentration).

Figure 4.1 describes the effect on TPC with respect to temperature and time, (1) temperature and solid-to-solvent ratio, (2) temperature and solvent concentration, (3) time and solid-to-solvent ratio, (4) time and solvent concentration, (5) and solid-to-solvent ratio and solvent concentration (6).

It is observed from the 3D plot that at initial time and temperature interaction TPC content is low. It increases as the temperature increases and then decreases. Again, as the time increases it increase and then decrease. TPC also increase with increase in interaction of the time and temperature and then again decrease (Figure 4.1(A)), initial point of temperature and solid-to-solvent ratio is minimum and as increasing in the interaction the TPC content increases (Figure 4.1(B)), it is not that much increase in solvent concentration but the interaction effect shows that TPC increase up to a certain level and then decrease (Figure 4.1(C)), time and solid-to-solvent ratio interaction shows that the maximum influence is of solid-to-solvent ratio. TPC increase in increasing of the solid-to-solvent ratio after the ratio 1:45 TPC decrease slightly (Figure 4.1(D)), time and solvent concentration interaction little increase in TPC and mostly decrease. Higher value obtains at the initial point (Figure 4.1(E)), solid-to-solvent ratio and solvent concentration plot describe the influence of solid-to-solid ratio has maximum influence. As increase in the ratio TPC increase at the end point it slightly decreases (Figure 4.1(F)).

Total phenolic content increases with the increase in the solvent concentration up to a 60% of the ethanol and then the TPC content decrease which is already describe by Sultana et al. (2009). Reason is TPC extraction is maximum in polar solvent (combination of water and ethanol) but not in the absolute ethanol. Minimum solid-to-solvent ratio 1:10 the solubility is less and at the ratio 1:40 solubility is maximum and again at higher stage 1:50 there excess amount of solvent so the extraction of the TPC increase and as the ratio increase and then decrease (Maran et al., 2013).

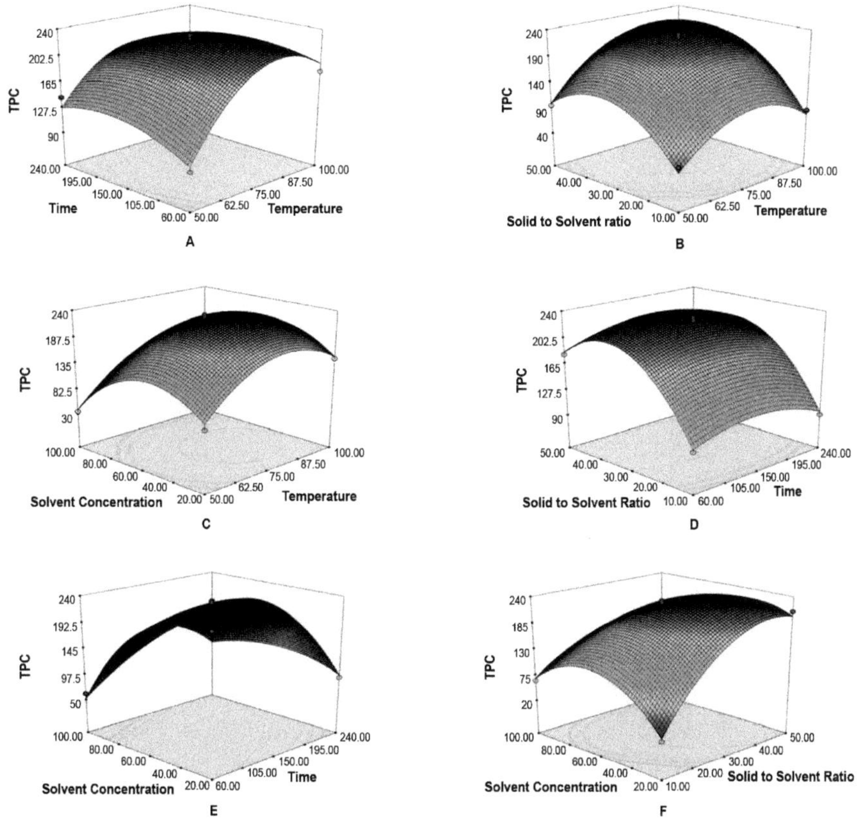

FIGURE 4.1 Response surface analysis for the TPC yield of Ceylon olive extract, Extracted by using conventional heat reflux extraction method with respect to temperature and time, (A) temperature and solid-to-solvent ratio, (B) temperature and solvent concentration, (C) time and solid-to-solvent ratio (D), time and solvent concentration, (E) and solid-to-solvent ratio and solvent concentration (F).

4.3.1.2.2 DPPH Radical Scavenging activity

The coefficient of the four independent variables as their linear, interaction, and quadratic function (Table 4.7) clearly describe that the linear function of solid-to-solvent ratio and interaction term of time and solvent concentration have the maximum positive effect on the DPPH activity of the extracted sample. The other positive effect terms are linear function of temperature, time, interaction term of temperature and solid-to-solvent ratio, temperature and solvent concentration, and time and solid-to-solvent ratio. Negative effect terms are linear function of solvent concentration, interaction term of solid-to-solvent ratio and solvent concentration, and all the quadratic term of all the four independent variables (temperature, time, solid-to-solvent ratio, and solvent concentration).

The RSM plots (Figure 4.2) with respect to DPPH scavenging activity are temperature and time, (1) temperature and solid-to-solvent ratio, (2) temperature and solvent concentration, (3) time and solid-to-solvent ratio, (4) time and solvent concentration, (5), and solid-to-solvent ratio and solvent concentration (6). Time has not much influence but the temperature influence is high on the DPPH radical scavenging activity. Increase as the temperature increase and decrease (Pompeu et al., 2009) (Figure 4.2(A)), temperature and solid-to-solvent interaction has much influence. Highly increase in increase in the interaction and then decrease. (Figure 4.2(B)), temperature and solvent concentration interaction has not much influence little increase at interaction value increase than decrease little (Figure 4.2(C)), In time and solid-to-solvent ratio, time has mostly same effect but DPPH scavenging activity increase as increase in solid-to-solvent ratio increase (Figure 4.2(D)), DPPH activity decrease with increase in time and solvent concentration but at higher temperature little increase observed and then decrease (Figure 4.2(E)), mostly increasing effect with increase in solid-to-solvent ratio increase. Interaction of solid-to-solvent ratio and solvent concentration helps to increase DPPH activity as the value of interaction increase. Finally, it decreases at higher value. (Figure 4.2(E)). As the ethanol concentration increase DPPH radical scavenging activity increases up to 50% of ethanol and then the DPPH scavenging activity decreases (Liyana-Parthirana and Shahidi, 2005).

4.3.1.3 OPTIMIZATION OF THE PROCESS VARIABLE FOR EXTRACTION OF TPC BY CONVENTIONAL HEAT REFLUX METHOD FROM CEYLON OLIVE

Workable optimum conditions of the extraction from Ceylon olive by conventional heat reflux extraction method were determined by the numerical multi response optimization technique. Optimization was done according to the maximum values of TPC and the DPPH radical scavenging activity (%) (Table 4.8). Desirability of the predicted result was one.

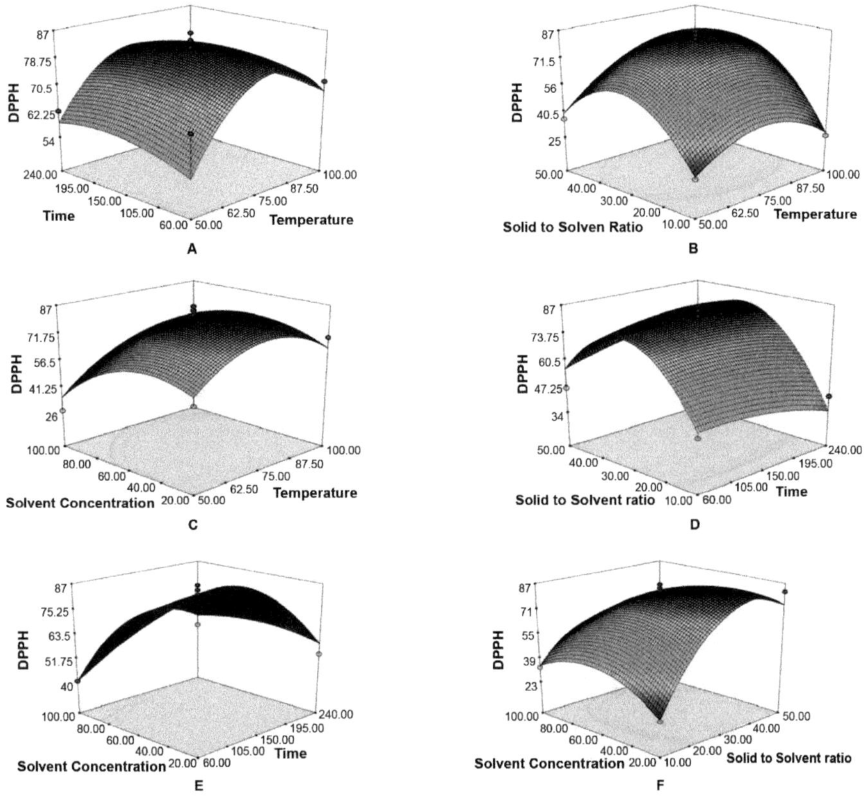

FIGURE 4.2 Response surface analysis for the DPPH radical scavenging activity of Ceylon olive extract, Extracted by using conventional heat reflux extraction method with respect to temperature and time, (A) temperature and solid-to-solvent ratio, (B) temperature and solvent concentration, (C) time and solid-to-solvent ratio, (D) time and solvent concentration, (E) and solid-to-solvent ratio and solvent concentration (F).

4.3.2 OPTIMIZATION OF EXTRACTION FROM CEYLON OLIVE BY MAE METHOD

4.3.2.1 MODEL FITTING

Microwave-assisted extraction optimization is done according to the responses (TPC and DPPH) value obtained from the total 29 setup of experiments done as per the experimental design. Table 4.9 represents the experimental design and their corresponding responses for TPC of the extract obtained from Ceylon olive by MAE. The least square technique was

TABLE 4.8 Optimum Condition of the Variable and Their Corresponding Predicted and Experimental Response Values of Ceylon Olive Extract by Conventional Heat Reflux Extraction Method

Real Process Variable				Response			
				Predicted		**Experimental**	
Temperature (X_1) (°C)	Time (X_2) (min)	Solid-to-Solvent Ratio (X_3) (W/V)	Solvent Concentration (X_4) (%)	TPC (mg GAE/100 g DW)	DPPH (% RSA)	TPC (mg GAE/100 g DW)	DPPH (% RSA)
80.55	133.34	1:37.20	43.23	245.562	86.4166	240.256	85.36

used to calculate the regression coefficients of the intercept, linear, quadratic, and interaction terms of the model (Zhang et al., 2013). These calculations are given in Table 4.10. It was shown that linear solid-to-solvent (X_4) and its quadratic parameter were highly significant at the level of $P < 0.01$, and all other linear and quadratic parameter are insignificant at level $P < 0.01$. The two interactions, irradiation time and solvent concentration, and solid-to-solvent ratio and solvent concentration were also significant at level $P < 0.01$. The final predictive equation (Eq. (4.5)) obtains the predicted value of TPC. Also, for the DPPH value as another response data according to the experimental design obtained are given in Table 4.9. This is also analyzed by a least-square technique to calculate the regression coefficient of intercept, linear, quadratic, and interaction terms. Table 4.11 represents the calculation. It is clearly observed that the only one linear term of solid-to-solvent ratio is significant at level of $P < 0.01$. All other linear, interaction, and quadratic terms are not significant at level of $P < 0.01$. Predictive equation for DPPH is given in Eq. (4.6).

The model for both the responses is significant at level of $P < 0.01$. The F-value of both the responses were 10.67861 and 6.046976. There was only 0.01% chance that this large "Model F-value" could occur due to noise. R^2 values for both the response were 0.9144 and 0.8581, respectively, for TPC and DPPH. It implies that the model was fit with the corresponding response obtained as per the experimental design. The adjusted R^2 is also close to 1 (Karazhiyan et al., 2011). CV% of both the response is low (10.20 and 12.54) which indicates that the deviation from the experimental data to predicted data is less. In general, the adequate precision which measure signal-to-noise ratio, value more than 4 is desirable and this experimental response shows the value higher than 4 (12.867 and 8.478) (Chen et al., 2012).

The polynomial equations developed through the design to correlate the relationship between the independent variable and the responses for MAE method are given in Eqs. (4.3) and (4.4).

$$\text{TPC} = +231.88 - 6.90\,X_1 + 18.30\,X_2 + 54.78\,X_3 - 10.42\,X_4 + 12.0\,X_1X_2 - 7.98\,X_1X_3 + 66.75\,X_1X_4 + 6.03\,X_2X_3 - 5.65\,X_2X_4 - 31.71\,X_3X_4 + 14.41\,X_1^2 - 1.69\,X_2^2 - 32.76\,X_3^2 - 9.80\,X_4^2 \quad (4.3)$$

$$\text{DPPH} = +74.73 + 0.24\,X_1 + 2.53\,X_2 + 12.18\,X_3 - 3.67\,X_4 + 3.40\,X_1X_2 - 5.99\,X_1X_3 + 12.80\,X_1X_4 + 0.71\,X_2X_3 + 6.56\,X_2X_4 - 8.47\,X_3X_4 - 1.61\,X_1^2 - 5.34\,X_2^2 - 15.97\,X_3^2 - 7.23\,X_4^2 \quad (4.4)$$

TABLE 4.9 Values of Experimental Data for the MAE of Ceylon Olive

Experiment No.	Real Process Variable				Experimental Response	
	Microwave Power. (X_1) (W)	Irradiation Time (X_2) (s)	Solid-to-Solvent Ratio (X_3) (mL/g)	Solvent Concentration (X_3) (%)	TPC (mg GAE/100 g DW)	DPPH (% RSA)
1	450	60	30	50	300.3264	74.48
2	450	30	20	50	212.217	69.15
3	450	30	30	60	228.33	58.17
4	600	90	20	60	299.6412	78.36
5	450	60	30	70	202.617	51.84
6	300	60	10	60	180.325	39.36
7	600	60	30	60	236.0439	69.88
8	450	30	20	70	186.504	48.02
9	300	90	20	60	244.7868	59.66
10	450	60	10	70	155.8203	46.36
11	450	90	10	60	135.1425	39.48
12	450	60	20	60	250.367	79.09
13	300	60	20	70	176.507	49.33
14	450	60	20	60	239.5123	72.95
15	450	60	20	60	215.427	81.19
16	450	60	20	60	233.546	76.45

TABLE 4.9 *(Continued)*

Experiment No.	Real Process Variable				Experimental Response	
	Microwave Power. (X_1) (W)	Irradiation Time (X_2) (s)	Solid-to-Solvent Ratio (X_3) (mL/g)	Solvent Concentration (X_3) (%)	TPC (mg GAE/100 g DW)	DPPH (% RSA)
17	600	60	20	50	143.649	49.07
18	450	60	10	50	126.6789	35.11
19	450	30	10	60	125.8218	41.36
20	600	60	20	70	285.9276	66.36
21	450	90	20	50	270.834	70.08
22	300	60	20	50	301.235	83.25
23	300	30	20	60	227.6448	64.85
24	600	30	20	60	234.5016	69.94
25	300	60	30	60	300.169	83.75
26	450	90	20	70	222.5022	75.19
27	450	60	20	60	220.5546	63.95
28	450	90	30	60	261.7569	59.11
29	600	60	10	60	148.1064	49.43

TABLE 4.10 Estimated Regression Coefficient and ANOVA for Response Surface Quadratic Model of TPC of Ceylon Olive Extract by MAE Method

Source	Estimated Coefficients	Sum of Squares	DF	*F* Value	*P*-value
Model		75033.25	14	10.67861	<0.0001
Intercept	231.88	–	–	–	–
X_1-Microwave power	−6.90	571.291	1	1.138273	0.3041
X_2-Irradiation Time	18.30	4020.305	1	8.010287	0.0134
X_3-solid-to-solvent Ratio	54.78	36008.9	1	71.7462	<0.0001
X_4-solvent concentration	−10.42	1303.379	1	2.596928	0.1294
X_1X_2	12.00	575.9424	1	1.147541	0.3022
X_1X_3	−7.98	254.5062	1	0.507093	0.4881
X_1X_4	66.75	17823.13	1	35.51183	<0.0001
X_2X_3	6.03	145.2772	1	0.289459	0.5990
X_2X_4	−5.65	127.9025	1	0.25484	0.6215
X_3X_4	−31.71	4022.781	1	8.01522	0.0133
X_1^2	14.41	1347.74	1	2.685315	0.1235
X_2^2	−1.69	18.57767	1	0.037015	0.8502
X_3^2	−32.76	6961.598	1	13.87069	0.0023
X_4^2	−9.80	623.0133	1	1.241328	0.2840
Lack of Fit		6224.736	10	3.105526	0.1430
Pure Error		801.7626	4		
R^2	0.9144				
Adjusted R^2	0.8287				
Predicted R^2	0.5478				
CV%	10.20				
Adequate Precision	12.867				

TABLE 4.11 Estimated Regression Coefficient and ANOVA for Response Surface Quadratic Model of DPPH of Ceylon Olive Extract by MAE Method

Source	Estimated Coefficients	Sum of Squares	DF	*F* Value	*P*-value Prob > *F*
Model		5155.442	14	6.046976	0.0009
Intercept	74.73				
X_1-microwave power	0.24	0.672133	1	0.011037	0.9178
X_2-irradiation time	2.53	76.96268	1	1.263806	0.2798
X_3-solid-to-solvent ratio	12.18	1779.498	1	29.22119	<0.0001
X_4-solvent concentration	−3.67	161.6268	1	2.654079	0.1256
X_1X_2	3.40	46.30803	1	0.760425	0.3979
X_1X_3	−5.99	143.2809	1	2.35282	0.1473
X_1X_4	12.80	655.616	1	10.76589	0.0055
X_2X_3	0.71	1.9881	1	0.032647	0.8592
X_2X_4	6.56	172.1344	1	2.826624	0.1149
X_3X_4	−8.47	287.133	1	4.71502	0.0476
X_1^2	−1.61	16.76318	1	0.275269	0.6080
X_2^2	−5.34	185.0584	1	3.03885	0.1032
X_3^2	−15.97	1654.596	1	27.17017	0.0001
X_4^2	−7.23	339.3098	1	5.571817	0.0333
Lack of fit		669.4889	10	1.462754	0.3808
Pure error		183.0763	4		
R^2	0.8581				
Adjusted R^2	0.7162				
Predicted R^2	0.3105				
CV%	12.54				
Adequate precision	8.478				

4.3.2.2 INFLUENCES OF INDEPENDENT VARIABLES ON THE RESPONSES FOR MAE OF CEYLON OLIVE

4.3.2.2.1 Total Phenolic Content

The effect of the independent variable is analyzed by 3D plot. The responses are kept at the Z-axis and two independent variables were in *X*-axis and

Y-axis. Interaction of two independent variables is analyzed in one graph and other factors remain constant at their center point.

From the equation it is shown that solid-to-solvent ratio has the maximum positive effect on the TPC. The other positive effect factors were linear function of irradiation time, interaction function of microwave power and irradiation time, microwave power and solvent concentration, Irradiation time and solid-to-solvent ratio, and quadratic function of microwave power. Negative effect functions were linear function of microwave power, solvent concentration, interaction term of microwave power and solid-to-solvent ratio, irradiation time and solvent concentration, solid-to-solvent ratio and solvent concentration, and quadratic terms of irradiation time, solid-to-solvent ratio, and solvent concentration.

The RSM plots analyze the effect of the independent variable on the TPC of the Ceylon olive extracted obtained by MAE process. These 3D RSM plots were microwave power and irradiation time, (1) microwave power and solid-to-solvent ratio, (2) microwave power and solvent concentration, (3) irradiation time and solid-to-solvent ratio, (4) irradiation time and solvent concentration, (5) and solid-to-solvent ratio and solvent concentration (6)

TPC content decreases with the increase in microwave power and it is maximum near 600 W (Ahmad and Langrish, 2012) (Figure 4.3(A)), microwave power and solvent ratio interaction that microwave power have mostly linear effect and increase with solid-to-solvent ratio (Maran et al., 2013) (Figure 4.3(B)). Microwave power and solvent concentration interaction shows that initially it is more TPC content and then decrease and again increase at the maximum value of both the parameter which is also described by Prasad et al. (2011) (Figure 4.3(C)). Irradiation time has linear effect and increase in TPC with increase in solid-to-solvent ratio (Figure 4.3(D)), TPC content increase with the irradiation time but solvent concentration does not has that much effect (Figure 4.3(E)), as the solid-to-solvent ratio increase TPC content increase and solvent concentration has mostly linear effect. (Figure 4.3(F)).

4.3.2.2.2 DPPH Radical Scavenging Activity

Optimization of the DPPH activity was done by the RSM plots generated according to the independent variable. DPPH activity is kept in the Z-axis and two independent variables are kept in the *X*-axis and *Y*-axis. Also rest are kept constant at their center level.

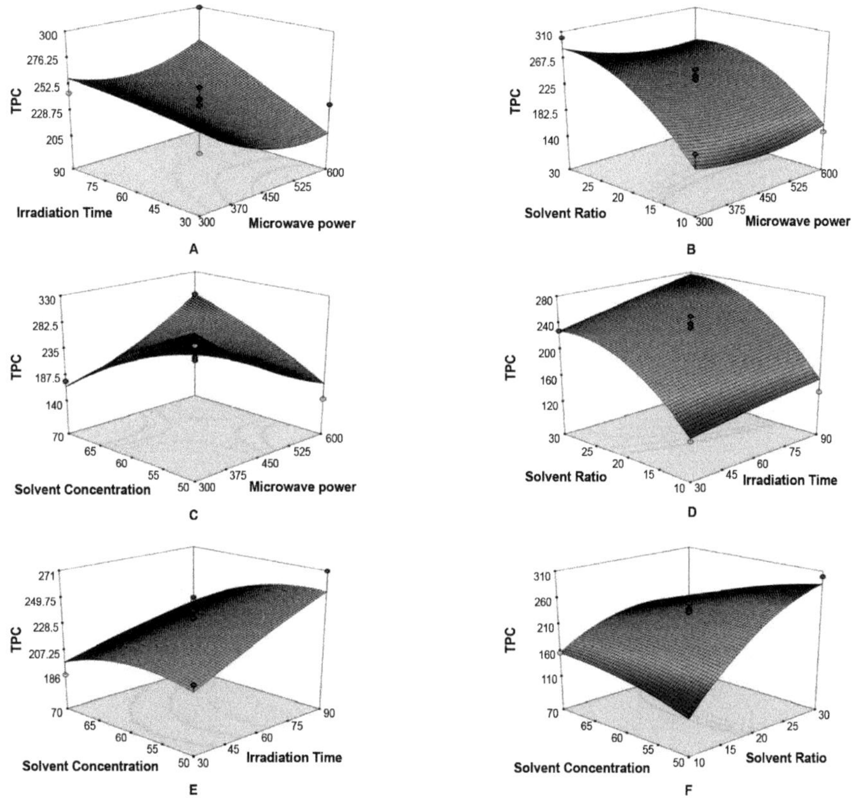

FIGURE 4.3 Response surface analysis for the TPC yield of Ceylon olive extract, Extracted by using MAE with respect to Microwave power and Irradiation time, (A) microwave power and solid-to-solvent ratio, (B) Microwave power and solvent concentration, (C) Irradiation time and solid-to-solvent ratio (D), irradiation time and solvent concentration, (E) and solid-to-solvent ratio and solvent concentration (F).

In equation the maximum positive effect found in the factors were linear function of solid-to-solvent ratio and solvent concentration. Maximum negative effect had linear function of microwave power and irradiation time. The other positive effect functions were interaction of microwave power and irradiation time, microwave power and solvent concentration, irradiation time and solid-to-solvent ratio, and Irradiation time and solvent concentration. The other negative effect functions were interaction term of microwave power and solid-to-solvent ratio, solid-to-solvent ratio and solvent concentration, and all the quadratic term of all four independent variables (microwave power, irradiation time, solid-to-solvent ratio, and solvent concentration).

These 3D plots were microwave power and irradiation time, (1) microwave power and solid-to-solvent ratio, (2) microwave power and solvent concentration, (3) irradiation time and solid-to-solvent ratio, (4) irradiation time and solvent concentration, (5) and solid-to-solvent ratio and solvent concentration (6).

Microwave Power has not much influence but the Irradiation time influence is high on the DPPH radical scavenging activity. Increase as the irradiation time increase and decrease (Figure 4.4(A)), microwave power and solid-to-solvent interaction has much influence. Highly increase in increase in the solid-to-solvent ratio increases and then decrease as the solid-to-solvent ratio decreases. A potential proxidant effect of the polyphenol is indicated by Dai and Mumper (2010) (Figure 4.4(B)), microwave power and solvent concentration interaction have not much influence little decrease at interaction value increase than increase little (Figure 4.4(C)), In microwave power and solid-to-solvent ratio, microwave power has mostly same effect but DPPH scavenging activity increase as increase in solid-to-solvent ratio (Figure 4.4(D)), DPPH activity increase with increase in irradiation time and solvent concentration but at higher temperature little increase observed and then decrease (Figure 4.4(E)), mostly increasing effect with increase in solid-to-solvent ratio increase. Interaction of solid-to-solvent ratio and solvent concentration helps to increase DPPH activity as the value of interaction increase. Finally, it decreases at higher value. (Figure 4.4(E)).

4.3.2.3 OPTIMIZATION OF THE PROCESS VARIABLE FOR EXTRACTION BY MAE METHOD FROM CEYLON OLIVE

Process variable optimization was done by the numerical multi response optimization technique. Main responses are TPC and DPPH radical scavenging activity. Maximum value of TPC and DPPH radical scavenging activity (%) was considered. The optimum level of the process variable at optimized condition was predicted by the software and shown in Table 4.12. The desirability of the optimization result was 0.990.

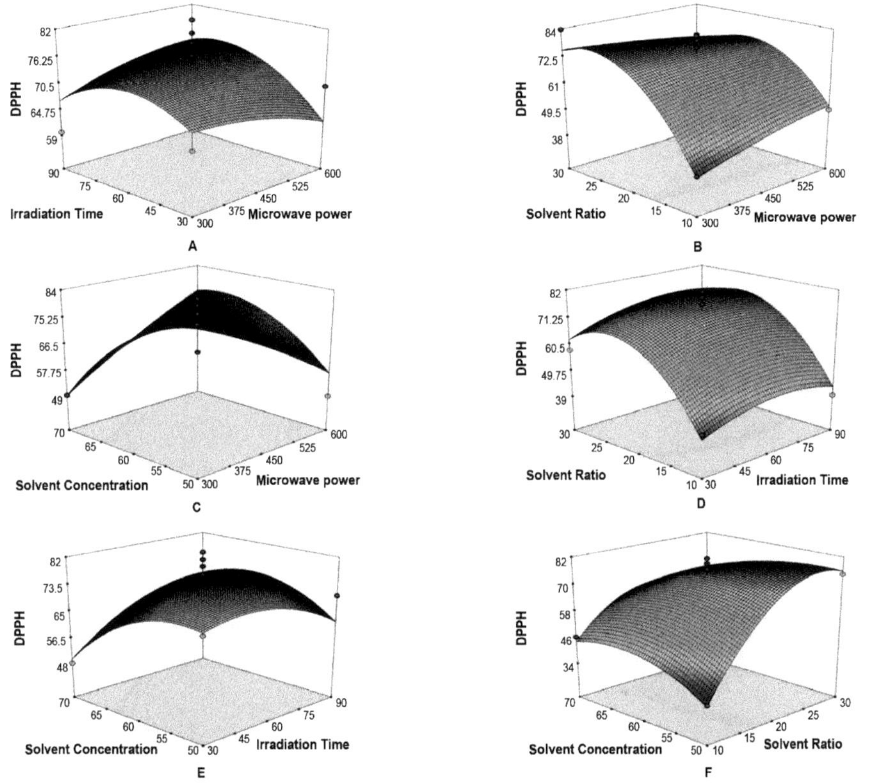

FIGURE 4.4 Response surface analysis for the DPPH radical scavenging activity of Ceylon olive extract, Extracted by using MAE with respect to microwave power and irradiation time, (A) microwave power and solid-to-solvent ratio, (B) microwave power and solvent concentration, (C) irradiation time and solid-to-solvent ratio, (D) irradiation time and solvent concentration, (E) and solid-to-solvent ratio and solvent concentration (F).

4.3.3 COMPARISON OF MAE WITH CONVENTIONAL HEAT REFLUX METHOD

Optimum condition of all the process variables for conventional heat reflux method and MAE method form the Ceylon olive are given in Table 4.13. It is clearly shown that at optimum condition the TPC content of the extract is more in the MAE process. The two main factors that are extraction time and the solid-to-solvent ratio are also less in case of the MAE method. Microwave-assisted extraction method takes only 90 s and 1:20 (approximately) solid-to-solvent ratio to obtain TPC of 315.247 mg GAE/100 g DW of sample. Whereas in conventional heat reflux extraction need 72 min

TABLE 4.12 Optimum Condition of the Variable and Their Corresponding Predicted and Experimental Response Values of Ceylon Olive Extract by MAE Method

Real Process Variable				Response			
				Predicted		Experimental	
Microwave Power (X_1) (W)	Irradiation Time (X_2) (s)	Solid-to-Solvent Ratio (X_3) (mL/g)	Solvent Concentration (X_3) (%)	TPC (mg GAE/100 g DW)	DPPH (% RSA)	TPC (mg GAE/100 g DW)	DPPH (% RSA)
300	58	26.19	58.05	297.086	82.0471	300.125	80.325

TABLE 4.13 Comparison of Extraction Yield of Phenolics and DPPH Scavenging Activity Ceylon Olive by Conventional Heat Reflux Method (CSE), and MAE Method

	Extraction Method	Temp (°C)	Extraction Time	Microwave Power (W)	Solid-to-Solvent Ratio (mL/g)	Ethanol Concentration (%)	TPC (mg GAE/100 g DW)	DPPH (RSA %)
Elaeocarpus serratus	CSE	80.55	133.34 min		37.20	43.23	240.256	85.36
	MAE		58 s	300 W	26.19	58.05	300.125	80.325

(approximately) time, 1:34 (approximately) solid-to-solvent ratio to obtain 239.256 mg GAE/100 g DW of sample. It means MAE is more effective. It is time and solvent saving process. So, we can say that the MAE process is a novel technique and most efficient technique for the extraction compare to the Conventional Heat Reflux System.

4.3.4 HIGH PERFORMANCE CHROMATOGRAPHY ANALYSIS

4.3.4.1 HPLC SPECTRA OF STANDARD

Standards spectra run for the comparison with the extracted sample are Gallic acid (Figure 4.5(A)) and Quercetine (Figure 4.5(B)). Gallic acid peak was found at the time of 9.86 min and the Quercetine standard UV spectra peak was found at the time of 28.43 min. The standards spectra are given in Figure 4.5.

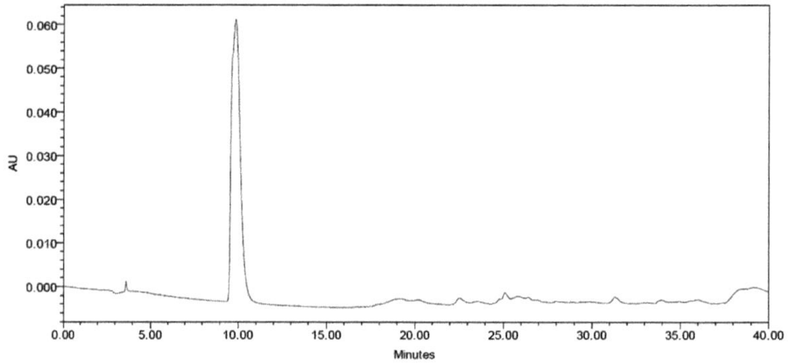

(A)

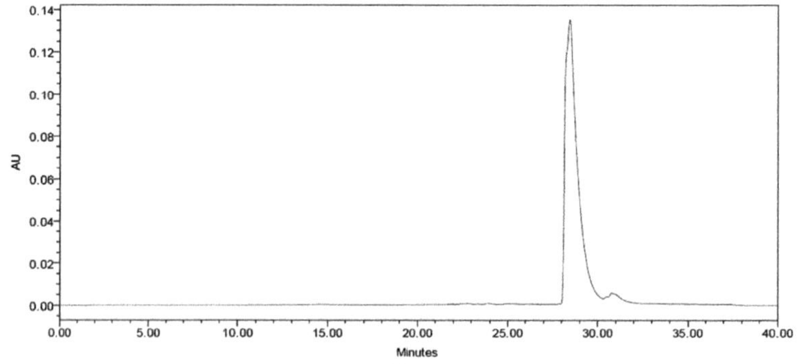

(B)

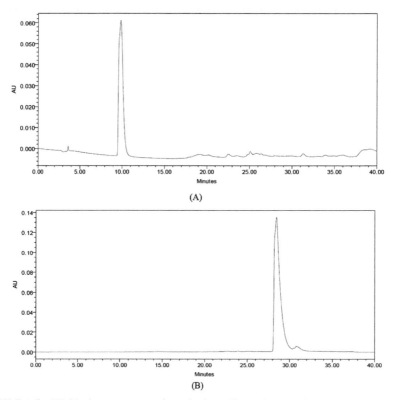

FIGURE 4.5 HPCL chromatogram of standards Gallic acid (A), and Quercetine (B).

4.3.4.2 HPLC OF CEYLON OLIVE EXTRACT

HPLC was done for the Ceylon olive fruit extract in methanol and MAE process for the TPC. And for the detection of the compound, UV detector with 280 nm and the analysis time of 40 min was used to detect the phenolic compound in both the conventional heat reflux extract and microwave-assisted extracted sample. Peaks are analyzed and compared with the standard run at 280 nm and 360 nm run for 40 min. It is found that the peak 1 (Figure 4.6(A)) which is at 9.933 min in conventional heat reflux extract and peak 1 (Figure 4.6(B)) (Romani et al., 2012) which is at 10.05 min in the microwave-assisted extract shows the presence of gallic acid. Detection is compared with the standard spectral data of gallic acid found at time 9.733333 min. One another compound is detected is quercetine in the conventional heat reflux extract at 28.733 min (peak 2, Figure 4.6(A)). It is compared with the standard spectral data for the quercetine found at 28.4333 min.

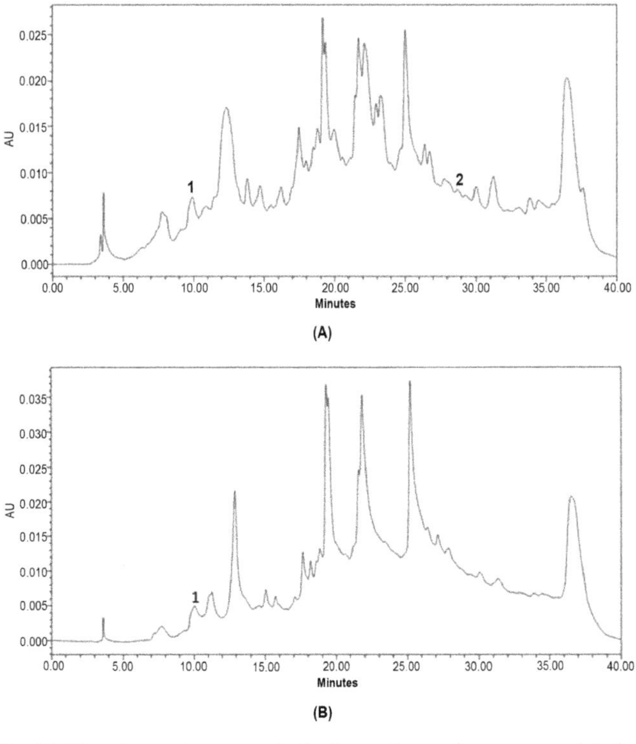

FIGURE 4.6 HPCL chromatogram of Ceylon olive microwave-assisted extract (A), Conventional Heat Reflux extract, (B) compound gallic acid (Peak 1), and quercetin (Peak 2) were found in the sample.

4.4 CONCLUSIONS

This study of the extraction of phenolic content from Ceylon olive was done by using conventional heat reflux and the MAE method. The influence of the four independent parameters (namely temperature, time, solid-to-solvent ratio, and solvent concentration) for the conventional heat reflux method was analyzed. Influences were checked with respect to the TPC and the DPPH radical scavenging activity. The temperature and the solid-to-solvent ratio have the maximum influence. The optimized conditions were temperature 80.55 °C, time 133.34 min, solid-to-solvent ratio 1:37 w/v, and solvent concentration 43.23% ethanol and the experimental value for the responses were TPC 240.256 mg GAE/100g DW and DPPH 85.36% RSA.

Microwave-assisted extraction of Ceylon olive was analyzed with the four independent variables, the independent variables were microwave power, irradiation time, solid-to-solvent ratio, and solvent concentration. Solvent-to-solid ratio and the irradiation time have the maximum influence on the MAE process. The optimized conditions for the process were microwave power 300 W, irradiation time 58 s, solid-to-solvent ratio 1:26.19, and solvent concentration 58.05% ethanol and the experimental response values were TPC 300.125 mg GAE/100 g DW, and DPPH 80.325% RSA. Among conventional heat reflux and MAE process, MAE methods had the maximum efficiency to extract the phenolic compound.

KEYWORDS

- **microwave-assisted extraction**
- **heat reflux extraction**
- **Ceylon olive**
- **phenolic compounds**
- **response surface methodology**
- **HPLC analysis**

REFERENCES

Ahmad, J., & Langrish, T. (2012). Optimisation of total phenolic acids extraction from mandarin peels using microwave energy: The importance of the Maillard reaction. *Journal of Food Engineering, 109*(1), 162–174.

Amorati, R., & Valgimigli, L. (2012). Modulation of the antioxidant activity of phenols by non-covalent interactions. *Organic and Biomolecular Chemistry*, *10*(21), 4147–4158.

Biesaga, M. (2011). Influence of extraction methods on stability of flavonoids. *Journal of Chromatography A*, *1218*(18), 2505–2512.

Bimakr, M., Rahman, R. A., Taip, F. S., Ganjloo, A., Salleh, L. M., Selamat, J., & Zaidul, I. S. M. (2011). Comparison of different extraction methods for the extraction of major bioactive flavonoid compounds from spearmint (*Mentha spicata* L.) leaves. *Food and Bioproducts Processing*, *89*(1), 67–72.

Liyana-Parthirana, C., Shahidi F. (2005). Optimisation of extraction of phenolic compounds from wheat using response surface methodology. Food Chemistry, *93*, 47–56

Chemat F, Abert-Vian M, Zill-e-Huma Y.-J. (2009) Microwave assisted separations: green chemistry in action. In: Pearlman JT (ed) Green chemistry research trends. Nova Science Publishers, *New York, pp 33–62.*

Chen, W., Wang, W., Zhang, H., & Huang, Q. (2012). Optimization of ultrasonicassisted extraction of water-soluble polysaccharides from Boletus edulis mycelia using response surface methodology. *Carbohydrate Polymers*, *87*(1), 614–619.

Dahmoune, F., Spigno, G., Moussi, K., Remini, H., Cherbal, A., & Madani, K. (2014). *Pistacia lentiscus* leaves as a source of phenolic compounds: Microwave-assisted extraction optimized and compared with ultrasound-assisted and conventional solvent extraction. *Industrial Crops and Products*, *61*, 31–40.

Dai, J., Mumper, R.J., 2010. Plant phenolics: extraction, analysis and their antioxidantand anticancer properties. *Molecules, 15*, 7313–7352.

Garcia-Salas, P., Morales-Soto, A., Segura-Carretero, A., & Fernández-Gutiérrez, A. (2010). Phenolic-compound-extraction systems for fruit and vegetable samples. *Molecules*, *15*(12), 8813–8826.

Huang, W. Y., Cai, Y. Z., & Zhang, Y. (2009). Natural phenolic compounds from medicinal herbs and dietary plants: potential use for cancer prevention. *Nutrition and Cancer*, *62*(1), 1–20.

Kalia, K., Sharma, K., Singh, H. P., & Singh, B. (2008). Effects of Extraction Methods on Phenolic Contents and Antioxidant Activity in Aerial Parts of Potentilla atrosanguinea Lodd. and Quantification of Its Phenolic Constituents by RP-HPLC. *Journal of Agricultural and Food Chemistry*, *56*(21), 10129–10134.

Karazhiyan, H., Razavi, S., & Phillips, G. O. (2011). Extraction optimization of a hydrocolloid extract from cress seed (Lepidium sativum) using response surface methodology. *Food Hydrocolloids*, *25*(5), 915–920.

Kaufmann, B., & Christen, P. (2002). Recent extraction techniques for natural products: microwave-assisted extraction and pressurised solvent extraction. *Phytochemical Analysis*, *13*(2), 105–113.

Khoddami, A., Wilkes, M. A., & Roberts, T. H. (2013). Techniques for analysis of plant phenolic compounds. *Molecules*, *18*(2), 2328–2375.

Li, H., Deng, Z., Wu, T., Liu, R., Loewen, S., & Tsao, R. (2012). Microwave-assisted extraction of phenolics with maximal antioxidant activities in tomatoes. *Food Chemistry*, *130*(4), 928–936.

Lu, Y., Ma, W., Hu, R., Dai, X., & Pan, Y. (2008). Ionic liquid-based microwave-assisted extraction of phenolic alkaloids from the medicinal plant *Nelumbo nucifera* Gaertn. *Journal of Chromatography A*, *1208*(1), 42–46.

Mandal, V., Mohan, Y., & Hemalatha, S. (2007). Microwave assisted extraction-an innovative and promising extraction tool for medicinal plant research. *Pharmacognosy Reviews*, *1*(1), 7.

Maran, J. P., Sivakumar, V., Thirugnanasambandham, K., & Sridhar, R. (2013). Optimization of microwave assisted extraction of pectin from orange peel. *Carbohydrate polymers*, *97*(2), 703–709.

Pfundstein, B., El Desouky, S. K., Hull, W. E., Haubner, R., Erben, G., & Owen, R. W. (2010). Polyphenolic compounds in the fruits of Egyptian medicinal plants (Terminalia bellerica, Terminalia chebula and Terminalia horrida): characterization, quantitation and determination of antioxidant capacities. *Phytochemistry*, *71*(10), 1132–1148.

Pompeu, D. R., Silva, E. M., & Rogez, H. (2009). Optimisation of the solvent extraction of phenolic antioxidants from fruits of Euterpe oleracea using Response Surface Methodology. *Bioresource Technology*, *100*(23), 6076–6082.

Prasad, K. N., Hassan, F. A., Yang, B., Kong, K. W., Ramanan, R. N., Azlan, A., & Ismail, A. (2011). Response surface optimisation for the extraction of phenolic compounds and antioxidant capacities of underutilised Mangifera pajang Kosterm. peels. *Food Chemistry*, *128*(4), 1121–1127.

Proestos, C., & Komaitis, M. (2008). Application of microwave-assisted extraction to the fast extraction of plant phenolic compounds. *LWT-Food Science and Technology*, *41*(4), 652–659.

Romani, A., Campo, M., & Pinelli, P. (2012). HPLC/DAD/ESI-MS analyses and anti-radical activity of hydrolyzable tannins from different vegetal species. *Food Chemistry*, *130*(1), 214–221.

Sultana, B., Anwar, F., & Ashraf, M. (2009). Effect of extraction solvent/technique on the antioxidant activity of selected medicinal plant extracts. *Molecules*, *14*(6), 2167–2180.

Zhang, G., Hu, M., He, L., Fu, P., Wang, L., & Zhou, J. (2013). Optimization of microwave-assisted enzymatic extraction of polyphenols from waste peanut shells and evaluation of its antioxidant and antibacterial activities in vitro. *Food and Bioproducts Processing*, *91*(2), 158–168.

CHAPTER 5

Surfactant-Mediated Ultrasound-Assisted Extraction of Phenolic Compounds from *Musa balbisiana Bracts*: Kinetic Study and Phytochemical Profiling

GIRIJA BRAHMA[1], KONA MONDAL[1], ABHINAV MISHRA[2], and AMIT BARON DAS[1,*]

[1]*Department of Food Engineering and Technology, Tezpur University, Napaam, Sonitpur, Assam 784028, India*

[2]*Department of Food Science and Technology, University of Georgia, Athens, GA 30602, USA*

[*]*Corresponding author. E-mail: amitbaran84@gmail.com*

ABSTRACT

In the present work, the surfactant-mediated ultrasound-assisted extraction of total phenolic content, anthocyanin, and antioxidant from *Musa balbisiana* bracts was modeled and optimized. The D-optimal design followed by response surface methodology was used to study the effects of independent variables on dependent variables. The optimum extraction conditions were, solvent concentration of 51.64%, surfactant (Tween-80) concentration of 278.54 ppm, the solvent-to-solid ratio of 13:1, extraction temperature of 38.59 °C, and sonication time of 21 min. At optimal condition, extraction kinetics was estimated and compared with other extraction process using nonexperimental two-parameter equation modified Peleg's model. The surfactant-mediated ultrasound-assisted extraction showed higher extraction rate and yield than ultrasound-assisted and conventional extraction process. The HPLC analysis showed that the extract contained very high amount of Cyanidin-3-glucoside (98.593 mg/L) and Peonidin-3-glucoside (7.491 mg/L).

5.1 INTRODUCTION

Banana (*Musa balbisiana*) is a popular fruit grown widely in tropical and subtropical regions in the world. Banana has valuable agronomic traits such as resistance to diseases and ability to thrive well in drier conditions (Schmidt et al., 2015). The inflorescence, which is one of the by-products of banana, is a unique structure that protrudes out and remains at the terminal part fruit bunch until the fruit matures (Padam et al., 2012). It is a leafy maroon cone with cream florets, which is consumed either cooked or raw by some ethnics of the Asian region. Previous studies revealed that by-products of banana have antibacterial (Mokbel and Hashinaga, 2005), antioxidants (Sulaiman et al., 2011), anticancer (Dahham et al., 2015), and antifungal (Ho et al., 2007) properties. The banana flower is rich in vitamins, flavonoids, and proteins, which has been used in traditional medicine to treat bronchitis, constipation, and ulcer problems (Bhaskar et al., 2011). The banana flower contains high number of phenolic compounds such as flavonoid, phenolic acids that act as a potent antioxidant agent (Loganayaki et al., 2010).

Recently, the use of by-products from fruits and vegetable is increasing steadily as a possible source of bioactive compounds. To separate these biologically active compounds from the by-products needs an effective and eco-friendly extraction technique. Several methods such as cold pressing, heating reflux, soxhlet, and solvent extraction have been widely used to extract bioactive components. However, all these traditional methods have some limitations such as longer processing time, higher environmental risk, lower efficiency, hydrolysis, oxidation, and ionization of the bioactive compounds. To surpass these limitations, ultrasound as an emerging technology is applied to extract bioactive compounds from plant sources (Sharma et al., 2015). The extraction of bioactive compounds with ultrasound can offer higher productivity in shorter time, reduced solvent consumption, lower temperature, and lower energy input (Sun et al., 2007).

Currently, expensive and toxic organic solvents, such as methanol and ethanol are used as extraction solvents. Therefore, a highly green, more accurate, and efficient alternative extraction solvent system is of great interest for the enhancement of recovery rate and health protection. Surfactant-mediated extraction has been recognized as a novel method to extract bioactive compounds (Sharma et al., 2015).

Surfactants are the amphiphilic molecules having hydrophobic and hydrophilic components. These moieties when present at a lower concentration get adsorbed onto the interface or the surface of the system hence altering the interfacial free energies (Sharma et al., 2015). Thus, surfactant can be solubilizing different types of hydrophilic and lipophilic compounds and the possibility of concentrating the solutes which result in high extraction efficiency and product recovery. Recent research has demonstrated the use of surfactant to separate phenolic compounds from different plant sources (Sharma et al., 2015; Sun et al., 2007). However, there is scanty of literature on surfactant-mediated ultrasound-assisted extraction of phenolic compounds. To our knowledge, extraction of phenolic compounds from inflorescence of banana has not been investigated.

Thus, in the present study, phenolic compounds (anthocyanin) were extracted from the bracts of *M. balbisiana* inflorescence using surfactant-mediated ultrasound-assisted extraction process. The parameters of the extraction process were optimized using response surface methodology (RSM). The extraction kinetics of surfactant-mediated ultrasound-assisted extraction process was investigated and compared with other extraction process. The phytochemical and antioxidant properties of the extract were also investigated.

5.2 MATERIALS AND METHODS

5.2.1 RAW MATERIAL

Fresh banana (bheem kol) inflorescence was obtained from the market of Tezpur, Napaam District, Assam, India. The bracts were separated from the inflorescence. The separated bracts were blanched before preparing a paste from it and the then paste was stored at 4 °C for extraction.

5.2.2 CHEMICAL REAGENTS

The tween-80, Folin–Ciocalteu reagent, and 2,2-diphenyl-1-picrylhydrazyl were purchased from Sigma-Aldrich (St. Louis, MO, USA). The ethanol, hydrochloric acid, sodium carbonate, methanol, sodium hydroxide, potassium hydroxide phthalate, and aluminum trichloride were purchased from Sisco Research Laboratories Pvt. Ltd (Mumbai, India).

5.2.3 EXTRACTION

5.2.3.1 SURFACTANT-MEDIATED ULTRASOUND-ASSISTED EXTRACTION

The surfactant-mediated ultrasound-assisted extraction was carried out in an ultrasonic water bath (RK 510 H, 35 kHz, 230 W, Bandelin Sonores). Five extraction variables, namely, extraction temperature (30–70 °C), sonication time (10–60 min), solvent (ethanol) concentration (10%–100%), surfactant (Tween-80) concentration (0–500 ppm), and solvent-to-solid ratio (10–40 mL/g) were considered as shown in Table 5.1. During extraction, the extractant pH was maintained acidic (pH 4) using HCl (0.1N). After extraction, the extract was passed through Whatman no. 2 filter and the filtrate was stored at −18 °C for further analysis.

5.2.3.2 ULTRASOUND-ASSISTED EXTRACTION (UAE)

The UAE was conducted under optimum condition of surfactant-mediated ultrasound-assisted extraction process. The experiment was conducted in an ultrasonic water bath (RK 510 H, 35 kHz, 230 W, Bandelin Sonores) having temperature and timer controller. After extraction, the extract was filtrated through Whatman no. 2 filter and was stored at −18 °C for further use.

5.2.3.3 CONVENTIONAL EXTRACTION

The conventional extraction of anthocyanin and total phenolic content (TPC) from banana bracts was carried out under optimal condition of surfactant-mediated ultrasound-assisted extraction process. The extraction was carried out in a time and temperature control water bath (OVFU O-SRWB2, India). The extraction was carried out with solvent concentration of 51.64%, solvent-to-solid ratio of 13:1, extraction temperature of 38.59 °C, and extraction time of 21 min. Then, the extract was filtrated and collected for feature analysis.

5.2.4 ANALYSIS

5.2.4.1 TOTAL PHENOLIC CONTENT

The TPC of the banana inflorescence extracts was determined using Folin–Ciocalteu assay (Deng et al., 2017) with some modifications in the process.

TABLE 5.1 Experimental Design for Extraction Process Based on D-Optimal Design

Run No.	Temperature (°C)	Time (min)	Solvent Concentration (%)	Surfactant (ppm)	Solvent/Solid (mL/g)	TPC (mg GAE/100 g)	Anthocyanin (mg C3G/ 100 g)	Antioxidant Activity (%)
1	49.69	35.26	10	500	10	53.27	0.406	39.42
2	70	35.4	56.78	241.52	25.58	37.72	0.816	30.24
3	51.02	60	55.15	0	25.06	39.54	0.909	33.27
4	51.02	60	55.15	0	25.06	37.54	0.827	31.09
5	70	60	10	500	40	23.63	0.08001	26.47
6	49.69	35.26	10	500	10	51.22	0.403	38.33
7	30	10	10	500	40	24	0.0834	10.59
8	30	60	10	500	17.6	51.09	0.1335	26.35
9	70	60	100	500	10	21.18	0.0208	21.64
10	53.98	10	10	266.08	23.67	40.57	1.201	27.35
11	30	60	100	0	10	49.09	0.269	30.66
12	30	10	100	500	40	31.36	0.484	21.51
13	70	10	100	0	10	49.9	0.886	33.39
14	70	60	100	0	40	16.36	0.0184	20.57
15	30	10	100	0	40	23.18	0.0951	18.51
16	30	10	100	500	10	44.09	0.897	60.03
17	30	11.56	12.81	0	40	19.54	0.0166	11.52
18	30	10	100	0	10	41.27	0.807	55.56
19	70	10	10	0	40	32.09	0.05	18.37
20	70	10	100	500	40	38.36	0.0651	24.52
21	30	60	10	0	40	38.18	0.0766	21.64
22	30	60	100	500	40	45.09	0.079	24.16
23	70	60	10	0	10	26.36	0.116	24.58
24	30	10	10	0	10	40.18	0.659	41.37
25	43.13	60	49.61	274.23	10	58.81	1.467	49.38
26	70	10	26.86	500	10	40.45	0.0333	41.21
27	43.13	60	49.61	274.23	10	57.07	1.401	47.61
28	70	60	100	500	40	19.36	0.0554	22.38
29	69.38	10	100	0	40	30.9	0.0971	22.22
30	70	35.4	56.78	241.52	25.58	36.13	0.974	28.67
31	53.98	10	10	266.08	23.67	42.34	1.17	25.77

Briefly, 20 μL of extract was taken and mixed with 1.58 mL of distilled water followed by 100 μL Folin–Ciocalteu reagent. After 8 min, 300 μL of Na_2CO_3 (20%) was added to the solution and vortexed immediately. The samples were then incubated for 30 min at 40 °C. The absorbance was measured at 765 nm in UV–Vis spectrophotometer (Spectra scan UV-2600, Thermo Fisher Scientific, Massachusetts, USA) and the phenolic content was expressed in mg gallic acid/100 g.

5.2.4.2 MONOMERIC ANTHOCYANIN

The anthocyanin content of the extracts was determined using the pH differential method (Celli et al., 2015). Absorbance was measured at 515 and 700 nm using a UV–Vis spectrophotometer (Spectra scan UV-2600, Thermo Fisher Scientific, Waltham, MA, USA). In order to obtain the absorbance (A) related to the total anthocyanins the following equation was used:

$$A = (A_{515} - A_{700})_{pH1.0} - (A_{515} - A_{700})_{\text{pH4.5}} \tag{5.1}$$

Considering the Beer–Lambert law the concentration of monomeric anthocyanin (mg/L) was calculated according to Eq. (5.2):

$$\text{Anthocyanin content}\left(\frac{\text{mg}}{L}\right) = \frac{\text{A}\times\text{MW}\times\text{DF}\times 1000}{\varepsilon\times 1} \tag{5.2}$$

where DF is dilution factor, MW is cyaniding-3-glucoside molecular weight (449.2), and ε is molar absorptivity (26,900).

5.2.4.3 2,2-DIPHENYL-1-PICRYLHYDRAZYL (DPPH) RADICAL SCAVENGING ACTIVITY

DPPH radical scavenging activity of banana inflorescence was evaluated according to Sharma et al. (2016). Initially, 100 μL of the extract was taken from each sample separately, and 1.4 mL of DPPH radical methanolic solution (10^{-4} M) was added. The solution was incubated for 30 min and then absorbance was measured at 517 nm using a UV–Vis spectrophotometer (Spectra scan UV-2600, Thermo Fisher Scientific, Waltham, MA, USA). The percentage of radical scavenging activity was calculated using the equation:

$$\text{Radical scavanging activity (\%)} = \frac{A_o - A_s}{A_o} \times 100 \quad (3)$$

where A_o is the absorbance of control blank and A_s is absorbance of sample extract.

5.2.4.4 CYANIDIN-3-GLUCOSIDE (C3G) AND PEONIDIN-3-GLUCOSIDE (P3G)

For determination of C3G and P3G in the extract, the sample was initially washed to remove nonpolar compounds with hexane in a separating funnel. For detection and estimation of C3G and P3G content in bracts extract, an HPLC Ultimate 3000 Liquid Chromatography Systems with a UV detector was used. For the estimation, a C_{18} having 4.6 mm diameter and 250 mm length was used. The mobile phase was a mixture of water, methanol, and formic acid with a ratio of 75:20:5 v/v. The flow rate was 0.5 mL/min. For analysis 20 μL of sample was injected at 25 °C and the wavelength was 530 nm. The C3G and P3G were identified with respect to standard and data were analyzed using Chameleon ver. 6.80 software.

5.2.5 DESIGN OF EXPERIMENT

In the present study, D-optimal design followed by RSM was applied to investigate the effect of independent variable on surfactant-mediated ultrasounds-assisted extraction process. In the D-optimal design five process parameters viz., extraction temperature (X_1), extraction time (X_2), solvent concentration (X_3), surfactant concentration (X_4), and solvent-to-solid ratio (X_5), were used as the independent variable. Total 31 experiments were conducted as shown in Table 5.1. After analyzing the variance of dependent parameters a polynomial quadratic equation was developed as follows:

$$y = \beta_0 + \sum_{i=1}^{n} \beta_i x_i + \sum_{i=1}^{n} \beta_{ii} x^2_{ii} + \sum_{i=1}^{n} \sum_{j=i+1}^{n} \beta_{ij} x_i x_J + \varepsilon \quad (5.4)$$

The coefficients of the polynomial equation were β_0 (constant), β_i (linear effects), β_{ii} (quadratic effects), and β_{ij} (interaction effects). x_i and x_j were the coded independent variables.

5.2.6 EXTRACTION KINETIC

The kinetics of the extraction process was investigated. The extraction kinetics was described by the Peleg's model as

$$C(t) = C_o + \frac{t}{K_1 + K_2 t} \tag{5.5}$$

where *C(t)* is the concentration of anthocyanin (mg C3G/100 g) at time *t*, *t* is the extraction time (min), C_0 is the initial concentration of TPC or anthocyanin at time $t = 0$, K_1 is Peleg's rate constant, and K_2 is Peleg's capacity constant. Since C_0 in all experimental runs was zero, Eq. (5.5) was used in the final form

$$C(t) = \frac{t}{K_1 + K_2 t} \tag{5.6}$$

5.2.7 STATISTICAL ANALYSIS

The D-optimal design was used for experiment through Design Expert 7.0 software (State-Ease lnc. Minneapolis, MN, USA). Significant thresholds at $p < 0.05$ were selected for all statistical tests.

5.3 RESULT AND DISCUSSION

5.3.1 MODELING AND RESPONSE SURFACE ANALYSIS FOR TPC

From the ANOVA analysis (Table 5.2), it was observed that the linear term in TPC model is highly significant ($p \geq 0.05$). In addition, some of the interaction terms and quadratic terms are also significant at level ($p \geq 0.05$). For TPC model (Eq. (5.7)), the coefficient of determination was (R^2) 0.98, which implied the adequacy of the applied regression model. The lack of fit was found to be insignificant, which signifies that the developed model can efficiently establish the relationship between dependent and independent parameter.

$$y_{TPC} = 5035 - 4.91x_1 - 054x_2 + 0.24x_3 + 1.26x_4 - 5.43x_5 - 7.92x_1x_2 - 0.044x_1x_3 - 1.47x_1x_4 + 1.76x_1x_5 \\ - 2.01x_2x_3 + 0.66x_2x_4 + 1.52x_2x_5 + 0.57x_3x_4 - 0.031x_3x_5 + 1.85x_4x_5 - 8.19x_1^2 - 4.83x_2^2 - 2.98x_3^2 - 3.88x_4^2 + 4.54_5^2 \quad (5.7)$$

To analyze the interactive effects of the independent variables and their collective interaction on the total phenolic content, three-dimensional response surface profiles of nonlinear regression model were plotted as shown in Figure 5.1. The effect of extraction temperature, solvent concentration, and their interaction is shown in Figure 5.1a. An increase in phenolic content in the extract was observed with increase in extraction temperature and solvent concentration. However, the reverse trend was observed after reaching a temperature of 45 °C and solvent concentration of 50%. These results are in agreement with the observation of Wong Paz et al. (2015). It was suggested that due to the presence of heat during extraction, the plant tissue became soft and thereby weaken the interactions between phenolic compounds, protein, and polysaccharide, leading to more phenolic compounds diffusion into the extractant. However, beyond 45 °C, there was a sharp decrease in TPC, which might be due to the oxidative degradation of TPC in the extract at elevated temperature. With the increasing solvent concentration, the polarity of the mixture may change, resulting in an increase in the solubility of phenolic compounds in the extractant and hence, there is an increase in the phenolic yield (Ghafoor et al., 2009). However, for the further increase in solvent concentration in the extractant, polarity changes such that the solubility of phenolic compounds decreases thereby decreasing the yield.

TABLE 5.2 ANOVA for Quadratic Model for Responses

Source	Total Phenolic Content (mg GAE/100 g)	Anthocyanin (mg/100 g)	DPPH Scavenging Activity (%)
Model	<0.0001	<0.0001	<0.0001
x_1-Temperature	<0.0001	0.0031	0.0001
x_2-Time	0.0377	0.0004	0.0002
x_3-Solvent	0.0252	0.0021	0.0095
x_4-Surfactant	0.0172	0.0386	0.0087
x_5-Solvent/Solid	<0.0001	0.0001	<0.0001
x_1x_2	<0.0001	0.0599	0.9057
x_1x_3	0.9285	0.2152	0.0035
x_1x_4	0.0112	0.0020	0.3638
x_1x_5	0.0043	0.0496	<0.0001

TABLE 5.2 *(Continued)*

Source	Total Phenolic Content (mg GAE/100 g)	Anthocyanin (mg/100 g)	DPPH Scavenging Activity (%)
x_2x_3	0.0015	0.0023	<0.0001
x_2x_4	0.185	0.3247	0.9283
x_2x_5	0.0082	0.0009	<0.0001
x_3x_4	0.2537	0.0218	0.0186
x_3x_5	0.9499	0.0045	0.5976
x_4x_5	0.003	0.0028	0.0639
x_1^2	<0.0001	<0.0001	0.0011
x_2^2	0.0048	0.2289	0.3673
x_3^2	0.0545	0.0028	<0.0001
x_4^2	0.0161	<0.0001	0.0637
x_5^2	0.0045	0.9832	<0.0001
Lack of Fit	0.0813	0.100	0.0728
R^2	0.98	0.97	0.99

The effect of sonication time and surfactant concentration on TPC is shown in Figure 5.1b. It was elucidated from Figure 5.1b that the TPC in the extract was increased with increase in the sonication time and surfactant concentration in the extractant but the trend was reversed after sonication time of 35 min and surfactant concentration of 280 ppm. Dranca and Oroian (2016) also observed a similar type of effect of time on the extraction of TPC from eggplant peel. This effect was understandable because it was already demonstrated that an extended extraction time enhances the mass transfer during extraction of phenolic compounds. During sonication, the proliferation of ultrasonic waves through the extractant and resulting cavitation exerted high shear forces on the plant material, which enhanced the eddy and internal diffusion. However, it was also reported that prolonged extraction had a detrimental effect on TPC. For prolong sonication, phenolic compounds may be volatile due to the heat generation during sonication (Tiwari et al., 2010) and some unwanted reactions namely enzymatic degradation, and oxidation. The extraction of TPC yield was increased up to 280 ppm of surfactant concentration in the extractant system. Similar observations were reported by Sharma et al. (2015) during extraction of TPC from fruit juice.

This may be due to surfactant-based assemblies tend to form a complex with both polar and nonpolar (hydrophobic and hydrophilic) components of TPC because of their amphiphilic nature and hence an enhance extraction was achieved (Rosen and Kunjappu, 2012). However, after 280 ppm concentration of Tween-80, the TPC yield was decreased which might be due to the increase of extractant viscosity that inhibits the diffusion of TPC from bracts to solvent (Sharma et al., 2015).

It can be observed (Figure 5.1c) that for the increase of solvent-to-solid ratio up to a certain limit there was no change in the TPC yield. With further increase in the solvent-to-solid ratio, there was a sharp decrease in the TPC yield. It may be attributed to the high solvent-to-solid ratio; the high amount of solvent in the extraction system would not change the driving force anymore as the mass transfer is more confined into the solid interior and thereby decrease the recovery of TPC (Zhang et al., 2008).

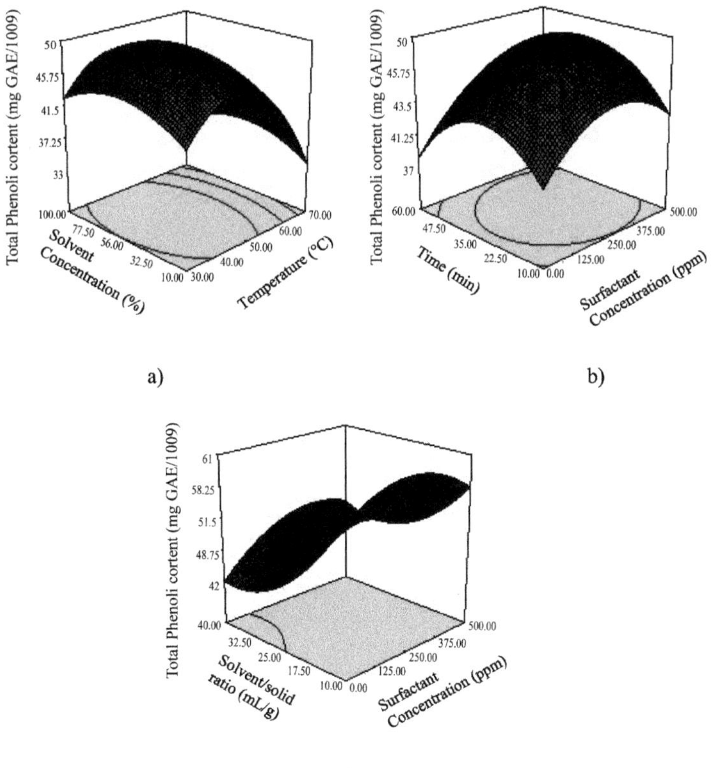

FIGURE 5.1 Response surface for the effect of process variables on total phenolic content.

5.3.2 MODELING AND RESPONSE SURFACE ANALYSIS OF ANTHOCYANIN CONTENT

The ANOVA (Table 5.2) of anthocyanin content demonstrated that model, linear, interaction, and quadratic terms are significant at level $p \geq 0.05$. The following predicted equation (Eq. (5.8)) was obtained after ANOVA at $p \geq 0.05$ significant level:

$$\begin{aligned} y_{an} &= 1.48 - 0.080x_1 - 0.10x_2 + 0.087x_3 - 0.047x_4 - 0.13x_5 + 0.044x_1x_2 + 0.028x_1x_3 + 0.088x_1x_4 \\ &+0.048x_1x_5 - 0.084x_2x_3 + 0.021x_2x_4 + 0.096x_2x_5 + 0.057x_3x_4 - 0.079x_3x_5 + 0.083x_4x_5 - 0.51x_1^2 \\ &+0.077x_2^2 - 0.24x_3^2 - 0.61x_4^2 + 1.19\times10^{-3}x_5^2 \end{aligned} \tag{5.8}$$

The coefficient of determination (R^2) for the model was 0.97, which suggested a high degree of correlation between the observed and predicted values. The lack of fit was found to be insignificant, signifies that the developed relationship was suitable for the experimental data and shows good correlations between variables. Thus, it can be assumed that the selected model can be used for the optimization of extraction process variables in terms of anthocyanin content in the extract.

Figure 5.2a shows the effects of solvent concentration and extraction temperature on anthocyanin content in extract. With the increase of solvent concentration up to 50%, a sharp increase in the anthocyanin yield from bracts was observed, whereas, after 50%, revers trend was observed, similar as TPC. For the polarity change of extractant the anthocyanin diffusivity from bracts effected and causes a reduction in anthocyanin extraction from the bracts (Mokrani and Madani, 2016). Likewise, for the increase in extraction temperature up to 45 °C, anthocyanin content in bracts extract was increased. It may be due to the diffusion of a higher number of phenolic compounds into the extractant during heating. During the heating, the banana bracts become soft and thereby interactions of phenol with other components in the banana bracts become weak, thus, more polyphenols are diffused into the solvent. However, increase in the temperature beyond 45 °C results in a decrease in the anthocyanin content in the extract. Anthocyanin is a heat liable bioactive compound, thereby at a higher temperature, anthocyanin undergoes the oxidation reaction and cause degradation (Ghafoor et al., 2009).

The effects of sonication time and surfactant concentration are shown in Figure 5.2b. Initially, the anthocyanin yield from bracts was constant and then a decreasing trend was observed with the increasing sonication time. The decrease in anthocyanin content is due to the degradation because of various sonochemical reactions such as free radical generation, polymerization/

depolymerization, and some other reactions, which can modify the chemical structure and thereby degrade the anthocyanin in the extract (Tiwari et al., 2009). From Figure 5.2b, it was observed that anthocyanin increases with increase in surfactant concentration until 280 ppm and decreases with further increase in surfactant concentration in extractant. The high hydrophilic–lipophilic balance value of Tween-80 alters the interfacial free energies of the interface and thereby increases the water solubility and dissolving capacity of anthocyanin. On the other hand, after a certain concentration of surfactant in extractant, due to the higher viscosity that creates a difficulty for transferability of the anthocyanin from banana bracts to surfactant phase. Figure 5.2c illustrates the effect of solvent-to-solid ratio on the extraction of anthocyanin from banana bracts. It was observed that there was a decreasing trend of anthocyanin due to increasing the solvent-to-solid ratio. Similar kinds of results were reported by Santos et al. (2010) during the extraction of antioxidant from Jabuticaba. Most likely, the higher amount of liquid would not change the driving force anymore as the mass transfer is more confined to the solid interior thereby decrease the anthocyanin recovery from banana bracts.

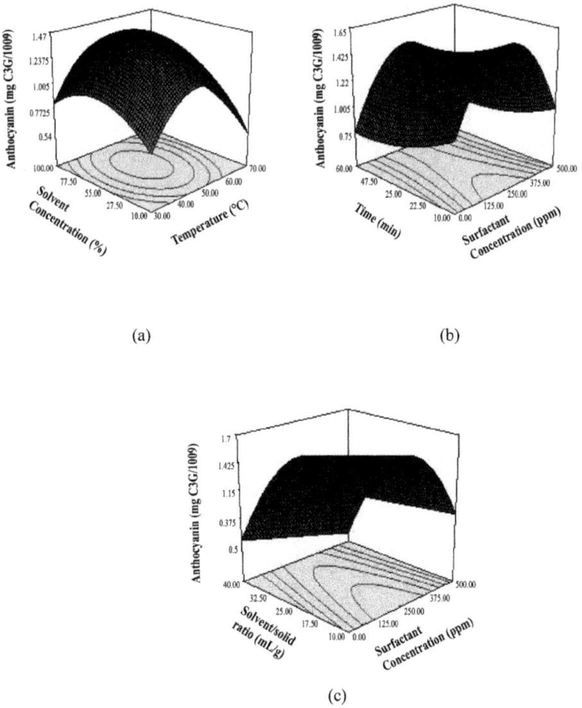

FIGURE 5.2 Response surface for the effect of process variables on anthocyanin content.

5.3.3 MODELING AND RESPONSE SURFACE ANALYSIS FOR 2, 2-DIPHENYL-1-PICRYLHYDRAZYL

From Table 5.2, it can be observed that DPPH activity demonstrated that developed model, and its linear, interaction, and quadratic terms are significant at level of $p \geq 0.05$. The following final predicted model (Eq. (5.9)) was obtained at level $p \geq 0.05$:

$$y_{anti} = 37.31 - 2.55x_1 - 2.28x_2 + 1.41x_3 + 1.34x_4 - 7.58x_5 + 0.053x_1x_2 - 1.69x_1x_3 - 0.42x_1x_4 + 4.76x_1x_5 - 2.98x_2x_3 - 0.040x_2x_4 + 4.97x_2x_5 + 1.24x_3x_4 - 0.25x_3x_5 - 0.92x_4x_5 - 5.14x_1^2 + 1.18x_2^2 - 10.54x_3^2 - 2.61x_4^2 + 7.75x_5^2 \quad (5.9)$$

From the determination of coefficient ($R^2 = 0.99$) for DPPH scavenging activity, it can be suggested that the developed model can efficiently correlate the dependent and independent variable. The lack of fit was found to be insignificant corresponding to pure error ($p \geq 0.05$) which implies that the develop model can efficiently depicted the relationship between variables.

The effects of solvent concentration and extraction temperature on antioxidant activity are shown in Figure 5.3a. For the increase in solvent concentration in the extractant up to 50%, a sharp increase in antioxidant activity was noticed and the further increase in solvent concentration resulted in a sharp decrease in antioxidant activity. These phenomena can be understood with the help of the relationship between extractant polarities with the solubility of the targeted compound. Antioxidant activity was observed to increase with the increase in extraction temperature and solvent concentration but when these reached to 45 °C and 50%, respectively, the activity of antioxidant was found to degrade abruptly. The increase in temperature up to a certain limit increases the TPC and anthocyanin content in the extract, thus there was an increase in antioxidant activity of the extract. Moreover, Ashokkumar et al. (2008) suggested that the antioxidant activity of components might increase because of the increase in the degree of hydroxylation of molecules due to radicals OH• formed in ultrasound processing. However, after 45 °C, due to the oxidation and other degradation, there was a decrease in TPC and anthocyanin, thereby a sharp decline of antioxidant activity of the bracts extract.

Figure 5.3b illustrated the effects of sonication time and surfactant concentration on antioxidant activity of bracts extract. Similar to anthocyanin, the antioxidant activity decreased with an increase of sonication time. The decrease in antioxidant activity is due to the degradation of TPC and anthocyanin because of various sonochemical reactions during the prolonged sonication time. From Figure 5.3b, it was also observed that with the increase in surfactant concentration until 280 ppm, antioxidant activity increased and decreased with further increase in surfactant concentration. Initially, tween 80 increased the water solubility and dissolving capacity of TPC and anthocyanin due to the hydrophilic–lipophilic activity and thereafter for the increase in extractant viscosity, the transferability of anthocyanin and TPC decreased, thus decreasing the antioxidant activity. From Figure 5.3c, it can be observed that with the increasing solvent-to-solid ratio, there was a decline in the antioxidant activity. It is due to decrease in TPC and anthocyanin content in the bract extract.

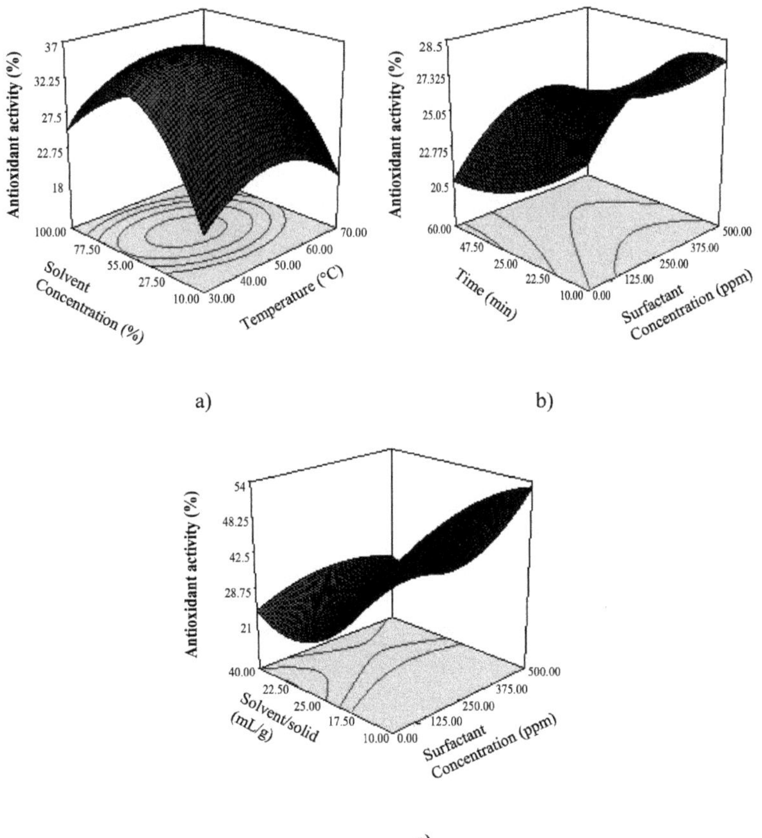

FIGURE 5.3 Response surface for the effect of process variables on DPPH scavenging activity.

5.3.4 OPTIMIZATION AND VALIDATION

In this step, the surfactant-mediated ultrasonic-assisted extraction conditions was optimized for high TPC, anthocyanin content, and DPPH scavenging activity of the extract. It was observed that the predicted model, namely, 7, 8, and 9 can efficiently predict the TPC, anthocyanin content, and DPPH scavenging activity of the surfactant-mediated ultrasonic-assisted extract from banana bracts. Thus, to optimize the surfactant-mediated ultrasonic-assisted extraction process variables, the predicted models were used. Optimum conditions were selected on the basis of highest desirability. The maximum desirability was 0.99 and the optimum condition was: solvent concentration of 51.64%, surfactant concentration of 278.54 ppm, the solvent-to-solid ratio of 13:1, extraction temperature of 38.59 °C, and sonication time of 21 min. For the optimum conditions, the extract contained 58.81 mg GAE/100 g total phenolic contents, 1.66 mg C3G/100 g anthocyanin content and antioxidant activity of 60.03%. For validation, the experiment was conducted at the optimal process conditions and it was observed that the extract contented 61.01 mg GAE/100 g TPC, 1.58 mg C3G/100 g anthocyanin and antioxidant activity of 63.15%, which does not have a significant difference with predicted results. Hence, the developed models were found to effectively optimize the extraction process.

5.3.5 EXTRACTION KINETIC AND COMPARISON

In the present study, Peleg's model was used to investigate the kinetics of surfactant-mediated ultrasound-assisted extraction of TPC, anthocyanin, and antioxidant from banana bracts. The Peleg's parameters (K_1 and K_2), and R^2 are shown in Table 5.3. In general, a higher K_1 value indicates a faster rate of the extraction process, while the lower K_2 value indicates maximum yield (Ghafoor et al., 2014). In surfactant-mediated ultrasound-assisted extraction Peleg's model showed a better fit to the experimental data in terms of R^2 value as 0.97, 0.96, and 0.98, respectively. From Table 5.3, it could also be implied that surfactant-mediated ultrasound-assisted extraction had the higher yield of TPC, anthocyanin, and antioxidant than the ultrasound-assisted and conventional extraction. From Table 5.3, it can be observed that the surfactant-mediated ultrasound-assisted extraction had highest K_1 and lowest K_2 which elucidated that surfactant-mediated ultrasound-assisted extraction had the higher rate of extraction and yield in terms of TPC and anthocyanin from banana bracts.

TABLE 5.3 Values of Peleg's Constants (K_1 and K_2) and Correlation Coefficient (R^2)

Extraction	Anthocyanin			TPC			Antioxidant		
	K_1	K_2	R^2	K_1	K_2	R^2	K_1	K_2	R^2
Surfactant-Mediated UAE	0.027	0.001	0.960	0.063	0.003	0.970	0.051	0.004	0.980
UAE	0.021	0.007	0.940	0.049	0.009	0.981	0.043	0.008	0.973
Conventional Extraction	0.014	0.015	0.933	0.025	0.019	0.961	0.028	0.018	0.935

In conventional extraction, the driving force is the concentration gradient and solubility of the targeted compound in extractant. Whereas in ultrasound-assisted extraction, due to the proliferation of ultrasonic waves and resulting cavitation, there is an enhancement of extraction yield. The ultrasonic wave produces high shear on plant material and thereby the mass transfer from banana bracts has been increased. The ultrasonic wave also causes breakdown of particle which increases the surface area further increasing mass transfer (Chemat et al., 2017). On the other side, the collapse of cavitation bubbles generates macroturbulence that causes interparticle collisions and agitation in microporous particles of the bracts, which further increases the eddy and internal diffusion (Chemat et al., 2017; Tao et al., 2014). In surfactant-mediated ultrasound-assisted extraction, in addition to the ultrasound effect, surfactant alters the interfacial free energies and thereby increases the water solubility and dissolving capacity of the targeted compounds, thus increasing the yield (Rosen and Kunjappu, 2012).

5.3.6 PHYTOCHEMICAL PROFILING

The phytochemical profile of the banana bracts extract was investigated in terms of total phenolic content, anthocyanin, cyaniding-3-glucoside, peonidin-3-glucoside, and antioxidant activity. The extract was contented 58.81 mg GAE/100 g of TPC, 1.612 mg C3G/100 g anthocyanin, and 60.3% antioxidant activity. The extract also contented very high amount of cyanidin-3-glucoside (98.593 mg/L) and peonidin-3-glucoside (7.491 mg/L). The present study confirms that banana bracts could be a potential source of natural phytochemical, and surfactant-assisted ultrasound extraction could be an effective method for the extraction of phenolic compounds from plant materials.

5.4 CONCLUSION

This research illustrated the effect of surfactant-mediated ultrasonic-assisted extraction of TPC, anthocyanin, and antioxidant from banana bracts. The effects of various independent parameters such as extraction temperature, extraction time, solvent concentration, surfactant concentration, and solvent-to-solid ratio on extraction were investigated. The extraction process was optimized using the D-optimal design and the optimum conditions were solvent concentration of 51.64%, surfactant concentration of 278.54 ppm, solvent-to-solid ratio of 13:1, extraction temperature of 38.59 °C, and sonication time of 21 min. The extraction kinetics was investigated using the modified Peleg's model and it was found that the surfactant-mediated ultrasonic-assisted extraction had the highest extraction rate and maximum yield than the conventional extraction processes. The phytochemical profile of the extract was analyzed and it had high amount of total phenolic content, anthocyanin, cyaniding-3-glucoside, and peonidin-3-glucoside. The present study confirms that banana bracts could be a potential source of natural phytochemical and surfactant-assisted ultrasound-extraction could be an effective method for the extraction of phenolic compounds from plant materials.

ACKNOWLEDGMENT

The authors are grateful to the Tezpur University, Tezpur, Assam for providing the analytical facility.

KEYWORDS

- ***Musa balbisiana***
- **surfactant**
- **Peleg's model**
- **total phenolic content**
- **antioxidant**

REFERENCES

Ashokkumar, M.; Sunartio, D.; Kentish, S.; Mawson, R.; Simons, L.; Vilkhu, K.; Versteeg, C., Modification of food ingredients by ultrasound to improve functionality: A preliminary study on a model system. *Innovative Food Science & Emerging Technologies* **2008**, *9* (2), 155–160.

Bhaskar, J. J.; Shobha, M. S.; Sambaiah, K.; Salimath, P. V., Beneficial effects of banana (Musa sp. var. elakki bale) flower and pseudostem on hyperglycemia and advanced glycation end-products (AGEs) in streptozotocin-induced diabetic rats. *Journal of Physiology and Biochemistry* **2011**, *67* (3), 415–425.

Celli, G. B.; Ghanem, A.; Brooks, M. S.-L., Optimization of ultrasound-assisted extraction of anthocyanins from haskap berries (*Lonicera caerulea* L.) using Response Surface Methodology. *Ultrasonics Sonochemistry* **2015**, *27*, 449–455.

Chemat, F.; Rombaut, N.; Sicaire, A. G.; Meullemiestre, A.; Fabiano-Tixier, A. S.; Abert-Vian, M., Ultrasound assisted extraction of food and natural products. Mechanisms, techniques, combinations, protocols and applications. A review. *Ultrasonics Sonochemistry* **2017**, *34*, 540–560.

Dahham, S. S.; Mohamad, T.; Tabana, Y. M.; Majid, A., Antioxidant activities and anticancer screening of extracts from banana fruit (Musa sapientum). *Academic Journal of Cancer Research* **2015**, *8*, 28–34.

Deng, J.; Xu, Z.; Xiang, C.; Liu, J.; Zhou, L.; Li, T.; Yang, Z.; Ding, C., Comparative evaluation of maceration and ultrasonic-assisted extraction of phenolic compounds from fresh olives. *Ultrasonics Sonochemistry* **2017**, *37*, 328–334.

Dranca, F.; Oroian, M., Optimization of ultrasound-assisted extraction of total monomeric anthocyanin (TMA) and total phenolic content (TPC) from eggplant (*Solanum melongena* L.) peel. *Ultrasonics Sonochemistry* **2016**, *31*, 637–646.

Ghafoor, K.; Choi, Y. H.; Jeon, J. Y.; Jo, I. H., Optimization of ultrasound-assisted extraction of phenolic compounds, antioxidants, and anthocyanins from grape (*Vitis vinifera*) seeds. *Journal of Agricultural and Food Chemistry* **2009**, *57* (11), 4988–4994.

Ghafoor, M.; Misra, N. N.; Mahadevan, K.; Tiwari, B. K., Ultrasound assisted hydration of navy beans (Phaseolus vulgaris). *Ultrasonics Sonochemistry* **2014**, *21* (1), 409–414.

Ho, V. S.; Wong, J. H.; Ng, T., A thaumatin-like antifungal protein from the emperor banana. *Peptides* **2007**, *28* (4), 760–766.

Loganayaki, N.; Rajendrakumaran, D.; Manian, S., Antioxidant capacity and phenolic content of different solvent extracts from banana (Musa paradisiaca) and mustai (Rivea hypocrateriformis). *Food Science and Biotechnology* **2010**, *19* (5), 1251–1258; (b) Bhaskar, J. J.; Chilkunda, N. D.; Salimath, P. V., Banana (Musa sp. var. elakki bale) flower and pseudostem: dietary fiber and associated antioxidant capacity. *Journal of Agricultural and Food Chemistry* **2011**, *60* (1), 427–432.

Mokbel, M. S.; Hashinaga, F., Antibacterial and antioxidant activities of banana (Musa, AAA cv. Cavendish) fruits peel. *American Journal of Biochemistry and Biotechnology* **2005**, *1* (3), 125–131.

Mokrani, A.; Madani, K., Effect of solvent, time and temperature on the extraction of phenolic compounds and antioxidant capacity of peach (Prunus persica L.) fruit. *Separation and Purification Technology* **2016**, *162*, 68–76.

Padam, B.; Tin, H.; Chye, F.; Abdullah, M., Antibacterial and Antioxidative Activities of the Various Solvent Extracts of Banana (Musa paradisiaca cv. Mysore) Inflorescences. *Journal of Biological Sciences* **2012**, *12* (2), 62–73.

Rosen, M. J.; Kunjappu, J. T., *Surfactants and Interfacial Phenomena*. John Wiley & Sons: 2012.

Santos, D. T.; Veggi, P. C.; Meireles, M. A. A., Extraction of antioxidant compounds from Jabuticaba (*Myrciaria cauliflora*) skins: Yield, composition and economical evaluation. *Journal of Food Engineering* **2010**, *101* (1), 23–31.

Schmidt, M. M.; Prestes, R. C.; Kubota, E. H.; Scapin, G.; Mazutti, M. A., Evaluation of antioxidant activity of extracts of banana inflorescences (Musa cavendishii). *CyTA—Journal of Food* **2015**, *13* (4), 498–505.

Sharma, P.; Ramchiary, M.; Samyor, D.; Das, A. B., Study on the phytochemical properties of pineapple fruit leather processed by extrusion cooking. *LWT—Food Science and Technology* **2016**, *72*, 534–543.

Sharma, S.; Kori, S.; Parmar, A., Surfactant mediated extraction of total phenolic contents (TPC) and antioxidants from fruits juices. *Food Chemistry* **2015**, *185*, 284–288.

Sulaiman, S. F.; Yusoff, N. A. M.; Eldeen, I. M.; Seow, E. M.; Sajak, A. A. B.; Supriatno; Ooi, K. L., Correlation between total phenolic and mineral contents with antioxidant activity of eight Malaysian bananas (Musa sp.). *Journal of Food Composition and Analysis* **2011**, *24* (1), 1–10.

Sun, C.; Xie, Y.; Tian, Q.; Liu, H., Separation of glycyrrhizic acid and liquiritin from licorice root by aqueous nonionic surfactant mediated extraction. *Colloids and Surfaces A: Physicochemical and Engineering Aspects* **2007**, *305* (1), 42–47.

Tao, Y.; Wu, D.; Zhang, Q.-A.; Sun, D.-W., Ultrasound-assisted extraction of phenolics from wine lees: Modeling, optimization and stability of extracts during storage. *Ultrasonics Sonochemistry* **2014**, *21* (2), 706–715.

Tiwari, B. K.; O'Donnell, C. P.; Cullen, P. J., Effect of sonication on retention of anthocyanins in blackberry juice. *Journal of Food Engineering* **2009**, *93* (2), 166–171.

Tiwari, B. K.; Patras, A.; Brunton, N.; Cullen, P. J.; O'Donnell, C. P., Effect of ultrasound processing on anthocyanins and color of red grape juice. *Ultrasonics Sonochemistry* **2010**, *17* (3), 598–604.

Wong Paz, J. E.; Muñiz Márquez, D. B.; Martínez Ávila, G. C. G.; Belmares Cerda, R. E.; Aguilar, C. N., Ultrasound-assisted extraction of polyphenols from native plants in the Mexican desert. *Ultrasonics Sonochemistry* **2015**, *22*, 474–481.

Zhang, Z.-S.; Wang, L.-J.; Li, D.; Jiao, S.-S.; Chen, X. D.; Mao, Z.-H., Ultrasound-assisted extraction of oil from flaxseed. *Separation and Purification Technology* **2008**, *62* (1), 192–198.

CHAPTER 6

Current Prospects of Bio-Based Nanostructured Materials in Food Safety and Preservation

TABLI GHOSH, KONA MONDAL, and VIMAL KATIYAR*

Department of Chemical Engineering, Indian Institute of Technology Guwahati, North Guwahati 781 039, Assam, India

**Corresponding author. E-mail: vkatiyar@iitg.ac.in*

ABSTRACT

Food-grade nanostructured materials are extensively used with several approaches for improved food properties in terms of quality and health benefits. The food-grade nanostructured materials generally include inorganic and organic materials, where the use of organic nanomaterials, such as polysaccharides, proteins, lipids, and others, has been increased for their bio-based resources. Food-grade nanostructured materials may offer improved food properties in terms of texture, color, flavor, nutrient contents, rheology, and others, which have to be critically monitored. The nanostructured materials are also used to develop packaging materials, in both primary and secondary packaging, for tailored properties with reduced waste. However, the food safety is measured in terms of migration properties, toxicological behavior of nanoparticle between package and food materials, as food safety is a major concern in protecting the packaged products throughout the life cycle. Among available, polysaccharide-based nanostructured materials, such as nanocellulose, nanochitosan, nanostarch, etc., are extensively used materials for tuned food properties. Considering the fact, this chapter focuses on bio-based nanostructured materials for various prospects in food application with added valuable properties of biodegradability, nontoxicity, biocompatibility, etc., which make them a potential candidate to be used in food-based applications. Further, the nanostructured materials are widely used as encapsulating agents, delivery of bioactive compounds, and provide protection in retaining bioactive agents. The modification in the available bio-based nanostructured materials can provide additional benefits over others. However, the food-grade nanoparticles can also have adverse effects on human health if consumed above the permissible limits, thus food safety of the available nanostructured materials is a critical concern to be considered for food application. In this chapter, the various rules and regulations available for the safe use of the addressed nanoparticle in relation to migration, toxicology of nanoparticles into food, etc., in relation to food and their packaging materials, will also be discussed.

6.1 INTRODUCTION

Nanostructured materials are the most engrossing outgrowth of nanotechnology which focuses mainly on the fabrication, characterization, and application of the emerging nanomaterials. In this context, the definition of nanotechnology implies the conception, control, and power of nanorange material having the dimension of 1–100 nm, which are embedded with unique phenomena for various applications (de Azeredo, 2009; Kumar et al., 2009; Neethirajan and Jayas, 2011; Sekhon, 2010). This particular scale range of nanomaterial is able to develop unique and functional properties, which are further helpful in several commercial applications. Indeed, the considerable benefits of this technology are recognized by various trades such as microelectronics, aerospace, pharmaceutical, agricultural, cosmetics, food products, etc. However, applications of nanotechnology toward the food industry are rather limited. Nevertheless, this technology can be utilized in the food sector for improved shelf life, quality, safety, nutritional benefits, and other physicochemical properties (Sozer and Kokini, 2009). Food is a complex biological system and its functions are generally controlled by several basic mechanisms and principles. Further, the discoveries and development achieved by the researchers would expect to have a good impact on the food industry. Nanosystems provide a new outgrowth to the researchers to obtain tailor-made properties at the nanoscale level for functioning in a valuable way for the targeted areas such as food safety and preservation, food packaging, food processing, and quality, food fortification, etc. (Figure 6.1). However, sometimes the nanostructured materials are incorporated intentionally into the food system for modifying optical, rheological, or flow properties. In addition, researchers and industry stakeholders have already recognized the potential uses of nanotechnology effectively in every segment of the food industry such as agriculture, food processing and food packaging as a carrier of nutrients. In the case of food processing, the impact of nanomaterials is used in encapsulating flavor and odor compounds of the food for enhancing the sensory appeal, as an agent of improving the textural property of food and are also utilized as gelation and viscosifying agent. Moreover, in food packaging, nanomaterials are predominantly used as antimicrobial agents' sensors for detecting pathogen and gas, UV-protection agents, etc. Along with the above functions, it can also be used as active agents for delivering nutraceuticals with better stability and bioavailability.

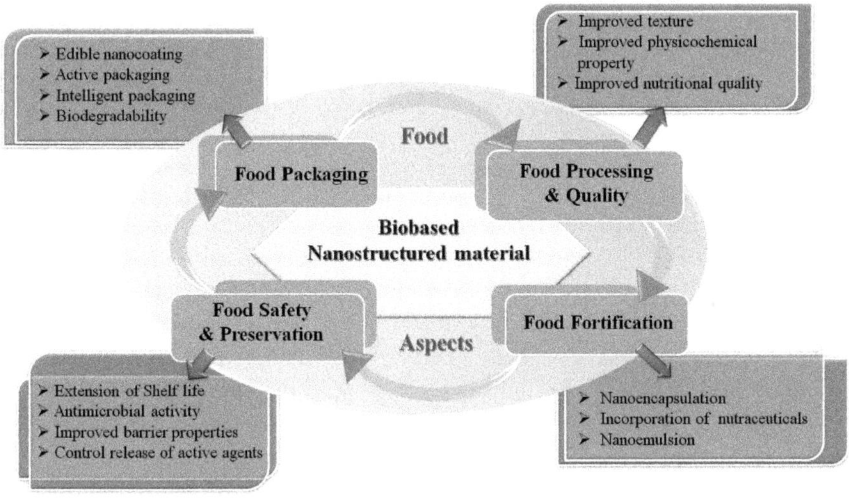

FIGURE 6.1 Schematic of applications of bio-based nanostructured material in food sector.

Interestingly, various nanostructured materials are found to present naturally in some food materials such as milk contain casein micelles, plant or animal cells contain certain organelles (Livney, 2010; Holt et al., 2003; Iwanaga et al., 2007) and, further, the nanostructured materials are formulated as a result of regular food processing including grinding, homogenization, cooking, etc. (Gupta et al., 2016). Thus, nanoparticles have been developed unintentionally as a natural consequence which aids their influence on processing conditions. In this context, it is important to determine the potential sources of various nanoscale materials in food or in other natural sources. Besides, the designed nanoscale materials are also developed intentionally to incorporate into the food for tailoring the inherent property, which makes it functional food for consumer benefits. However, the restriction of usage of these materials is mainly due to safety concerns after consuming the food. Moreover, the mechanism of their action is not completely discovered, and sometimes the added nanoparticles into the food or packaging material may leach out and transfer to another side of packaging material. These designed nanostructured materials are basically nanoparticles where the size, shape, composition, and interfacial properties are developed to achieve functional attributes. Moreover, the aim of developing these nanoscale materials is to utilize as a delivering agent for nutraceuticals, colors, flavors, preservatives. In addition, the potential renewable origin-based bio-nanostructured materials are also able to improve the texture, appearance, and stability of the food.

Based on the above discussion, this chapter mainly focuses on the brief overview of bio-based origin of the nanostructured material and their development processes, property analysis, and application into the food sector. Along with this, the various safety concern, available rules, and regulations relating to the nanomaterials will be discussed as they are susceptible to cause health issues. However, the possibility of showing toxicity is caused by these nanostructured materials either in a chronic or acute way. Although the application of newly developed nanomaterials in food is likely to be consumed at a low level over an extended period of time, where the occurrence of the adverse effect is less. Furthermore, nanostructured material can also alter human health by damaging the microbial cell, which generally affects the human gastrointestinal tract. Despite these limitations, the benefits of nanotechnology are promising for the development of novel functional materials, and product processing at micro and nanoscale level, where the design of materials, methods, and instrumentation are very crucial for food safety and biosecurity.

6.2 AVAILABILITY AND CATEGORIES OF FOOD-GRADE NANOSTRUCTURED MATERIALS

The categorization of nanomaterials generally depends on some factors such as the dimension of the nanomaterial, origin of the nanostructured material (bio-based or others), and their compositions (organic or inorganic) (Figure 6.2). In this regard, based on the dimension three main categories of nanostructured materials are available, such as nanoparticles, nanotubes, and nanolayers. Further, particulates are also referred to as equiaxed or isodimensional nanoparticles or nanogranules or nanocrystals (e.g., cellulose, silica) and nanospherical (e.g., metal oxides such as TiO_2, ZnO), when the three dimensions of the specified nanostructured materials are in nanometer scale. Again, particulates are also recognized as nanotubes or nanofibers, nanowhiskers, nanorods (e.g., carbon nanotubes (CNTs), cellulose whiskers, and titania nanotubes (TiNTs)), where two dimensions are in nanometer range and the third dimension is larger, forming an elongated structure. The particulates which are characterized based on dimension in nanometer scale, are also referred as nanolayers or nanosheets, which have a thickness in the range of one to few nanometer and length of 100–1000 nm (e.g., clay (layered silicates), layered double hydroxides) (Jordan et al., 2005; Chiu and Lin, 2012). Furthermore, the bio-based nanostructured materials are fabricated from organic compounds such as polysaccharide, protein, and lipids, where the substances are extracted from natural sources including marine and agriculture. However, the sources also have an impact on the nature of the nanostructured material in terms of safety and application. Moreover, nanotechnology is seeking an interest in utilizing food grade or edible nanomaterial for imparting bioactive compounds such as vitamins, nutrients, and nutraceuticals into the food product (McClements, 2015a; Yang et al., 2015; Kim et al., 2015). The characteristic properties of nanomaterials including higher optical clarity, greater stability to aggregation and gravitational separation, and enhanced bioavailability make them more advantageous compared to other techniques for delivering functional compounds into the biomatrix (McClements and Rao, 2011; Sagalowicz and Leser, 2010). However, these significant characteristic properties have been achieved due to the smaller size of the nanomaterial. Further, higher optical clarity and better aggregation stability have obtained due to smaller dimensions of particles, which help to reduce attractive forces between colloidal particles more rapidly than the repulsive forces (McClements, 2002; Mcclements, 2011). The proper stability for sedimentation is occurred due to the action of weak gravitational force and the influence of Brownian motion (Mason et al., 2006). The bioavailability of encapsulated bioactive components can be enhanced as digestive enzymes are able to hydrolyze smaller particles more rapidly than larger particles in the gastrointestinal tract (GIT). Moreover, food-grade nanomaterials can be fabricated from various kinds of organic food biomolecules such as lipids, proteins, carbohydrates, and other materials (Mcclements, 2011; Fathi et al., 2014; Wan et al., 2015). Considerably, the composition of nanoparticle determines the physicochemical, functional attributes, and gastrointestinal fate of nanomaterial into the biomatrix.

Therefore, it is significant to identify the most suitable type of nanoparticle for a specific application. Preferably, the nanoparticles should be fabricated using simple, reproducible, and inexpensive methods that can easily be scaled up for commercial applications. In this context, it would be beneficial for consumers if the nanostructured materials could be fabricated from eco-friendly substances, such as natural lipid, protein, and carbohydrate.

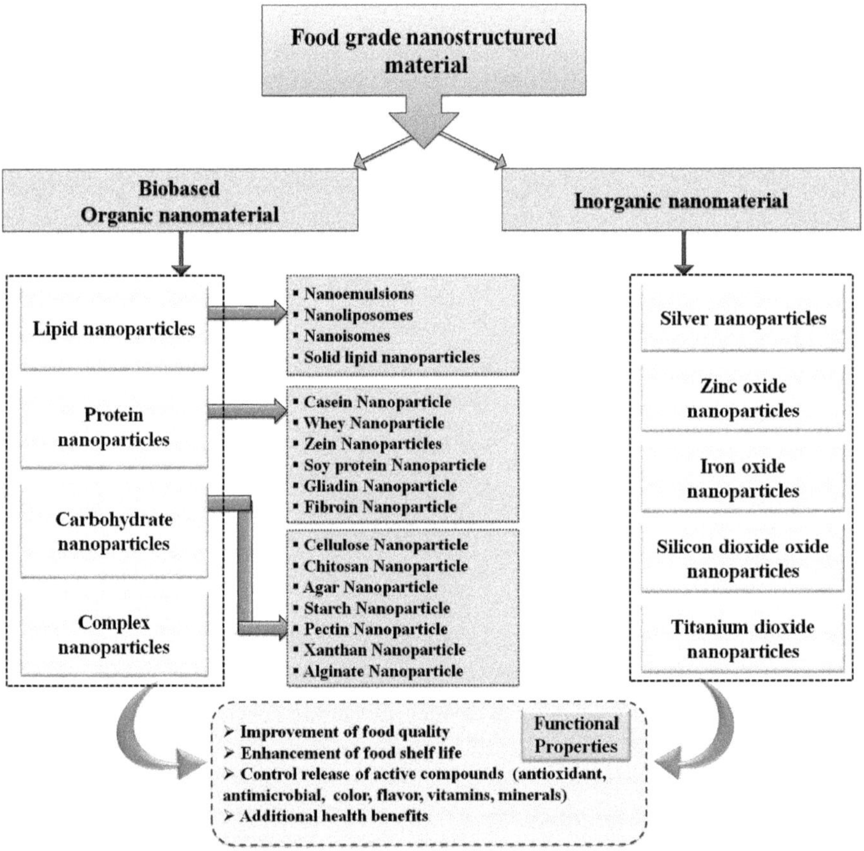

FIGURE 6.2 Classification of food-grade nanostructured materials based on their origin.

6.2.1 ORGANIC FOOD-GRADE NANOSTRUCTURED MATERIALS

Lipids, proteins, and carbohydrates are predominantly utilized as organic substances for the fabrication of nanostructured materials. Due to the growing concern of environmental safety and the scarcity of fossil resources, the utilization of natural renewable material is one of the emerging novel ways of reducing waste disposal problems. Organic substances can be extracted from natural resources and further converted into nanostructured material. Several materials are available for the extraction of these organic substances including agricultural waste, biomass resources. Generally, these organic substances are present in nature as solid, liquid, and semi-solid state. Moreover, both spherical and nonspherical (nanofibers) type of organic nanostructured materials are used in the food. It is noteworthy to mention that organic nanostructured materials in food are less toxic than inorganic nanomaterials in food system as human digestive enzymes are able to hydrolyze them within the human GIT. On the contrary, there are some circumstances where they could be able to show toxicity. Furthermore, several advantages of organic nanomaterials in different areas of food are listed in Table 6.1.

6.2.1.1 LIPID NANOSTRUCTURED PARTICLES

Lipid nanoparticles are mostly used as delivery agents to protect and encapsulate the hydrophobic bioactive agents, which retains color, flavors, antioxidants, nutrients, nutraceuticals, etc. (Livney, 2015; McClements, 2015b; Shin et al., 2015). Generally, it consists of a lipid core which is surrounded and stabilized by a surface-active material. For stabilizing the lipid, single or a mixture of surfactants can also be used such as lecithin and polysorbates. Sometimes, bile salts, butanol as a co-surfactant may be added into the formulation. The fabrication and processing of lipid nanostructured particles are different, where the final particle properties depend on various factors. However, homogenization and cooling are generally used for the particle formulation. Many commercial food items consist of lipid nanoparticles such as soft drinks, fortified waters, fruit juices, and dairy drinks (Piorkowski and McClements, 2014). Moreover, the use of these nanostructured materials is advantageous for enhanced bioavailability and functional property of encapsulated components, which also provide physical stability to the product. In general, triacylglycerols, diacylglycerols, monoacylglycerols, free fatty acids, and phospholipids are the constituent of food-grade lipid nanoparticles. However, different forms of lipid nanoparticles can be developed such as nanoemulsions, nanocapsules, nanoliposomes, nanoisomes and solid lipid particle, for applying into the food products (Fathi et al., 2012).

6.2.1.1.1 Nanoemulsions

The particle size of nanoemulsion is in the range of 20 to 200 nm with suitable thermodynamic stability (Delmas et al., 2011). The emulsion is generally formed in the presence of surfactant, where two immiscible liquids develop transparent emulsions, where high- and low-energy methods are generally used to produce these nanoemulsions (Yang et al., 2012). The common methods used for low energy is membrane emulsification, spontaneous emulsification, solvent substitution, reverse emulsion, and reverse phase. In the case of high energy, mechanical methods are used such as colloid mills, high-pressure, and ultrasound homogenizer (Date et al., 2010, Jaiswal et al., 2015). In addition, the industry prefers high-energy methods due to the formation of adequate, controlled size, and high-quality emulsions. Further, nanoemulsions are mainly utilized for the development of functional drinks where vitamins, minerals, and other bio-actives are supplied, which also helps in controlling the functional compounds in the developed functional drinks (Rao and McClements, 2012; Lane et al., 2014). The release of functional compounds from nanoemulsion also depends on several factors including heat, pH, sound frequencies, and other stimulants. Further, the antimicrobial activity of nanoemulsion is helpful for decontaminating the food equipment, packaging materials.

6.2.1.1.2 Nanoliposomes

Another form of lipid nanostructured material is liposome which is also used to make functional foods. It is a colloidal particle of lipid molecules, which accumulates in the form of organized bilayer membranes after reacting with water. Liposomes have the capacity to encapsulate both hydrophilic and hydrophobic substances. It can encapsulate hydrophilic active compounds inside the membrane and hydrophobic substances between the membranes. However, nanoliposomes are less used in the food industry due to its unstable structure.

6.2.1.1.3 Nano-niosomes

This form of lipid nanostructured material is used to transfer nutraceuticals. Nano-niosomes are generally formed based on nonionic surfactants in an aqueous environment, which enables to overlap both types of hydrophilic and hydrophobic nutraceuticals. Nano-niosomes are more advantageous in terms of higher stability, low cost, and biocompatibility than liposomes; however, both are similar in structure (Moghassemi and Hadjizadeh, 2014).

6.2.1.1.4 Solid Lipid Nanoparticles (SNLs)

This form of nanoparticle has received great attention in the food industry due to the properties of high stability and having control over the release of components that are entrapped within the lipid matrix. It is more efficient than liquid nanoparticle due to the mentioned properties. On the contrary, this nanoparticle has a tendency of aggregation, which limits their use in commercial application though it can be overcome by incorporating specific surfactants that can modify polymorphic transitions or by selecting a lipid in which alpha and beta polymorphic transition is slow. This form of lipid nanoparticle differs from others. The size range of this particle is within the range of 40 to 1000 nm. Moreover, SLNs can be formed by substituting emulsified oil phase (oil in water) with solid oil or a mixture of solid SLNs which can also be formed through dispersing 0.1–30% solid lipid in the liquid phase, and 0.5%–5% of surfactants may be used if necessary (Üner and Yener, 2007).

Moreover, the fabricated lipid nanoparticles are optically transparent, which is considered as one of the most desirable properties in food and beverages. Food contains various types of lipid nanoparticles including micelles, vesicles, oil droplets, and fat crystals with varying dimensions, structures, and compositions. Additionally, the electrical characteristics of these particles depend on the type of molecules present at their surfaces. However, their susceptibility towards hydrolysis is based on the functioning of the digestive enzymes lipase and phospholipase presents at GIT (McClements and Xiao, 2012). Due to the high specific surface area, the lipid nanoparticles are able to hydrolyze rapidly, which indicates that they are directly absorbed by the GIT fluid. Though, they are able to absorb and accumulate directly in the intestinal cells and other organs, but, their ability to enhance the bioavailability of bioactive compounds which are hydrophobic in nature may cause undesirable effects. In contrast, some lipid nanoparticles are not digested by the enzymes in the GIT such as terpenes and hydrocarbons, which are preferably found in some flavor, essential, and mineral oils. Another reason includes the utilization of coated material, which inhibits the digestive enzymes from hydrolyzing the oil droplets through creating an interfacial layer in encapsulated lipids (McClements, 2010). Thus, an indigestible nanoparticle of lipid may be absorbed directly in the body. Moreover, the potential toxicity of these nanoparticles after absorption has not been established so far.

6.2.1.2 PROTEIN NANOPARTICLES

Protein nanoparticles are a good candidate for releasing drugs and various beneficial components into the food. It has plenty of advantages including enhancement of drug release, reduction of toxicity, and improvement of bioavailability and also provides better formulation opportunities. These particles are also able to show better action at the minimum dose and help to decrease the drug resistance in the body. However, the most common protein nanoparticle is casein micelles in bovine milk, which is the small clusters of casein molecule and calcium phosphate ions (Livney, 2010; Holt et al., 2003). Further, this nanostructured material, casein micelle is naturally present in milk as a delivering agent of nutrients like protein and minerals. More recently, after observing the properties of this nanomaterial, much interest has been raised for developing a variety of protein nanostructured material for food application. Although a little concern for potential toxicity exists for long-term consumption of protein nanoparticles. In addition, other protein nanoparticles used in food are whey, zein, gliadin, gelatine, and soy protein particles, where their size varies from few nanometers to hundreds of nanometers. Similar to lipid nanoparticles, they are also able to encapsulate, protect, deliver bioactive compounds including color, flavor, preservative, vitamins, minerals, and nutraceuticals.

6.2.1.3 CARBOHYDRATE NANOPARTICLES

Carbohydrates are the natural biopolymer, which is extracted from natural resource (i.e., animal and plant) including agricultural waste. Some of the common polysaccharides are cellulose, starch, chitosan, alginate, pectin, and xanthan. Based on their chemical nature, polysaccharides can be digestible and indigestible inside the human body. However, nanoparticles can be developed from these biopolymers by cleaving the natural glycosidic linkage presents between the monomers. After breaking the linkages within polymer chains, various biopolymers, including cellulose, starch, and chitosan, are fabricated. Further, breaking down of the polymer chain is done in two different ways using chemical and mechanical processes. The chemical process deals with the application of temperature, changing in pH, application of enzyme, acid, and ionic liquids, whereas the mechanical process includes ultrasonication, ball milling, homogenization, etc.

The shape and size of the carbohydrate nanoparticles vary depending on the source of raw material and process of fabrication. In this context, cellulose nanoparticles have different morphology, such as cellulose nanocrystals, cellulose nanofibers, cellulose nanorod, and cellulose nanowhiskers, can be obtained based on the various processing conditions. Although, all these morphological structures are in the nanometer range. Furthermore, these nanomaterials can be utilized in different sector of food including food preservation, packaging, edible, active, and intelligent packaging by releasing nutraceuticals, vitamins, minerals, antimicrobial agents, anticarcinogenic agent, antidiabetic agents, which helps to improve the quality and texture of the food along with extended shelf life. However, it has been known that cellulose cannot be digested by human GIT enzymes therefore utilizing this carbohydrate nanoparticle in primary and secondary packaging would be advantageous as it has a tendency to improve the mechanical and barrier properties of packaging material. Also, it can be used as a delivering agent through the release of beneficial compounds such as inorganic particles, vitamins, minerals, antimicrobial agents, etc. In addition, cellulose nanoparticles are hydrophilic in nature (George and Sabapathi, 2015) and require mixing with other polymers or biopolymers for developing packaging material. In this regard, for complete mixing and development of good properties of composites, incorporation of inorganic compounds has been done, which helps to modify the surface property of this nanomaterial. Moreover, incorporation of iron particles onto cellulose nanocrystals and its composites has shown better thermomechanical, induced electrical and magnetic properties, which additionally can be useful for electronics products, such as fabrication of sensors, where this nanocomposite can be used as laminated films into food packaging. However, other carbohydrate nanoparticles, such as chitosan nanoparticles, help to improve the textural property, when applied as edible coating material (Muxika et al., 2017).

Further, in the case of starch, depending on the type, some are hydrolyzed by amylases originated from small intestine and mouth, and some starches are also resistant to hydrolyze by digestive enzymes in the GIT.

Since, polysaccharides are used to fabricate carbohydrate nanoparticles, which are not digested by the GIT enzymes, mostly known as dietary fibers. However, they can be digested at lower GIT when fermented by enzymes released by the microbiota. Considerably, carbohydrate nanoparticles which are completely digested by the GIT enzymes do not exhibit toxicity as they produce simple sugar as a conventional form of carbohydrates. However, indigestible carbohydrate nanoparticles show adverse health effects. Carbohydrate nanoparticles are also capable to provide bioavailability and bioactivity of encapsulated substances.

6.2.1.4 COMPLEX NANOPARTICLES

Complex nanoparticles are the combination of different ingredients including carbohydrate, protein, and lipid, which are fabricated for potential food applications (Joye et al., 2014). Coacervates are an example of this type of nanoparticle, which is made by electrostatic interaction of oppositely charged proteins and polysaccharides. In this case, lipid nanoparticles are coated by nanolaminated layers which can be formed by successive electrostatic precipitation of oppositely charged biopolymers over the surfaces of particle. Another example of this complex nanoparticle is a mixture of protein and polysaccharide which carries an opposite charge and the complex formation is driven by the attractive electrostatic interactions between the two biopolymers. The developed complexes may be soluble or insoluble in nature depending on the strength of the interactions, the balance of negative and positive charges, and their distribution on the biopolymers, molecular weight, and biopolymer flexibility under the processing conditions (e.g., pH, ionic strength, concentrations of biopolymers). In addition, complexes are soluble in nature if the overall charges are not neutralized. Furthermore, when proteins carry a positive charge at a pH below its isoelectric point, the complex makes the strongest attraction between proteins and anionic polysaccharides (e.g., carrageenan, pectins). Complexes can also be formed between chitosan (anionic polysaccharide) and whey protein isolate and the strongest interactions occur at pH above the isoelectric point. These complex nanoparticles give rise to different textured food. However, complex nanoparticles can show adverse health effects when indigested by the GIT enzymes. Since the laminated layers also have pronounced effects on the GIT fate of ingested particles.

6.2.2 INORGANIC FOOD-GRADE NANOSTRUCTURED MATERIALS

The main concern in food processing is to protect against foodborne illnesses, which causes a significant problem in public health worldwide. One of the possible ways to prevent foodborne diseases occurring from food through the development of active food packaging. Furthermore, the foodborne diseases can be prevented by the antimicrobial agents, releasing from the active food packaging material. In general, active packaging is developed by the addition of functional bioactive compounds into the packaging material itself and by releasing these compounds from the packaging material into the food. In this context, inorganic food-grade nanostructured materials can also aid the beneficial effects through releasing the biocide into the food directly or to the space around the food. Several inorganic nanoparticles used in food mostly include iron oxide, titanium dioxide, silicon dioxide, silver as listed in Table 6.2 (Pietroiusti et al., 2016). The shape and sizes of these nanoparticles vary depending on the origin and the process of fabrication. These nanoparticles are either crystalline or amorphous solids at room temperature and also vary to disintegrate into different solutions depending on the processing conditions such as pH and ionic strength. Moreover, the chemical reactivity of inorganic nanomaterials has an impact on GIT fate and toxicity.

6.2.2.1 SILVER NANOPARTICLES

Among other metal particles, silver nanoparticles (AgNPs) have the most effective bactericidal property against bacteria, yeast, fungi, and viruses (Martínez-Abad et al., 2012; AshaRani et al., 2009). Due to this reason, they are used in food sectors for a different type of application. This nanoparticle has been used as antimicrobial agents in food and packaging material (Hajipour et al., 2012; Gaillet and Rouanet, 2015; Pulit-Prociak et al., 2015). Further, AgNPs have shown better antimicrobial activity than metallic silver due to their large surface area which provides sufficient surface area for optimum attachment with the microorganism (Toker et al., 2013). Moreover, they exhibit low volatility and stability at high temperatures (Youssef and Abdel-Aziz, 2013). AgNPs can be incorporated in different matrices such as polymers and stabilizing agents (citrates and long-chain alcohols) (Toker et al., 2013), through different strategies including coating, or can directly be incorporated in the synthesis process (Martínez-Abad et al., 2012). However, the use of AgNPs as antimicrobial agents in food packaging is the most common technology, though the concern is associated with the potential risk of silver ions after ingestion. Considerably, there is a tendency of migrating the silver ions from packaging material into food and beverages. Some US companies have used AgNPs as an antimicrobial agent in food packaging containers such as Kinetic Go Green basic nanosilver food storage container, Oso fresh food storage container, and Fresher Longer™ Plastic Storage bags (Echegoyen and Nerín, 2013).

Moreover, the European Food Safety Authority (EFSA) Panel on Food Additives and Nutrient Sources stated that products containing AgNPs in food packaging and food supplements are not allowed in the EU if crosses permissible limits (Bumbudsanpharoke and Ko, 2015). The recommended upper limit of silver migration from packing should not exceed 0.05 mg/L in water and 0.05 mg/kg in food by ESFA. They also recommend for some in vitro tests including genotoxicity, absorption, distribution, metabolism, and excretion by the manufacturers before commercialization in 2011 (EFSA Scientific Committee, 2011). Additionally, the United States Food and Drug Administration (USFDA) also recommend that manufacturers need to prepare a toxicological profile for each container with nanomaterials. Further, one estimation suggested that adults can consume silver within the range of 20–80 μg/day in which a small fraction of nanoparticle is permissible (Fröhlich and Fröhlich, 2016). However, very limited information is available about the potential toxicity of the AgNPs ingested with food, where some studies have reported to provide no toxicity for AgNPs and others reported appreciable toxicity. GIT may be susceptible to the AgNPs induced toxicity but the adverse effect of this nanoparticle on GIT remains indecisive. Furthermore, animal studies have been conducted to check the effect of AgNPs after ingestion and it has been found that this particle can accumulate in various organs such as liver, kidneys, spleen, stomach, and small intestine (Gaillet and Rouanet, 2015; Hendrickson et al., 2016; Kim et al., 2008). This outcome indicates that this nanoparticle can be absorbed by the GIT. Furthermore, the majority of the ingested material excretes through faces or urine, where a small quantity still accumulates inside the tissues (Hendrickson et al., 2016). Finally, no toxicity has been found after oral gavage in this particular study with the dose level of 2000 and 250 mg/kg body weight for single and multiple doses. Other studies on animals such as rats and mice revealed that no toxic effects of ingestion of AgNPs over a 28 days period at 30,300 and 1000 mg/kg day ingestion but slight liver damage occurred at the highest dose level (Kim et al., 2015). Another study reported the adverse effect of this nanoparticle over kidney and liver after multiple time ingestion (Park et al., 2010, Garcia et al., 2016). Thus, it can be stated that the consumption of high levels of AgNPs may cause adverse health effects. However, there is a need of further study for establishing the appropriate safe limit of AgNPs and long term chronic toxicity test must be conducted using the dose levels of nanoparticles that is more similar to the human consumption rate. Moreover, cell culture studies have been done in a few studies and they reported that AgNPs may promote cytotoxicity through various mechanisms (Chen et al., 2016; Georgantzopoulou et al., 2016). This study indicates the effect of the particles depending on the size and nature of the coating material on their surfaces. Moreover, one of the most important factors of AgNPs for contributing toxicity is their ability to generate reactive oxygen species (ROS). The generation of ROS helps to promote oxidative stress, which causes damage to cell membranes, organelles, and the nucleus (Sharma et al., 2014). Further, biochemical functions and the nature of gut microbiota may alter due to the strong antimicrobial activity of this nanoparticle. On the contrary, the cell culture test is not much relevant as the dose level used is much higher than normal.

6.2.2.2 ZINC OXIDE NANOPARTICLES

Zinc is a trace element essential for maintaining human health. Zinc oxide (ZnO) nanoparticle is an inorganic substance that has a wide application in pharmaceutical, cosmetic, food, rubber, commodity chemical, painting, ceramic, glass industries, etc. Recently, ZnO is considered as Generally Recognized as Safe (GRAS) compound by the USFDA and also used as a food additive. The development of nanotechnology led to fabricate nanomaterials with unique properties for using antimicrobial agents. ZnO nanoparticles have shown good antimicrobial activities, has a potential application in food preservations. Thus, it can be utilized as a food additive in food packaging to prevent contamination and also as ultraviolet (UV) light absorbers to protect sensitive foods from UV light exposure (EFSA Panel on Food Contact Materials et al., 2016). The activity of ZnO nanoparticles depends on the size, where smaller size provides the higher activity. Its direct contact with microbial cell wall may result in the disruption of bacterial cell integrity, and liberation of antimicrobial ions Zn^{2+} ions, and generation of ROS that damage key cellular components, which leads to cytotoxicity (Sirelkhatim et al., 2015). This mechanism could lead to adverse health effects in human body, if this nanoparticle ingested in an inadequate quantity. It has been observed that ZnO is the most effective antimicrobial agent for inhibiting the growth of *Staphylococcus aureus* when compared to MgO and CaO. Further the application of ZnO nanoparticles in powder, film, PVP capped, and coating showed effects against *L. monocytogenes* and *Salmonella enteritidis* in liquid egg white and in culture media (Jin et al., 2009). This nanoparticle is also very much effective against foodborne pathogens including *Bacillus cereus, Escherichia coli, S. aureus,* and *S. enterdis*. In the case of active food packaging, films with ZnO nanoparticles may lead to good antimicrobial properties. It has been reported that a single oral dose of ZnO nanoparticles is responsible for occurring hepatic injury, kidney toxicity, and lung damage in animals (Esmaeillou et al., 2013). Another study reported about the nontoxicity but after mixing with ascorbic acid it became toxic (Wang et al., 2014). This study suggests that it is important to measure the impact of specific food components on the toxicity of this type of nanoparticle. Further, the aggregation of this particle in aqueous solution has an impact on GIT fate and toxicity. These aggregates are much bigger than individual particles depending on their size and structure. Thus, studies need to be done for selecting an appropriate dose level for safe human consumption in the long term.

6.2.2.3 IRON OXIDE NANOPARTICLES

Iron oxide (Fe_2O_3) nanoparticles may be utilized in food material as colorants and as a source of bioavailable iron (Raspopov et al., 2011; Zimmermann and Hilty, 2011; Wu et al., 2014; Hilty et al., 2010). In the United States, the use of iron oxide as a coloring agent is limited only up to 0.1 wt% in sausages as part of casings. One survey has estimated that the intake of iron oxide from these products is 450 μg/day. The intake of iron oxide is higher for consumers who take iron-fortified functional foods or supplements (Fulgoni et al., 2011). In addition, iron intake from fortified food or supplements does not provide iron oxide or iron nanoparticles to the consumer. The range of iron in fortified food is 10–23 mg/day and dietary supplement is 10–32 mg/day (Fulgoni et al., 2011). Moreover, the toxicity of iron oxide particles may alter due to their varying size, shape, and different crystalline form (Patil et al., 2015). However, the potential toxicity of these nanoparticles depends on their mechanism and ability to generate ROS (Wu et al., 2014). Several studies on animals have been done to check the toxicity from this nanoparticle. The outcome of those studies is negative and no accumulation of toxic compound observed during taking oral doses at higher (250–1000 mg/kg body weight) and lower rate (3 mg/kg body weight) (Hilty et al., 2010).

6.2.2.4 TITANIUM DIOXIDE NANOPARTICLES

Titanium dioxide (TiO_2) has been extensively used in the food industry as functional ingredients, such as color additive, and provides optical properties, such as increased lightness and brightness (Lichtenstein et al., 2015). Further, TiO_2 nanoparticles possess significant properties including high stability, relatively low-cost production (Anwar et al., 2010) long-lasting, safe, antimicrobial agent (Cerrada et al., 2008), optical and high refractive index rating, dielectrical, catalytical property, and UV protecting agent, which allows its industrial applications such as pigments, fillers, catalyst supports, and photocatalysts (Mahshid et al., 2007). The light-scattering properties of TiO_2 depend on the size of the particles and the optimized size of the particle is in the range of 100–300 nm, which helps to increase their light scattering property (Jovanović, 2015). However, TiO_2 particles have different particle size variations and the most significant proportion of particles is within 100 nm which allows them to be considered as nanoparticles (Warheit et al., 2015). The mean diameter of commercially available food-grade TiO_2 powder has found to be about 110 nm in which more than 36% of the particle is more than 100 nm (Weir et al., 2012). The estimated dietary exposure of humans to TiO_2 has been reported to be up to 1.1 and 2.2 mg/kg body weight/day in UK and US, respectively (Weir et al., 2012). Additionally, consuming one piece of chewing gum can result in an intake of 1.5–5.1 mg of TiO_2 nanoparticles (Chen et al., 2013). The consumption of this nanoparticle is higher in children than adults as children consume more candies, gums, desserts, and beverages.

Further, TiO_2 nanoparticles may vary in their sizes, shapes, crystal form, interfacial properties, and aggregation states, which have an influence on GIT fate and toxicity. Additionally, TiO_2 can exhibit both crystalline and amorphous forms. The amorphous form of TiO_2 is photocatalytically inactive (Watson et al., 2004). TiO_2 have three crystalline structures such as anatase, rutile, and brookite (Anwar et al., 2010), where anatase is most stable at sizes less than 11 nm, brookite at size ranges between 11 and 35 nm, rutile at sizes greater than 35 nm (Lin, 2006). The most commonly occurring polymorphic forms are anatase and rutile (Chen, 2009) with different crystal packing and physicochemical properties (Yang et al., 2014). Moreover, depending on the source, surface composition of food-grade TiO_2 particles may vary and X-ray analysis also proved the different levels of presence of phosphorous, aluminum, and silica content depending on their origin (Yang et al., 2014).

The cytotoxicity of TiO_2 nanoparticles after ingestion has been studied and the result indicates that the accumulation of this nanoparticle in tissues of mammals and other vertebrates with limited elimination rate (Jovanović, 2015). Further, single oral doses of this nanoparticle have shown their accumulation in liver, spleen, kidney, and lung tissues of mice which led to hepatic injury, nephrotoxicity, and myocardial damage (Wang et al., 2014). In the case of a higher dosage, it adversely affects the liver, blood, and immune system (Duan et al., 2010). Subsequently, it also affects intestine by induced inflammatory cytokine production, T-cell proliferation, hypertrophy, and hyperplasia in the mucosal epithelium (Nogueira et al., 2012, Bu et al., 2010). In contrary few studies have reported the nontoxicity of the TiO_2 nanoparticle and elimination of this particle through urine and faces at a lower dose. Additionally, this contradiction on the toxic effect of TiO_2 nanoparticle depends on some factors such as the oral dose, crystal form, particle size, aggregation state, and surface characteristics of the nanoparticles. In case of cell culture study, the cytotoxicity depends on the particle characteristics, such as dose, size, crystal form, and surface coating (Tada-Oikawa et al., 2016; Song et al., 2015; Brun et al., 2014; Gerloff et al., 2009). Nevertheless, the anatase phase of TiO_2 nanoparticles has shown more toxicity than rutile polymorph due to their higher photocatalytic activity (Dorier et al., 2015).

6.2.2.5 SILICON DIOXIDE NANOPARTICLES

Silicon dioxide (SiO_2) particles are usually amorphous solid spheres. The particle range of food-grade SiO_2 is between 100 and 1000 nm ranges. This nanoparticle can be used in powdered food products as an anticaking agent for enhancing flow properties such as salts, icing sugar, spices, dried milk, and dry mixes (Dekkers et al., 2011; Peters et al., 2012). In addition, the intake of SiO_2 is around 20–50 mg/day/person (Fröhlich and Fröhlich, 2016). In the case of commercial SiO_2 nanoparticle, the diameter range is between 10 and 50 nm, although these nanoparticles often exist in the form of larger clusters (100 to 1000 nm). Similar to other nanoparticles, cell culture, and animal feeding studies indicate that high levels of SiO_2 nanoparticles accumulate in the liver that may have an adverse effect on health. This nanoparticle has shown improved tensile strength and water resistance properties when incorporated for the development of nanocomposite films (Tang et al., 2009).

6.3 TRAITS OF BIO-BASED NANOSTRUCTURED MATERIALS IN FOOD PRODUCTS

Food industries are constantly looking for novel technologies to improve the nutritional value, texture, taste, flavor, shelf life, and safety of their food products. Further, the challenges for easy transportation of consumer goods and packaging include legislation, global markets, longer shelf life, convenience, safer, and healthier food, environmental concerns, authenticity, and food waste (Realini and Marcos, 2014). However, the bioactive compounds showed irrelevant or unsatisfactory therapeutic effects when orally administered in a free state, where they were rapidly degraded losing their bioactivity (Raspopov et al., 2011). The concept of using nanostructure material has come as a substitute to solve the above-mentioned restrictions. Moreover, the dimension of this emerging nanostructured material is in the nanometer scale, which aids to increase the specific surface. Due to their small size, they impart unique properties and the particles with less than 100 nm diameters are currently used in various applications, including food production to improve food texture, control release of antimicrobial, antioxidants, and other beneficial agents. In addition, the origin of this nanostructured material may be bio- or nonbio based. The purpose of using bio-based sources is due to specific properties such as bioactivity, bioavailability, biocompatibility, biodegradability, and nontoxicity. Besides, environmental concerns also enhance and stimulate the use of renewable resources for producing economically convenient applications to maintain and improve life quality. Thus, utilizing bio-based sources for developing protein, lipid, and carbohydrate nanostructured material has encountered much interest in the food sector. However, the hurdle associated with waste disposal of food plastic packaging material can also be replaced using biopolymers from renewable sources or through utilizing agricultural wastes. Furthermore, the above-stated limitation helps to increase the use of bio-based nanostructured material in the food sector.

6.3.1 BIOCOMPATIBILITY, BIODEGRADABILITY, AND NONTOXICITY

The term biocompatibility is defined as the ability of a biomaterial to perform its desired function without eliciting any undesirable or systemic effects in the recipient. In general, biocompatibility describes the property of materials being compatible with living tissue and does not produce any toxic or harmful response when exposed to other media. In this text, biocompatibility is one of the excellent properties of bio-based nanostructured material. Thus, there is a less chance of occurrence of toxicity in the food product incorporated with nanostructured material that means after coming in contact with food matrix the nanomaterial property will not change or they may not produce any toxic element into the consumer body when come in contact with the human body fluid.

Biodegradability is another beneficial property of nanostructured material which refers to the ability of organic substances and materials to be broken down into simpler substances through the action of enzymes/microorganisms when come in contact with biological media. Biodegradation is a natural phenomenon of earth. Disposal of plastic waste generating from food packaging is a worldwide problem due to early use of conventional plastics and these are either nonbiodegradable or takes a very long time to degrade. In recent trends, people have started using biopolymers for making packaging films, and incorporation of nanostructured material into the film will aid in faster and easier degradation.

Food products must contain substances that are nontoxic in nature. The term toxicity refers to the degree in which a chemical substance or a particular mixture of substances can damage cell. When the toxicity affects a cell it is known as cytotoxicity and when it affects the liver is known as hepatotoxicity. The probability of occurring toxicity in nanostructured material is less compared to inorganic nanoparticles.

6.3.2 MORPHOLOGY, STRUCTURE, AND DIMENSIONAL STABILITY

The physicochemical properties of nanostructure material (organic/inorganic) depend on their morphology, structure, and dimensional stability. Various forms of morphology have been obtained from bio-based nanostructured materials for example cellulose. Cellulose has different forms of nanostructured particle depending on their morphology, structure, and dimension including cellulose nanocrystals, nanowhiskers, nanorod, nanotube, nanofiber, nanosphere, etc. (Shak et al., 2018). Generally, cellulose nanocrystals have rod, needle, spherical shaped morphology depending on their origin and processing conditions. Further, nanowhiskers are also a form of cellulose nanocrystals with a high aspect ratio and rod-like morphology and a transverse dimension of as small as 3 to 30 nm (Eichhorn, 2011). The dimension of food nanoparticle varies from a few nanometers like surfactant micelles to a few hundred nanometers including protein, lipid, and carbohydrate nanoparticles depending on the origin and processing parameters used fabrication (Yada et al., 2014). The toxicity and GIT fate is much influenced by the dimension of nanostructured material under various mechanisms (Bellmann et al., 2015; Ensign et al., 2012; Fröhlich and Roblegg, 2012). Additionally, the tendency of smaller nanoparticles getting absorbed is faster compare to bigger nanoparticles. Since the surface area of the smaller particle is more and this helps to react with GIT fluids components including digestive enzymes, phospholipids, bile salts, or mineral oils. Further, smaller nanoparticle first absorbs at the upper GIT tract and those do not absorb or digested, trying to get absorbed at lower GIT. Also, the penetration and the transportation of nanoparticles at mucus layer coating epithelium cells depend on the size of the nanoparticle. The smaller size particle gets easily absorbed and transfer faster through different junctions.

6.3.3 AGGREGATION STATE AND INTERFACIAL PROPERTY

Nanoparticles also have a tendency of agglomeration or aggregation through forming bigger clusters due to many environmental factors including pH, ionic strength, interactions between particles and media substances, and mechanical forces. The formed cluster is also varying in size, morphology, and strength. The dimension of the nanoparticle cluster is much bigger than individual nanoparticle and these clusters have various effects on GIT fate. Moreover, nanoparticles inside the clusters are held together by physical forces, such as van der Waals, electrostatic, hydrogen bonding, and hydrophobic interactions. Furthermore, the interfacial characteristics of food-grade nanostructured material also have an influence on the GIT fate and may be responsible for occurring adverse health effects (Magnuson et al., 2011; Powell et al., 2010). This nanoparticle in food and in the GIT fluid is surrounded by coating material. The coating material obtained by the adsorptions of the substances from the biological system over the nanoparticle. This coating material has an influence on the developed characteristics of nanoparticles including electrical charge, hydrophobicity, thickness, digestibility, and chemical reactivity. Further, these surface properties will help to determine the behavior of the nanoparticle in the GIT. Additionally, these properties also allow us to determine the penetration ability of nanoparticles through biological barriers, their interaction with other components within the gut, and their aggregation states.

6.3.4 TAILOR-MADE FOOD PROPERTIES

The incorporation of nanostructured particles into the food matrix has shown improved properties in various food products. These nanoparticles present in the food matrix as a dispersed form that also may vary depending on their compositions, structure, and properties. Since food is a complex material in nature containing various constituents including carbohydrates, proteins, fats, minerals, and vitamins and these are assembled into different structural features such as bulk phases, droplets, particles, bubbles, and networks. Further, various processing techniques also have been subjected to produce food products such as mixing, heating, cooling, and many other treatments. Therefore, after incorporation of nanoparticles into the food, physicochemical, and structural property of these nanomaterials may be changed and consequently the properties of food. Moreover, changes in the property play a major role in determining GIT fate and toxicity.

Furthermore, it has been reported that lipid nanoparticles in the form of nanoemulsion edible coating on different foods have shown extended shelf life, antioxidant capacity, antimicrobial, and enzyme inhibition property (Table 6.1). The use of nanostructured lipid carriers in an edible coating of food shows various properties including antioxidant, anti-inflammatory, structural stability, and reduction in cholesterol level. However, a simple need for these nanostructured materials and their mechanism through packaging materials toward the food for better functionality has shown in Figure 6.3. Additionally, incorporation of carbohydrate nanoparticle, such as cellulose nanofiber, cellulose nanocrystals in biopolymer matrix including polylactic acid, polycaprolactone, and other biodegrdabale polymers provide improved packaging properties. Other way of using protein nanostructured material such as use of zein nanofibers and curcumin as an edible coating on apple provides improved shelf life and antimicrobial activity. Similarly, chitosan nanoparticle has been used to coat table grapes and the study has shown improved physicochemical, microbiological qualities, which further indicating the use of nanoparticle as food preservatives. Further, incorporation of silicon dioxide nanoparticle, titanium oxide nanoparticle, AgNPs, and zinc nanoparticle in biopolymer matrix for edible coating application has shown various improved properties such as decreasing respiration rate, maintenance of firmness, antimicrobial activity, flavor restoration with extended shelf life. Moreover, silk fibroin nanoparticle also has shown the improved water vapor and oxygen barrier properties, better freshness of food, and extension of shelf life while storing at room temperature, when used as an edible coating material.

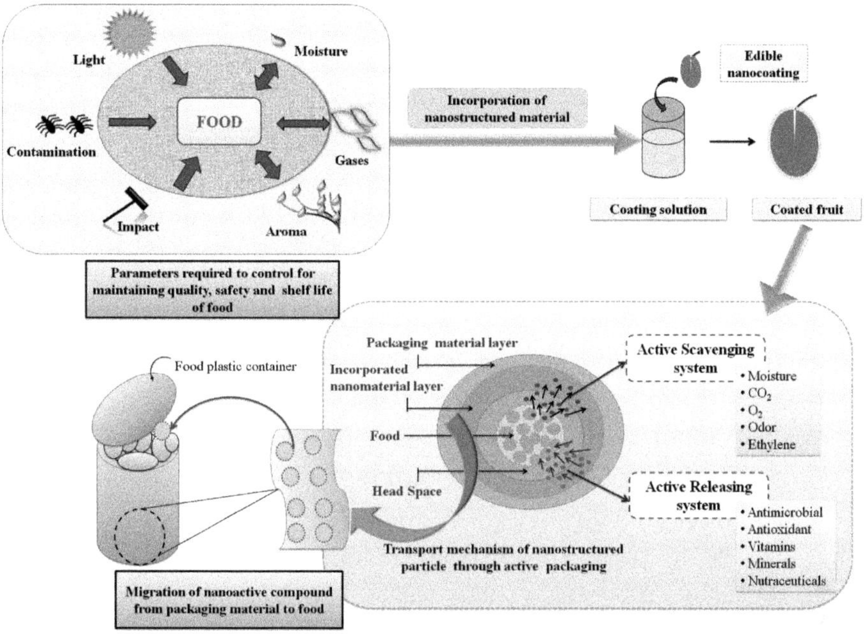

FIGURE 6.3 General need of nanostructured material and their mechanism of action through packaging toward food.

6.3.5 NANOPARTICLES IN FOOD SAFETY AND PRESERVATION

As mentioned earlier, the rapidly growing nanotechnology has a wide impact on the food sector, where the most interesting outcome of this technology is nanomaterials and their uses in the food system. However, the advancement of nanotechnology in the food industry needs more concerned research before commercializing food products with nanoparticles for the consumer since it may cause unhealthy attributes. Further, the food industry will only accept the novel nanomaterials when the issues related to safety are addressed (Ghosh, 2018; te Kulve et al., 2013). Various researches are still going on for establishing the permissible limit of using nanomaterials in food for consumption. Various authorized scientific committees and agencies are working on the safety issues of nanomaterials for food and animal, which will be discussed in the later sections. Interestingly, one guidance document has provided relating to the handling of risk associated with certain food-related nanotechnology. They discussed the engineered nanomaterials and their uses in food additives, enzymes, flavorings, food contact materials, novel foods, food supplements, feed additives, pesticides, etc. Further, the USFDA has made the draft guidance for the industrial use of nanomaterials in animal feed (Commissioner, 2019). However, more research is required to determine the actual impact of nanomaterials in food and on human health to ensure public safety and the public communication for the safe use of such materials along with the food (Handy and Shaw, 2007; Aschberger et al., 2015). Though, no such worldwide protocols for testing the toxicity for the nanostructured materials are available; however, the protocols are still under development stage by some of the organizations such as the International Alliance for Nano Environment, Human Health and Safety Harmonization (Maynard et al., 2006) and the US National Research Council (Council, 2007).

6.4 APPLICATION OF BIO-BASED NANOSTRUCTURED MATERIALS IN FOOD SAFETY AND PRESERVATION

As mentioned earlier, the bio-based nanostructured materials are extensively used in everyday life, where nanotechnology makes life easier to benefit the entire society. Bio-based nanostructured materials are deliberated as a promising candidate to supply tailor-made food properties with respect to preservation, protection, and safety throughout the product lifecycle. The nanostructured materials are significantly used to modify the available technologies for food protection particularly in the area of active packaging, smart packaging, edible nanocoating, etc. The significance of edible, active, smart/intelligent, and food functionalization are represented in Figure 6.4, where the nanomaterials play a remarkable role in various emerging sectors. The materials are very essential in many areas for their tunable properties in terms of biodegradability, biocompatibility, nontoxicity, dimensional stability, structure, surface properties, etc. Further, the effective nanomaterials in terms of functionality can be formulated using various functionalizers, plasticizers, cross-linkers, etc. The main functions of the emerging nanomaterials as food functionalization include modifications of texture, aesthetic property, barrier properties, mechanical properties, thermal stability, etc. However, the outgrowth areas of the bio-based nanostructured materials include food functionalization, food packaging as active, smart, edible packaging, etc. as mentioned earlier. To date, various categories of bio-based nanostructured materials are used as nanofillers for the fabrication of composite-based packaging materials such as biodegradable packaging materials, edible packaging, and others. Based on this, the section discusses the inclusion of various bio-based nanostructured materials in food safety and preservation *via* various approaches.

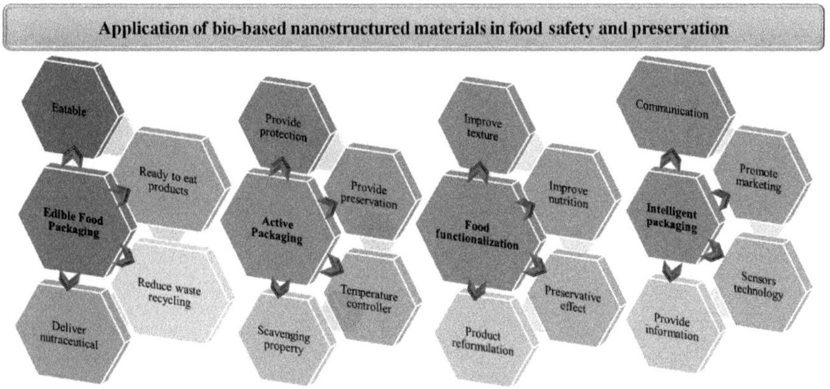

FIGURE 6.4 Application of bio-based nanostructured materials in food.

6.4.1 EDIBLE FOOD PACKAGING

Edible food packaging is an emerging class of packaging that is produced worldwide which is eatable in nature and available in different forms. The two categories of edible food packaging generally include edible films and coatings. Edible films are obtained as a thin layer of edible materials which are applied on food products or can be used as sandwich materials. Whereas, edible coatings are considered as a postharvest technique where the eatable materials are applied to food products *via* dipping or spraying (Figure 6.5). However, modernization in the world shows a great need for natural products. The postharvest management systems, such as drying, freezing, and addition of preservatives, generally change the food quality. In this regard, the world is in a need of natural, fresh, and ready to eat products, which can be obtained through edible food packaging. The materials for edible food packaging generally involve polysaccharides, proteins, fats, or lipids. However, among available polysaccharide materials, cellulose, chitosan, and starch are extensively used materials for edible packaging. Thus, a brief discussion of the available materials for edible food packaging has been detailed in the below section.

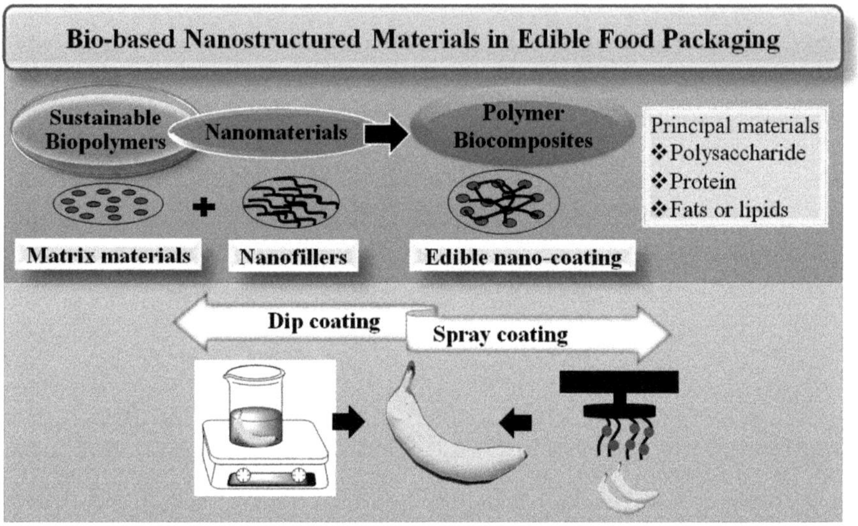

FIGURE 6.5 Bio-based nanostructured materials in edible food packaging with processes used for edible coating.

6.4.1.1 EDIBLE NANOCOATING

The elements of edible nano-coatings generally include biopolymer-based nanoparticles such as nanocrystals, nanofibers, nanowhiskers, SNLs, nanoemulsions, nanoencapsulation, nanocarriers, etc. The nanocellulose is considered as a promising candidate to be used as a material for edible nano-coating. The utilization of magnetic cellulose nanofiber reinforced chitosan provides a candidate for an improved shelf life of cut pineapple fruits (Ghosh et al., 2019). The magnetic cellulose nanofiber is considered as a nanofiller material that can be produced *via* a single-step coprecipitation method, where the addition of the nanofiller can act as a media for iron fortification to the human body. The magnetic cellulose nanofiber dispersed chitosan-based edible nanocoating materials provide improved mechanical, thermal, and texture properties. In this regard, the safety of the materials can be analyzed through the ICPMS technique, where the iron content in the edible nano-coating materials is compared with the available food materials, minimum daily intake, and others (Ghosh et al., 2019). Further, the gelatin-based edible coating with cellulose nanofibers and glycerol can increase the wetting property of coating formulation. The gelatin-based coating formulation has more wetting property on banana epicarp than eggplant epicarps (Andrade et al., 2014). The chitosan nanoparticles and chitosan also act as an active coating for fish fingers in reducing the microbial count at −18 °C (freezing storage) (Abdou et al., 2012). A report suggests that the application of chitosan and nanochitosan coating is also very effective for silver carp fillets at 4 °C (refrigerated storage) (Ramezani et al., 2015). Considerably, nanochitosan provides more antimicrobial activity than chitosan for refrigerated storage of silver carp fillets (Ramezani et al., 2015). Further, the chitosan and chitosan nanoparticle based-edible coating enhance the shelf life of fish fingers up to 6 months due to improving the microbiological properties of fish fingers. The use of chitosan nanoparticle-based edible nanocoating on fresh-cut apples provides a control activity of microbial count (Pilon et al., 2015). Thus, chitosan nanoparticles are very efficient in maintaining the food properties due to their antimicrobial property, which helps in reducing the microbial growth in fruits produces, seafood, meat, and meat products, etc. Though chitosan as the edible coating material is widely used, chitosan nanoparticles provide a significant potential in improving the product quality as edible coating materials. The utilization of various concentrations of chitosan nanoparticles as an edible coating on tomato, chilly, and brinjal helps to improve the product quality by decreasing the weight loss, where chitosan nanoparticle provides significant antifungal and antioxidant activity (Divya et al., 2018). The shrimp-based chitosan coating also helps in improving the product life of strawberry fruit, where the processes of chitosan extraction provide a significant effect in maintaining the quality of strawberry fruit (Benhabiles et al., 2013). The utilization of chitosan nanoparticle-based coatings in improving the quality of whiteleg shrimp, *Litopenaeus vannamei* are very effective at 4 °C in comparison to carboxymethyl chitosan (Wang et al., 2015). Considerably, chitosan nanoparticles with orange and pomegranate peel extract are very helpful in improving the storage life of silver carp at refrigerated storage, where pomegranate peel extract is stronger than orange peel extract (Zarei et al., 2015). Further, silk fibroin, a protein material is found to be very effective in improving the life of perishable fruit products (Marelli et al., 2016). The use of SNL, xanthan gum, polyethylene glycol as edible coating materials helps to improve the shelf life of guava, where the formulation provides a significant improvement in the product life of guava (Zambrano-Zaragoza et al., 2013). The edible coatings based on xanthan gum with various concentrations of SNLs are very significant as edible coating materials on guava fruits due to their proper distribution of nanoparticles in the coating materials. The edible coating is very effective for showing less changes in textural, color, and physiochemical properties (Zambrano-Zaragoza et al., 2013). The application of citrus essential oil loaded nanoemulsion based on chitosan nanoparticle fabricated by emulsion-ionic gelation technique has a significant potential for preserving seafood products (Wu et al., 2016). Further, chitosan nanoparticle with citrus essential oil-based edible coating is more efficient for reducing microbial count and oxidation (lipid) for silvery pomfret. Considerably, nanolaminate coatings are also used for enhanced product life, where nanolaminate-based coating is considered as a promising technique for maintaining the life of food products (Flores-López et al., 2016).

6.4.1.2 EDIBLE FILMS

Similarly, cellulose nanofibers are also utilized with mango puree as nanocomposite-based edible films, where cellulose nanofibers are considered very effective for obtaining tunable packaging properties (Azeredo et al., 2009). Further, the cotton linter cellulose nanofibril-reinforced sodium carboxymethyl cellulose improves the water barrier and tensile properties of carboxymethyl cellulose (Oun and Rhim, 2015). The edible films with cellulose nanofibril incorporated carboxymethyl cellulose form a smooth and flexible edible film with improved properties (Oun and Rhim, 2015). Alginate-acerola puree incorporated cellulose whisker can also be used as edible films, where the inclusion of cellulose whiskers helps to improve the tensile strength, elastic modulus, and water vapor barrier properties (Azeredo et al., 2012). The edible films based on pectin and crystalline nanocellulose can provide improved mechanical properties of the biocomposite films (Chaichi et al., 2017). The addition of pectin with crystalline nanocellulose has the capability of offering improved water barrier properties. The edible films based on pectin with 5% nanocrystalline cellulose can improve the tensile strength up to 84% (Chaichi et al., 2017). The development of polyelectrolyte films utilizing chitosan, olive oil incorporating cellulose nanocrystals is also available as edible films with significant properties (Pereda et al., 2014). The addition of cellulose nanocrystals and olive oil to chitosan-based films helps in reducing the water vapor permeability with improved tensile strength (Pereda et al., 2014). In this way, various biopolymer-based nanoparticles are used as an ingredient for edible films.

6.4.2 ACTIVE PACKAGING

Active packaging is a category of the packaging system, which involves the addition of certain components in the packaging materials or within the package to modify the conditions of packaged food products actively, causing an improvement in the food storage condition by enhancing the shelf life of food products (Figure 6.6). The mechanism of action of such a packaging system is already depicted in the earlier section as shown in Figure 6.3. However, the various active packaging systems are shown in Figure 6.6, where the activity of active packaging involves retaining the food components, eliminating the food components, releasing various components to food systems, and other activities. In such kind of packaging system, the product, packaging materials, and the environment interact with each other maintaining product quality. The mechanism of this specific packaging system involves the changes in the chemical, physical, biological conditions of the product actively for the packaging by maintaining quality, microbiological stability, etc. during the interactions between the internal atmosphere and food products. The main principle behind the packaging system involves enhancing the storage life of packaged food products by maintaining the quality during the interactions between the inside environment and food products. The utilization of active packaging is done through maintaining various changes in the food products such as microbial changes, chemical changes, physical changes, etc. The physiological processes in fresh fruits and vegetables generally involve the respiration kinetics of fresh fruits and vegetables, where the fast respiration rate causes ripening following degradation in the fruit products. Whereas, the chemical processes in the fresh fruits and vegetables occur due to changes in the various food components such as pectin, fat oxidation, etc. Further, the microbiological changes in the food products generally involve the growth of microorganisms due to the favorable condition, where the microbes can be of many types such as psychrophilic, mesophilic, thermophilic, and others. However, all the changes in the food products are interrelated with each other. In this regard, the active packaging is developed and can be controlled with the aid of various active compounds. The main active packaging includes oxygen scavengers, moisture absorber, and controller, generation of carbon dioxide, generation of ethanol, migration of antimicrobial compounds, etc. (Figure 6.6).

As mentioned earlier, active packaging is developed by incorporating some components in the food packaging, where the components are active agents having effective properties of antioxidant, antimicrobial properties against microbes, vapor removal, O_2/CO_2/ethylene scavengers. In this regards, the active packaging materials can be developed through incorporating active filler materials such as silver, TiO_2 nanoparticle, zinc oxide, lignin nanoparticle, chitosan nanoparticle, magnesium oxide, clove oil, polyphenols (sources: tea, coffee, plant extract, fruit extract, etc.) in the composite materials etc. Among available, chitosan nanoparticles are widely used as a sustainable source for developing active packaging, where the nanoparticles of chitosan have more effective antimicrobial properties than chitosan particles. Further, lignin is the most available plant-derived polymers after cellulose, it is the nonsugar part of the polymer found in plant cell walls. On the other hand, polyphenols are the nonvolatile secondary metabolites of plants, which is considered as a potential candidate for the development of antimicrobial packaging materials. Interestingly, various inorganic nanomaterials are widely used for active packaging. In this regard, silver nanoparticles are extensively used as a component in active packaging for the bactericidal effect. Silver nanoparticles can be incorporated into cellulose for developing active packaging with a bactericidal effect against some bacteria (De Moura et al., 2012). Active packaging is also developed using antioxidant compounds such as phenols, flavonoids, ascorbic acids, etc. The loading of eugenol to chitosan nanoparticle can be used as antioxidants for the development of active food packaging (Woranuch and Yoksan, 2013). Chitosan and zinc oxide-based biocomposites are also used as active packaging, which shows the reduction of viability of *E. coli, S. enterica,* and *S. aureus* by 99.9% (Al-Naamani et al., 2016). The edible coating-based on zinc oxide reinforced chitosan-based nanocomposites can be applied on polyethylene films, which provide the antimicrobial properties by developing active food packaging material. Further, a report suggested that binary and ternary polymer-based films based on polyvinyl alcohol/chitosan reinforced with lignin nanoparticles also provide antioxidant and antimicrobial properties, developing an active packaging material (Yang et al., 2016). The nanocomposite films with whey protein isolate, cellulose nanofiber, TiO_2 nanoparticle, rosemary essential oil are considered as an active packaging (Sani et al., 2017). The specific nanocomposite films are significant in preserving the sensory qualities of lamb meat. Further, the microbiological study in terms of total viable count, *Pseudomonas spp. Count, Enterobacteriaceae count,* Lactic acid bacteria show that the nanocomposite films are very effective in reducing the microbial count of the films. The used oxygen scavengers for active modified atmosphere packaging include sachets, labels, trays, films, etc. As shown in Figure 6.6, the active packaging systems include the use of oxygen scavengers, ethylene absorbers, carbon dioxide emitters, moisture scavengers, etc. The use of polyolefin-based nanocomposite films having iron modified kaolinite is also used as active food packaging materials (Busolo and Lagaron, 2012), where iron-containing kaolinite has a property of oxygen scavenging property. The oxygen scavenging-based active packaging can be used for oxygen sensitive food products. Interestingly, endospore-forming bacteria, iron-based component, entrapped microorganism, phenolic compounds, etc., are generally used as oxygen scavengers for active food packaging materials.

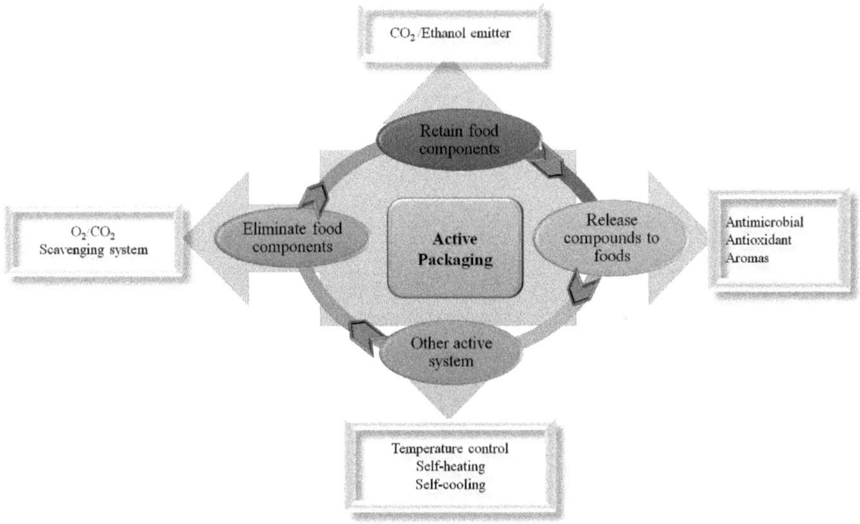

FIGURE 6.6 Various active packaging systems.

6.4.3 *INTELLIGENT PACKAGING*

As mentioned earlier, intelligent packaging provides information relating to food quality during storage and transportation. Further, intelligent or smart packaging generally communicates with the customer *via* promoting the products. Unlike active packaging, smart packaging systems do not involve in extending the shelf life, rather, this specific kind of packaging provides information regarding the packaged food products. The smart/intelligent packaging system generally consists of sensors, indicators, tags/barcodes, holograms, and thermos chromic inks (Sohail et al., 2018). The various technologies used in intelligent packaging (Figure 6.7) are: (1) sensors such as gas sensor (O_2/CO_2 sensor), biosensors, fluorescence-based oxygen sensors, allergen sensor, microbial growth sensors, pathogens, and contaminant sensors, ripeness indicator; (2) indicators such as microbial growth/freshness indicator, time–temperature indicator, gas leakage indicator, pH indicator, integrity indicator, radio frequency identification, physical shock indicator, etc.

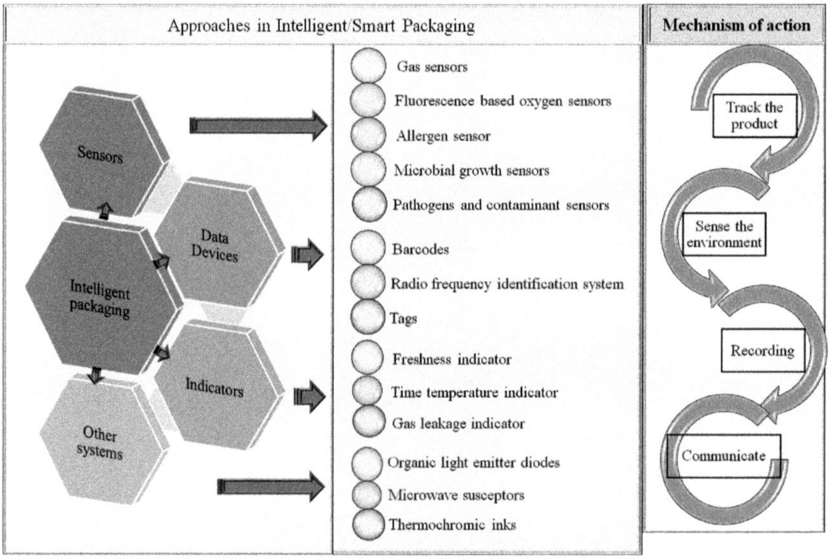

FIGURE 6.7 Approaches and mechanism of intelligent packaging systems.

In addition, the development of bacterial cellulose nanofibers and red cabbage extract containing anthocyanin-based colorimetric pH indicator are utilized as smart packaging for food products (Pourjavaher et al., 2017). The specified smart label can be used as a visual indicator for pH variations during stored packaged food products. A report suggests that nanoengineering of microfibrillated cellulose (MFC) films *via* inkjet-printing of silver nanoparticles can be used in smart packaging, where MFC films can act as a substrate for inkjet printing of electrically conductive structures and multilayer electronic structures (Chinga-Carrasco et al., 2012). The smart packaging can be developed based on bacterial cellulose nanofibers modified by conductive polypyrrole and zinc oxide nanoparticles (Pirsa et al., 2018). Similarly, silver nanoparticles and bacterial cellulose nanopaper are modulated when volatile compounds are released during food spoilage, which delivers a new outgrowth for smart packaging especially gas sensing related areas (Heli et al., 2016). The biodegradable films based on active chitosan/poly-vinyl alcohol with anthocyanin are used for developing smart packaging, where the formulations act as a time–temperature indicator (Pereira et al., 2015). The changes in the color of the developed films provide a simple and cost-effective film that will indicate the chemical changes of the food products such as changes in the pH of milk. Moreover, the water-resistant UV-activated oxygen indicators are formulated using alginate as a coating polymer redox dye, sacrificial electron donor, UV-absorbing semiconducting photocatalyst, where the indicator can be successfully photo-bleached and can regain the color in the oxygen environment (Vu and Won, 2013). Besides, there are many available intelligent packaging such as gravure-printed color indicators (helps in monitoring kimchi fermentation), myoglobin-based indicators (detect the freshness of unmarinated broiler cuts), carrageenan-based colorimetric oxygen indicator, agar/potato starch/anthocyanin extract-based pH indicator, etc. (Hong, 2002; Smolander et al., 2002; Vu and Won, 2014; Choi et al., 2017).

6.4.4 NANOMATERIALS FOR FOOD FUNCTIONALIZATION AND PRESERVATION

The available polymers such as polysaccharides, proteins, and lipids are extensively used materials for food functionalization. Further, the specified materials are very effective in improving and preserving the food texture in terms of firmness, crispiness, color, etc. Among available polysaccharides, cellulose is a versatile material to be used for food functionalization. Cellulose, a kind of indigestible carbohydrate, is widely known as dietary fiber, which is a potential candidate to be used as food additives for food functionalization for its various properties such as chemical, physical, surface properties. Further, cellulose is classified as GRAS food materials by Food and Drug Administration. The general form of cellulose used are cellulose and its derivatives such as carboxymethyl cellulose, hydroxymethyl cellulose, microcrystalline cellulose, nanocellulose, bacterial cellulose, etc. The addition of cellulose in food has many benefits such as an increase in bulkiness, an increase in fiber content, less impact on flavor. Thus, cellulose is also added to solid and liquid food items, when the increase in other fiber content may provide undesirable texture. The use of cellulose materials in food products has many other benefits such as calorie reducing agent, thickening agent. Considerably, cellulose is used to reduce the cost of the product by replacing flours, sugar, lipids, oil, etc. Cellulose aids bulkiness without adding calorie intake to the body. Among available microbial-derived cellulose-based products, one of the widely available products is Nata de coco, which is a jelly-like food product developed *via* fermentation of coconut water. Nata de coco is considered as a healthy food product as dietary fiber is one of the components which provide no-calorie compared to other dessert products. Cellulose-based nanomaterials are utilized for the development of gravies, puddings, soups, desserts, liquid drinks, etc. Nanocellulose is a remarkable candidate for food functionalization for modifying the rheological behavior of food products. Considerably, nanocellulose is used as a stabilizing agent, and a functional ingredient, where nanocellulose has a high surface area, high aspect ratio, better rheological behavior, no cytotoxic, and genotoxic properties, which make it a potential candidate to be used in food functionalization (Serpa et al., 2016).

Moreover, chitosan and chitosan nanoparticle are considered as a food supplement to improve food quality in terms of texture. In this regard, the bioevaluation of crustacean and fungal nanochitosan show that there are no signs of inflammation, fibrosis, or cirrhosis, which provide the safety of the chitosan (Darwesh et al., 2018). Interestingly, the chitosan glucose complex is considered as a modified form of chitosan prepared through heating chitosan with a glucose unit, which is a promising candidate for food preservation (Kanatt et al., 2008). Further, the addition of chitosan glucose complex to lamb meat can increase the shelf life by more than 14 days during chilled storage, which is due to the antioxidant and antimicrobial properties making them a promising candidate for preserving various food formulations. Considerably, nanochitosan is extensively used for food functionalization, where the addition of ascorbic acid-soluble nanochitosan into milk can provide improved functionality (Seo et al., 2011). Further, commercial products of protein supplemented food products are widely available, which are generally available as nanohydrogels, nanoemulsions, and nanoparticles for food fortification.

Based on this, the available bionanostructured materials are extensively used for food preservation. The nanomaterials are involved in the preparation of all types of packaging materials such as primary, secondary, and tertiary packaging materials. As discussed, primary packaging materials, such as edible nanocoating/edible films, are prepared with the various combinations of organic and inorganic nanomaterials. The properties of edible nanocoating are tuned with the aid of various modifications such as nanocellulose, nanochitosan, nanostarch, and others. Further, the secondary packaging materials in the form of blown films can be prepared with the aid of biocomposite-based films such as nanocellulose dispersed polylactic acid films, functionalized chitosan dispersed polylactic acid-based films, nanostarch dispersed polylactic acid, and other combinations. In this way, the various packaging materials are extensively available for the targeted application for an improved shelf life of food products.

6.5 IMPACT OF USING BIO-BASED NANOSTRUCTURED MATERIALS IN SAFETY ANALYSIS

As discussed in the previous section, bio-based nanostructured materials are used in the food sector due to their nontoxic, biodegradable, biocompatible nature. However, there are many available rules and regulations which need to be accompanied for their safe use in the food products. The nutritional requirement of food products should be within the permissible limit, otherwise, many food nutrients can cause adverse health effects if crosses the permissible limits. Further, food packaging is consisting of nanostructured materials, where the nanomaterials can migrate to food products or the components of food product can migrate to packaging materials. In this way, the packaging materials should be tested with many food simulants for ensuring the safety of the food products. Moreover, the components of food packaging materials may react with each other forming various other components. For example, some of the packaging materials in marine condition are harmful to marine life. The packaging components many times develop different hazardous components when come in contact with water or soil. It is well known that some of the conventional polymers destroy soil fertility if left untreated. Thus, a discussion has been made on the available packaging rules and regulations, their migration behavior and toxicological behavior for their safe use.

6.5.1 AVAILABLE RULES AND REGULATIONS

The utilization of food packaging materials should be done under various rules and regulations for their safe use as mentioned below:

- Standards of Weights and Measure Act, 1976
- Standards of Weight and Measures (Packaged Commodities) Rules, 1977 (SWMA)

- Prevention of Food Adulteration Act, 1954
- Prevention of Food Adulteration Rules, 1955 and its first amendment, 2003
- Fruit Products Order, 1955
- Meat Food Products Order, 1973 (MFPO)
- Edible Oil Packaging Order, 1998
- Agmark Rules.

Besides, the available approaches for food packaging sectors are Hazard Analysis Critical Control Point (HACCP), Good Manufacturing Practices (GMP), Good Hygienic Practices (GHP), USFDA, International Organization for Standardization (ISO), Food and Agriculture Organization, Food Safety and Standards Authority of India, etc. Further, food packaging systems should include quality management systems, track and trace systems, etc. Considerably, the materials of food packaging need to be analyzed following various rules and regulations, where the packaging elements should be tested for their toxicity levels, carcinogens content, etc.

Among available approaches, HACCP is generally used for hazard identification in food-based industries. In early days, HACCP was applicable to food processing; however, the HACCP study has been associated with the food packaging processes. The seven steps in HACCP analysis are hazard analysis, CCP identification, establishing critical limits, monitoring, corrective actions, verification, record keeping, and documentation. In the HACCP concept, Critical Control Points (CCP) are determined for chemical, physical, and biological agents, which can be prevented. The chemical hazards include the use of sanitizers, inks; the biological hazards include living organisms, bacteria, parasites, etc., and physical hazards include foreign bodies, such as broken glass, metals, etc. The critical control points for the packaging industry include allergens which include labeling or packaging decoration. The packaging-based industries should be availed with various CCPs for food packaging materials for the proper safety of the food products when packed in the packaging materials. Generally, HACCP-based systems are executed based on some prerequisite programs such as GMP, preventive maintenance program, supplier management system, cleaning and sanitation program, documentation, and Recordkeeping, etc. Further, GMP and GHP are also associated with food contact materials and packaging. According to GMP/GHP for food packaging materials, the contamination, damage should be prevented with the aid of packaging materials. The packaging materials should be able to withstand the handling, storage, and transportation of the packaged products. As discussed earlier, bio-based materials are nontoxic and create no harm to the food materials and environment, thus the food materials are safe for consumption, which is covered under GHP/GMP practices. Further, the GMP program is associated with the safety issues of food contact materials, products, control processes, the establishment of quality systems. However, GMP and FDA regulations deliver very specific guidelines, whereas FDA does not detail the GMP for packaging and other food contact materials. Among available ISO standards, ISO 22000 is related to food safety management systems, where ISO/TS 22002-4:2013 deals with the prerequisite programmers on food safety (Part 4: Food packaging manufacturing). The Codex Alimentarius Commission plays a role in developing the international food standards, code of practice, guidelines to be used in food, production, and safety for international trade. It also provides the general standards for the labeling of prepackaged food products. In this way, there are available various rules and regulations, which specifies the various food packaging requirement for food safety management, food contact materials. The food-based industries should be based on the available rules and regulations such as HACCP, ISO, DDA, GMP/GHP for safe production, and deliberations of food materials.

6.5.2 MIGRATION PROPERTY

Migration in food packaging is generally related to the presence of chemical food contaminants (nanomaterials in food packaging), where the transfer of food contacts materials to food products is generally known as migration. However, the components of food materials can also transfer to packaging materials. The migration in food materials is dependent on many factors such as storage temperature, time, type of food products, categories of packaging materials, properties of migrants, etc. The plastic-based packaging materials are defined as noninert packaging materials, where migration is a problem. In plastic-based packaging, migration can held from outside of the packaging films, and further, the nanostructured materials can also transfer directly to food materials from inside. In this regard, the developed packaging materials are tested for their migration properties in terms of overall migration behavior and specific migration behavior. The overall migration limit is defined as the total mass of migrants that is allowed to pass to food products. On the other hand, a specific migration limit is considered as the mass of specific migrants that may come in contact with the packaging materials.

Specific migration limit is generally measured to find the effect of specific packaging components which may harm human health. The newly developed biocomposite-based packaging materials are tested for overall and specific migration behavior using various food simulants such as ethanol, acetic acid, water, and others. According to European Union legislation, the overall migration limit for plastic is (1) 60 mg/kg of food or food simulants, or (2) 10 mg dm^{-2} (based on a contact area) (Bradley et al., 2009). The biocomposite-based packaging materials are generally consisting of bionanostructured materials, which should have limiting migration behavior when in contact with food simulants. For the measurement of specific migration behavior, the obtained migrant can be tested using nuclear magnetic resonance (NMR), gel permeation chromatography, and others. The packaging materials after the migration test can be tested for their various packaging properties such as barrier, mechanical, thermal properties to check the effect on food contact materials to the packaging properties. NMR is a kind of spectroscopy technique that is used to determine the content, purity, and molecular structure of a sample. The technique can be used to study the molecular conformation and physical properties such as solubility, diffusion, phase changes, conformational exchange, etc. Further, the thermal stability of films can be studied using thermogravimetric analysis, differential scanning calorimetry to know the effect of food simulants on thermal properties. Considerably, oil-based food products have a tendency to leak in the packaging materials, which may be harmful to consumers. Thus, the packaging materials should be tested for various food products to avoid food-related health problems. Further, the migration study should be done at various storage temperature and various risk assessment analyses are executed for the safe use of packaging materials.

6.5.3 TOXICOLOGY STUDY AND NANOBIO EFFECTS

The food products in their whole lifecycle include processing, packaging, storage, and transportation, where different elements can enter in the food chain in various ways. The general food-related contaminants include food additives in the packaging, microbes, environmental agents, toxic agents, pesticide residues, etc. Interestingly, the additives present in food packaging materials and food contact materials are a source of food contamination. Thus, proper assessment of food products and related packaging materials is a serious concern to provide health effective food products, where various toxicological study and nanobio effects are studied for the safety of the food products. In the conventional packaging materials, various organic chemicals, such as phthalates, are added to polyvinyl chloride and further, bisphenol A is added to polyethylene terephthalate (PET) bottles for improved properties, where phthalates can enter into the food products creating many health issues. Thus the development of bio-based packaging materials can be a solution to this problem, where the addition of bionanomaterials can improve the food properties. However, the nanobio effects are defined as the interactions between nanomaterials and biological entities such as polysaccharides, proteins, peptides, cells, human, etc. The nano-bio effects with protein, peptide, and DNA play a critical role in cellular activity. The toxicology study includes mutagenicity, oral toxicity, reproductive toxicity, chronic toxicity, carcinogenicity, cytotoxicity, endocrine disruption potential, etc. The toxicity of the specific compound is tested at various levels, as the toxic compounds provide a negative impact above the permissible limits. The in vitro cytotoxicity bioassays include cellular damage, cell proliferation, cell viability, effects on oxidative metabolism test, etc.

6.6 CURRENT PROSPECTS IN FOOD SAFETY AND PRESERVATION

Food safety and preservation processes play a very crucial role in delivering healthy food products over a year. However, seasonal availability, perishability of food products may restrict in obtaining our favorite food. The global demand for food, especially nutritious food is rising due to various health benefits and a busy lifestyle. In this regard, the various postharvest techniques are used for an improved product life of food products such as drying, freezing, edible nanocoating, edible films, etc. The application of various postharvest techniques for food preservation may change the nutritional quality, whereas the application of edible food packaging improves food quality. It is noteworthy to mention that the application of edible nanocoating or films can provide ready to eat food products with other health beneficial nutrients. As discussed earlier, the development of active and intelligent packaging can provide information regarding packaged food conditions. The food-related problems generally include foodborne illnesses and foodborne intoxication, which are the real cause of food-related diseases. The foodborne intoxication is caused due to the consumption of food containing toxin materials, where the toxins present in the food create health issues. So, the food products should be preserved using a promising approach with good packaging material. The bionanostructured materials are extensively utilized for the fabrication of green composites-based food packaging materials and can provide better safety as it contains

TABLE 6.1 Studies of the Impact of Bio-based Nanostructured Material in Different Aspects of Food

Bio-based Nanostructured Material	Biopolymer Matrix	Applications	Findings	References
Zein nanofibers	Thymol/γ-Cyclodextrin	Food packaging (apples coated by electro spinning)	Antimicrobial activity Enhancement of shelf life of Apple	(Aytac et al., 2017)
Cellulose nanofibers	Fish Gelatin/Palmitic acid	Edible coating	Improved water vapor barrier and mechanical strength	(Wang et al., 2017)
Magnetic cellulose nanofibers	Chitosan	Edible coating (pineapple)	Improved mechanical strength and texture	(Ghosh et al., 2019)
Nanocrystalline cellulose	Chitosan	Food packaging	Improved tensile and swelling property	(Tian et al., 2017)
Cellulose nanospheres, nanorods, nanofibers	Polylactic acid	Food packaging	Improved mechanical and barrier properties	(Yu et al., 2017)
Cellulose nanocrystals	Polylactic acid	Food packaging	Improved thermomechanical, rheological and barrier properties	(Dhar et al., 2017)
Cellulose nanocrystals	Gelatin	Edible coating (Fresh Strawberry)	Enhanced shelf life Minimization of post-harvest loss	(Fakhouri et al., 2014)
Soy protein nanoparticle	–	Release of nutraceuticals (curcumin) in food and biomedical application	Good stability and encapsulation capability Delivery agent of drug or nutraceuticals	(Teng et al., 2012)
Chitosan triphosphate nanoparticle	–	Release of tea catechins (Pharmaceutical and food application)	Control release of tea catechins in vitro study has been done	(Hu et al., 2008)
Soy protein nanoparticle	–	Stabilizer in food and bio sector	The aggregation of soyprotein nanoparticles acts as pickering stabilizer for oil in water emulsions	(Liu et al., 2013)

TABLE 6.1 *(Continued)*

Bio-based Nanostructured Material	Biopolymer Matrix	Applications	Findings	References
Chitosan nanoparticles	Cellulose	Edible films	Improved thermal, mechanical and water vapor barrier properties	(de Moura et al., 2011)
Chitosan nanoparticles	Corn flour	Edible coating (Fish finger)	Enhanced antimicrobial activity during frozen storage of food	(Osheba et al., 2013)
Chitosan nanoparticles	Tara gum	Food packaging	Improved mechanical, physiochemical and barrier properties	(Antoniou et al., 2015)
Chitosan nanoparticles	Fish gelatin	Edible packaging	Improved mechanical and barrier properties Better transparency	(Hosseini et al., 2015)
Silk nano discs	Polylactic acid	Food packaging	Improved thermomechanical and barrier properties	(Patwa et al., 2018)
Nanoemulsion (Lemongrass essential oil)	Chitosan	Edible coating (Grape berry)	Nanoemulsion showed better antimicrobial activity and retention of antioxidant capacity	(Oh et al., 2017)
Nanoemulsion (Clove bud and oregano essential oils)	Methylcellulose	Edible coating (Sliced bread)	Antimicrobial activity and shelf life extender	(Otoni et al., 2014)
Solid lipid nanoparticle	Carnauba wax and candelilla wax	Edible coating (Guava)	Suppression of senescence of several food products	(Zambrano-Zaragoza et al., 2013)

TABLE 6.2 Application of Inorganic Nanomaterial and their Functional Properties in Several Food System

Inorganic Nanostructured Material	Functionality	Polymer Matrix	Application	Key Findings	References
Silver nanoparticles	Retention of volatile compounds	CMC/guar gum	Edible coating (Kinnow fruit)	Preserve food aroma and flavor Extension of shelf life	(Shah et al., 2016)
Silver nanoparticles	Antimicrobial agent	Cellulose	Kiwi, melon juices Poultry and beef samples	99.9% reduction of total viable count of bacteria and yeast 99.0% reduction of total viable count of lactic acid bacteria	(Lloret, et al., 2012)
Silver nanoparticles	Antimicrobial agent	Pullulan	Turkey deli meat	Inhibited *L. monocytogenes*, *S. aureus* over two weeks of refrigerated storage	(Sharoba et al., 2013)
Silicon dioxide nanoparticles	Toughening agent	Soy protein isolate	Packaging	Improved mechanical properties	(Ai et al., 2007)
Siliconoxide nanoparticles	Extension of shelf life	Soy protein isolate	Edible coating (Apple)	Improved gas permeability and tensile strength	(Liu et al., 2017)
Titanium dioxide nanoparticles	Antimicrobial activity	Whey protein isolate/ cellulose nanofiber	Edible film (Lamb Meat)	Effective against gram positive, spoilage, and pathogenic bacteria Improved organoleptic properties Enhancement of shelf life	(Alizadeh Sani et al., 2017)
Zinc oxide nanoparticles	Antimicrobial activity	Carboxymethyl cellulose	Pomegranate	Improved antioxidant activity higher vitamin C and anthocyanin content Enhancement of storage life	(Koushesh Saba et al., 2017)

no hazardous toxic compounds, unlike conventional plastic materials. Moreover, the cellulose, starch, chitosan, silk-based nanostructured materials are extensively utilized for the development of biocomposite-based packaging materials, which are considered as a potential candidate for food packaging materials. The biodegradable packaging is biocompatible, biodegradable, nontoxic, and creates no harm to the environment, which increases its potential use throughout the world.

6.7 CONCLUSION

The available bio-based nanomaterials are considered as promising agents in the food sector especially in edible food packaging (edible nanocoatings and edible films), intelligent packaging, active packaging, and food functionalization. Both the organic and inorganic food-grade nanomaterials are utilized for the development of effective packaging materials. The bio-based nanostructured mediated packaging materials also combine with inorganic nanomaterials for improved effectiveness in terms of packaging properties and activity such as active or smart packaging. The inorganic food-grade nanomaterials play a significant role in the fabrication of smart or intelligent packaging materials. Considerably, the bio-based materials are extensively used for the preparation of biodegradable packaging materials for reduced plastic-based waste and associated carbon footprint. Further, the safety of bionanostructured materials can be measured following available rules and regulations such as European Union legislation, HACCP, GHP, GMP, etc. The migration and toxicological behavior of packaging materials play a very crucial role in selecting the particular packaging materials for specific food items. In this way, the various available bionanostructured materials are used for food preservation considering the available safety rules.

KEYWORDS

- **nanostructured materials**
- **modification**
- **application**
- **safety**
- **preservation**

REFERENCES

Abdou, E. S.; Osheba, A. S.; Sorour, M. A. Effect of Chitosan and Chitosan-nanoparticles as Active Coating on microbiological characteristics of fish fingers. *Int. J. Appl. Sci. Technol.* 2012, 2.

Ai, F.; Zheng, H.; Wei, M.; Huang, J. Soy Protein Plastics Reinforced and Toughened by SiO_2 Nanoparticles. *J. Appl. Polym. Sci.* 2007, 105, 1597–1604.

Alizadeh Sani, M.; Ehsani, A.; Hashemi, M. Whey Protein Isolate/Cellulose Nanofibre/TiO2 Nanoparticle/Rosemary Essential Oil Nanocomposite Film: Its Effect on Microbial and Sensory Quality of Lamb Meat and Growth of Common Foodborne Pathogenic Bacteria during Refrigeration. *Int. J. Food Microbiol.* 2017, 251, 8–14.

Al-Naamani, L.; Dobretsov, S.; Dutta, J. Chitosan-zinc Oxide Nanoparticle Composite Coating for Active Food Packaging Applications. *Inn. Food Sci. Emer. Technol.* 2016, 38, 231–237.

Andrade, R.; Skurtys, O.; Osorio, F.; Zuluaga, R.; Gañán, P.; Castro, C. Wettability of Gelatin Coating Formulations Containing Cellulose Nanofibers on Banana and Eggplant Epicarps. *LWT-Food Sci. Technol.* 2014, 58, 158–165.

Antoniou, J.; Liu, F.; Majeed, H.; Zhong, F. Characterization of Tara Gum Edible Films Incorporated with Bulk Chitosan and Chitosan Nanoparticles: A Comparative Study. *Food Hydrocoll.* 2015, 44, 309–319.

Anwar, N. S.; Kassim, A.; Lim, H.; A Zakarya, S.; Ming, H. Synthesis of Titanium Dioxide Nanoparticles via Sucrose Ester Micelle-Mediated Hydrothermal Processing Route (Sintesis Nanozarah Titanium Dioksida Melalui Kaedah Misel Ester Sukrosa Dalam Proses Hidroterma). *Sains Malays.* 2010, 39, 261–265.

Aschberger, K.; Gottardo, S.; Amenta, V.; Arena, M.; Moniz, F.B.; Bouwmeester, H.; Brandhoff, P.; Mech, A.; Pesudo, L.Q.; Rauscher, H.; Schoonjans, R.; Vettori, M.V.; Peters, R. Nanomaterials in Food - Current and Future Applications and Regulatory Aspects. *J. Phys. Conf. Ser.* 2015, 617, 012032.

Asha Rani, P. V.; Low Kah Mun, G.; Hande, M.P.; Valiyaveettil, S. Cytotoxicity and Genotoxicity of Silver Nanoparticles in Human Cells. *ACS Nano.* 2009, 3, 279–290.

Aytac, Z.; Ipek, S.; Durgun, E.; Tekinay, T.; Uyar, T. Antibacterial Electrospun Zein Nanofibrous Web Encapsulating Thymol/Cyclodextrin-Inclusion Complex for Food Packaging. *Food Chem.* 2017, 233, 117–124.

Azeredo, H. M.; Mattoso, L. H. C.; Wood, D.; Williams, T. G.; Avena-Bustillos, R. J.; McHugh, T. H. Nanocomposite Edible Films from Mango Puree Reinforced with Cellulose Nanofibers. *J. Food Sci.* 2009,74, N31–N35.

Azeredo, H. M.; Miranda, K. W.; Rosa, M. F.; Nascimento, D. M.; de Moura, M. R. Edible Films from Alginate-Acerola Puree Reinforced with Cellulose Whiskers. *LWT — Food Sci. Technol.* 2012, 46, 294–297.

Bellmann, S.; Carlander, D.; Fasano, A.; Momcilovic, D.; Scimeca, J. A.; Waldman, W.J.; Gombau, L.; Tsytsikova, L.; Canady, R.; Pereira, D. I. A.; Lefebvre, D.E. Mammalian Gastrointestinal Tract Parameters Modulating the Integrity, Surface Properties, and Absorption of Food-Relevant Nanomaterials. *Wiley Interdiscip. Rev. Nanomed. Nanobiotechnol.* 2015, 7, 609–622.

Benhabiles, M. S.; Drouiche, N.; Lounici, H.; Pauss, A.; Mameri, N. Effect of Shrimp Chitosan Coatings as Affected by Chitosan Extraction Processes on Postharvest Quality of Strawberry. *J. Food Meas. Charact.* 2013, 7, 215–221.

Bradley, E. L.; Castle, L.; Jickells, S. M.; Mountfort, K. A.; Read, W. A. Use of Overall Migration Methodology to Test for Food-contact Substances with Specific Migration Limits. *Food Addit. Contam.* 2009, 26, 574–582.

Brun, E.; Barreau, F.; Veronesi, G.; Fayard, B.; Sorieul, S.; Chanéac, C.; Carapito, C.; Rabilloud, T.; Mabondzo, A.; Herlin-Boime, N.; Carrière, M. Titanium Dioxide Nanoparticle Impact and Translocation through Ex Vivo, in Vivo and in Vitro Gut Epithelia. *Part. Fibre Toxicol.* 2014, 11, 13.

Bu, Q.; Yan, G.; Deng, P.; Peng, F.; Lin, H.; Xu, Y.; Cao, Z.; Zhou, T.; Xue, A.; Wang, Y.; Cen, X.; Zhao, Y.-L. NMR-Based Metabonomic Study of the Sub-Acute Toxicity of Titanium Dioxide Nanoparticles in Rats after Oral Administration. *Nanotechnology*. 2010, 21, 125105.

Bumbudsanpharoke, N.; Ko, S. Nano-Food Packaging: An Overview of Market, Migration Research, and Safety Regulations. *J. Food Sci.* 2015, 80, R910–R923.

Busolo, M. A.; Lagaron, J. M. Oxygen Scavenging Polyolefin Nanocomposite Films Containing an Iron Modified Kaolinite of Interest in Active Food Packaging Applications. *Inn. Food Sci. Emer. Technol.* 2012, 16, 211–217.

Cerrada, M. L.; Serrano, C.; Sánchez-Chaves, M.; Fernández-García, M.; Fernández-Martín, F.; de Andrés, A.; Riobóo, R. J. J.; Kubacka, A.; Ferrer, M.; Fernández-García, M. Self-Sterilized EVOH-TiO_2 Nanocomposites: Interface Effects on Biocidal Properties. *Adv. Funct. Mater.* 2008, 18, 1949–1960.

Chaichi, M.; Hashemi, M.; Badii, F.; Mohammadi, A. Preparation and Characterization of a Novel Bionanocomposite Edible Film Based on Pectin and Crystalline Nanocellulose. *Carbohydr. Polym.* 2017, *157*, 167–175.

Chen, N.; Song, Z.-M.; Tang, H.; Xi, W.-S.; Cao, A.; Liu, Y.; Wang, H. Toxicological Effects of Caco-2 Cells Following Short-Term and Long-Term Exposure to Ag Nanoparticles. *Int. J. Mol. Sci.* 2016, 17, 974.

Chen, X. Titanium Dioxide Nanomaterials and Their Energy Applications. *Chinese J. Catal.* 2009, 30, 839–851.

Chen, X.-X.; Cheng, B.; Yang, Y.-X.; Cao, A.; Liu, J.-H.; Du, L.-J.; Liu, Y.; Zhao, Y.; Wang, H. Characterization and Preliminary Toxicity Assay of Nano-Titanium Dioxide Additive in Sugar-Coated Chewing Gum. *Small.* 2013, 9, 1765–1774.

Chinga-Carrasco, G.; Tobjörk, D.; Österbacka, R. Inkjet-printed Silver Nanoparticles on Nano-engineered Cellulose Films for Electrically Conducting Structures and Organic Transistors: Concept and Challenges. *J. Nanopart. Res.* 2012, 14, 1213.

Chiu, C.-W.; Lin, J.-J. Self-Assembly Behavior of Polymer-Assisted Clays. *Prog. Polym. Sci.* 2012, 37, 406–444.

Choi, I.; Lee, J. Y.; Lacroix, M.; Han, J. Intelligent pH Indicator Film Composed of Agar/Potato Starch and Anthocyanin Extracts from Purple Sweet Potato. *Food Chem.* 2017, 218, 122–128.

Commissioner, O. of the. Nanotechnology Guidance Documents. *FDA*, 2019.

Darwesh, O. M.; Sultan, Y. Y.; Seif, M. M.; Marrez, D. A. Bio-evaluation of Crustacean and Fungal Nano-chitosan for Applying as Food Ingredient. *Toxicol. Rep.* 2018, 5, 348–356.

Date, A. A.; Desai, N.; Dixit, R.; Nagarsenker, M. Self-Nanoemulsifying Drug Delivery Systems: Formulation Insights, Applications and Advances. *Nanomedicine* 2010, 5, 1595–1616.

de Azeredo, H. M. C. Nanocomposites for Food Packaging Applications. *Food Res. Int.* 2009, 42, 1240–1253.

de Moura, M. R.; Lorevice, M. V.; Mattoso, L. H. C.; Zucolotto, V. Highly Stable, Edible Cellulose Films Incorporating Chitosan Nanoparticles. *J. Food Sci.* 2011, 76, N25–N29.

de Moura, M. R.; Mattoso, L. H.; Zucolotto, V. Development of Cellulose-based Bactericidal Nanocomposites Containing Silver Nanoparticles and their Use as Active Food Packaging. *J. Food Eng.* 2012, 109, 520–524.

Dekkers, S.; Krystek, P.; Peters, R.J.B.; Lankveld, D. P. K.; Bokkers, B. G. H.; van Hoeven-Arentzen, P. H.; Bouwmeester, H.; Oomen, A. G. Presence and Risks of Nanosilica in Food Products. *Nanotoxicology* 2011, 5, 393–405.

Delmas, T.; Piraux, H.; Couffin, A.-C.; Texier, I.; Vinet, F.; Poulin, P.; Cates, M.E.; Bibette, J. How to Prepare and Stabilize Very Small Nanoemulsions. *Langmuir* 2011, 27, 1683–1692.

Dhar, P.; Gaur, S. S.; Soundararajan, N.; Gupta, A.; Bhasney, S. M.; Milli, M.; Kumar, A.; Katiyar, V. Reactive Extrusion of Polylactic Acid/Cellulose Nanocrystal Films for Food Packaging Applications: Influence of Filler Type on Thermomechanical, Rheological, and Barrier Properties. *Ind. Eng. Chem. Res.* 2017, 56, 4718–4735.

Divya, K.; Smitha, V.; Jisha, M. S. Antifungal, Antioxidant and Cytotoxic Activities of Chitosan Nanoparticles and its Use as an Edible Coating on Vegetables. *Int. J. Biol. Macromol.* 2018,114, 572–577.

Dorier, M.; Brun, E.; Veronesi, G.; Barreau, F.; Pernet-Gallay, K.; Desvergne, C.; Rabilloud, T.; Carapito, C.; Herlin-Boime, N.; Carrière, M. Impact of Anatase and Rutile Titanium Dioxide Nanoparticles on Uptake Carriers and Efflux Pumps in Caco-2 Gut Epithelial Cells. *Nanoscale* 2015, 7, 7352–7360.

Duan, Y.; Liu, J.; Ma, L.; Li, N.; Liu, H.; Wang, J.; Zheng, L.; Liu, C.; Wang, X.; Zhao, X.; Yan, J.; Wang, S.; Wang, H.; Zhang, X.; Hong, F. Toxicological Characteristics of Nanoparticulate Anatase Titanium Dioxide in Mice. *Biomaterials* 2010, 31, 894–899.

Echegoyen, Y.; Nerín, C. Nanoparticle Release from Nano-Silver Antimicrobial Food Containers. *Food Chem. Toxicol.* 2013, 62, 16–22.

EFSA Panel on Food Contact Materials, Enzymes, Flavourings and Processing Aids (CEF) Safety Assessment of the Substance Zinc Oxide, Nanoparticles, for Use in Food Contact Materials. *EFSA J.* 2016, 14, 4408.

EFSA Scientific Committee. Guidance on the Risk Assessment of the Application of Nanoscience and Nanotechnologies in the Food and Feed Chain. *EFSA J.* 2011, 9, 2140.

Eichhorn, S. J. Cellulose Nanowhiskers: Promising Materials for Advanced Applications. *Soft Matter*, 2011, 7, 303–315.

Ensign, L. M.; Cone, R.; Hanes, J. Oral Drug Delivery with Polymeric Nanoparticles: The Gastrointestinal Mucus Barriers. *Adv. Drug Deliv. Rev.* 2012, 64, 557–570.

Esmaeillou, M.; Moharamnejad, M.; Hsankhani, R.; Tehrani, A.A.; Maadi, H. Toxicity of ZnO Nanoparticles in Healthy Adult Mice. *Environ. Toxicol. Pharmacol.* 2013, 35, 67–71.

Fakhouri, F. M.; Casari, A. C. A.; Mariano, M.; Yamashita, F.; Mei, L. H. I.; Soldi, V.; Martelli, S. M. Effect of a Gelatin-Based Edible Coating Containing Cellulose Nanocrystals (CNC) on the Quality and Nutrient Retention of Fresh Strawberries During Storage. *IOP Conf. Ser.: Mater. Sci. Eng.* 2014, 64, 012024.

Fathi, M.; Martín, Á.; McClements, D.J. Nanoencapsulation of Food Ingredients Using Carbohydrate Based Delivery Systems. *Trends Food Sci Technol.* 2014, 39, 18–39.

Fathi, M.; Mozafari, M.R.; Mohebbi, M. Nanoencapsulation of Food Ingredients Using Lipid Based Delivery Systems. *Trends Food Sci. Technol.* 2012, 23, 13–27.

Flores-López, M. L.; Cerqueira, M. A.; de Rodríguez, D. J.; Vicente, A. A. Perspectives on Utilization of Edible Coatings and Nano-laminate Coatings for Extension of Postharvest Storage of Fruits and Vegetables. *Food Eng. Rev.* 2016, 8, 292–305.
Fröhlich, E. E.; Fröhlich, E. Cytotoxicity of Nanoparticles Contained in Food on Intestinal Cells and the Gut Microbiota. *Int. J. Mol. Sci.* 2016, 17, 509.
Fröhlich, E.; Roblegg, E. Models for Oral Uptake of Nanoparticles in Consumer Products. *Toxicology*, 2012, 291, 10–17.
Fulgoni, V. L.; Keast, D. R.; Bailey, R. L.; Dwyer, J. Foods, Fortificants, and Supplements: Where Do Americans Get Their Nutrients? *J. Nutr.* 2011, 141, 1847–1854.
Gaillet, S.; Rouanet, J.-M. Silver Nanoparticles: Their Potential Toxic Effects after Oral Exposure and Underlying Mechanisms—A Review. *Food Chem. Toxicol.* 2015, 77, 58–63.
Garcia, T.; Lafuente, D.; Blanco, J.; Sánchez, D.J.; Sirvent, J. J.; Domingo, J. L.; Gómez, M. Oral Subchronic Exposure to Silver Nanoparticles in Rats. *Food Chem. Toxicol.* 2016, 92, 177–187.
Georgantzopoulou, A.; Serchi, T.; Cambier, S.; Leclercq, C. C.; Renaut, J.; Shao, J.; Kruszewski, M.; Lentzen, E.; Grysan, P.; Eswara, S.; Audinot, J.-N.; Contal, S.; Ziebel, J.; Guignard, C.; Hoffmann, L.; Murk, A. J.; Gutleb, A. C. Effects of Silver Nanoparticles and Ions on a Co-Culture Model for the Gastrointestinal Epithelium. *Part. Fibre Toxicol.* 2016, 13, 9.
George, J.; Sabapathi, S. Cellulose Nanocrystals: Synthesis, Functional Properties, and Applications. *Nanotechnol. Sci. Appl.* 2015, 8, 45–54.
Gerloff, K.; Albrecht, C.; Boots, A.W.; Förster, I.; Schins, R.P.F. Cytotoxicity and Oxidative DNA Damage by Nanoparticles in Human Intestinal Caco-2 Cells. *Nanotoxicology*.2009, 3, 355–364.
Ghosh, S. *Nanomaterials Safety: Toxicity and Health Hazards*; Walter de Gruyter GmbH & Co KG, 2018.
Ghosh, T.; Teramoto, Y.; Katiyar, V. Influence of Non-Toxic Magnetic Cellulose Nanofibers on Chitosan based Edible Nanocoating: A Candidate for Improved Mechanical, Thermal, Optical, and Texture Properties. *J. Agric. Food Chem.* 2019, 67, 4289–4299.
Gupta, A.; Eral, H. B.; Hatton, T. A.; Doyle, P. S. Nanoemulsions: Formation, Properties and Applications. *Soft Matter*. 2016, 12, 2826–2841.
Hajipour, M. J.; Fromm, K.M.; Akbar Ashkarran, A.; Jimenez de Aberasturi, D.; de Larramendi, I. R.; Rojo, T.; Serpooshan, V.; Parak, W. J.; Mahmoudi, M. Antibacterial Properties of Nanoparticles. *Trends Biotechnol*. 2012, 30, 499–511.
Handy, R. D.; Shaw, B. J. Toxic Effects of Nanoparticles and Nanomaterials: Implications for Public Health, Risk Assessment and the Public Perception of Nanotechnology. *Health Risk Soc*. 2007, 9, 125–144.
Heli, B.; Morales-Narváez, E.; Golmohammadi, H.; Ajji, A.; Merkoçi, A. Modulation of Population Density and Size of Silver Nanoparticles Embedded in Bacterial Cellulose Via Ammonia Exposure: Visual Detection of Volatile Compounds in a Piece of Plasmonic Nanopaper. *Nanoscale* 2016, 8, 7984–7991.
Hendrickson, O. D.; Klochkov, S. G.; Novikova, O. V.; Bravova, I. M.; Shevtsova, E. F.; Safenkova, I. V.; Zherdev, A. V.; Bachurin, S. O.; Dzantiev, B. B. Toxicity of Nanosilver in Intragastric Studies: Biodistribution and Metabolic Effects. *Toxicol. Lett.* 2016, 241, 184–192.
Hilty, F. M.; Arnold, M.; Hilbe, M.; Teleki, A.; Knijnenburg, J. T. N.; Ehrensperger, F.; Hurrell, R. F.; Pratsinis, S. E.; Langhans, W.; Zimmermann, M. B. Iron from Nanocompounds

Containing Iron and Zinc is Highly Bioavailable in Rats without Tissue Accumulation. *Nat. Nanotechnol.* 2010, 5, 374–380.

Holt, C.; de Kruif, C. G.; Tuinier, R.; Timmins, P.A. Substructure of Bovine Casein Micelles by Small-Angle X-Ray and Neutron Scattering. *Colloids Surf. A.*2003, 213, 275–284.

Hong, S. I. Gravure-printed Colour Indicators for Monitoring Kimchi Fermentation as a Novel Intelligent Packaging. *Packag. Technol. Sci. Int. J.* 2002, 15, 155–160.

Hosseini, S. F.; Rezaei, M.; Zandi, M.; Farahmandghavi, F. Fabrication of Bio-Nanocomposite Films Based on Fish Gelatin Reinforced with Chitosan Nanoparticles. *Food Hydrocoll.* 2015, 44, 172–182.

Hu, B.; Pan, C.; Sun, Y.; Hou, Z.; Ye, H.; Zeng, X. Optimization of Fabrication Parameters to Produce Chitosan-Tripolyphosphate Nanoparticles for Delivery of Tea Catechins. *J. Agric. Food Chem.* 2008, 56, 7451–7458.

Iwanaga, D.; Gray, D. A.; Fisk, I. D.; Decker, E. A.; Weiss, J.; McClements, D. J. Extraction and Characterization of Oil Bodies from Soy Beans: A Natural Source of Pre-Emulsified Soybean Oil. *J. Agric. Food Chem.* 2007, 55, 8711–8716.

Jaiswal, M.; Dudhe, R.; Sharma, P. K. Nanoemulsion: An Advanced Mode of Drug Delivery System. *3 Biotech.* 2015, 5, 123–127.

Jin, T.; Sun, D.; Su, J. Y.; Zhang, H.; Sue, H.-J. Antimicrobial Efficacy of Zinc Oxide Quantum Dots against Listeria Monocytogenes, Salmonella Enteritidis, and *Escherichia coli* O157:H7. *J. Food Sci.* 2009, 74, M46–M52.

Jordan, J.; Jacob, K.I.; Tannenbaum, R.; Sharaf, M.A.; Jasiuk, I. Experimental Trends in Polymer Nanocomposites—a Review. *Mater. Sci. Eng. A.* 2005, 393, 1–11.

Jovanović, B. Critical Review of Public Health Regulations of Titanium Dioxide, a Human Food Additive. *Integr. Environ. Assess Manag.* 2015, 11, 10–20.

Joye, I. J.; Davidov-Pardo, G.; McClements, D. J. Nanotechnology for Increased Micronutrient Bioavailability. *Trends Food Sci. Technol.* 2014, 40, 168–182.

Kanatt, S. R.; Chander, R.; Sharma, A. Chitosan glucose complex—A novel food preservative. *Food Chem.* 2008, 106, 521–528.

Kim, H.-Y.; Park, S.S.; Lim, S.-T. Preparation, Characterization and Utilization of Starch Nanoparticles. *Colloid Surface B.* 2015, 126, 607–620.

Kim, Y. S.; Kim, J. S.; Cho, H. S.; Rha, D. S.; Kim, J. M.; Park, J. D.; Choi, B.S.; Lim, R.; Chang, H. K.; Chung, Y. H.; Kwon, I. H.; Jeong, J.; Han, B. S.; Yu, I. J. Twenty-Eight-Day Oral Toxicity, Genotoxicity, and Gender-Related Tissue Distribution of Silver Nanoparticles in Sprague-Dawley Rats. *Inhal. Toxicol.* 2008, 20, 575–583.

Koushesh Saba, M.; Amini, R. Nano-ZnO/Carboxymethyl Cellulose-Based Active Coating Impact on Ready-to-Use Pomegranate during Cold Storage. *Food Chem.* 2017, 232, 721–726.

Kumar, A. P.; Depan, D.; Singh Tomer, N.; Singh, R. P. Nanoscale Particles for Polymer Degradation and Stabilization—Trends and Future Perspectives. *Prog. Polym. Sci.* 2009, 34, 479–515.

Lane, K. E.; Li, W.; Smith, C.; Derbyshire, E. The Bioavailability of an Omega-3-Rich Algal Oil Is Improved by Nanoemulsion Technology Using Yogurt as a Food Vehicle. *Int. J. Food Sci. Technol.*2014, 49, 1264–1271.

Lichtenstein, D.; Ebmeyer, J.; Knappe, P.; Juling, S.; Böhmert, L.; Selve, S.; Niemann, B.; Braeuning, A.; Thünemann, A. F.; Lampen, A. Impact of Food Components during in Vitro Digestion of Silver Nanoparticles on Cellular Uptake and Cytotoxicity in Intestinal Cells. *Biol. Chem.* 2015, 396, 1255–1264.

Lin, F. Preparation and Characterization of Polymer TiO_2 Nanocomposites via In-Situ Polymerization. 2006.
Liu, F.; Tang, C.-H. Soy Protein Nanoparticle Aggregates as Pickering Stabilizers for Oil-in-Water Emulsions. *J. Agric. Food Chem.* 2013, 61, 8888–8898.
Liu, R.; Liu, D.; Liu, Y.; Song, Y.; Wu, T.; Zhang, M. Using Soy Protein SiOx Nanocomposite Film Coating to Extend the Shelf Life of Apple Fruit. *Int. J. Food Sci. Technol.*2017, 52, 2018–2030.
Livney, Y. D. Milk Proteins as Vehicles for Bioactives. *Curr. Opin. Colloid Interface Sci.* 2010, 15, 73–83.
Livney, Y. D. Nanostructured Delivery Systems in Food: Latest Developments and Potential Future Directions. *Curr. Opin. Food Sci.* 2015, 3, 125–135.
Lloret, E.; Picouet, P.; Fernández, A. Matrix Effects on the Antimicrobial Capacity of Silver Based Nanocomposite Absorbing Materials. *LWT—Food Sci Technol.* 2012, 49, 333–338.
Magnuson, B. A.; Jonaitis, T. S.; Card, J. W. A Brief Review of the Occurrence, Use, and Safety of Food-Related Nanomaterials. *J. Food Sci.* 2011, 76, R126–R133.
Mahshid, S.; Askari, M.; Ghamsari, M.S. Synthesis of TiO2 Nanoparticles by Hydrolysis and Peptization of Titanium Isopropoxide Solution. *J. Mater.* 2007, 189, 296–300.
Marelli, B.; Brenckle, M. A.; Kaplan, D. L.; Omenetto, F. G. Silk Fibroin as Edible Coating for Perishable Food Preservation. *Sci. Rep.* 2016, 6, 25263.
Martínez-Abad, A.; Lagaron, J. M.; Ocio, M.J. Development and Characterization of Silver-Based Antimicrobial Ethylene–Vinyl Alcohol Copolymer (EVOH) Films for Food-Packaging Applications. *J. Agric. Food Chem.* 2012, 60, 5350–5359.
Mason, T. G.; Wilking, J.N.; Meleson, K.; Chang, C. B.; Graves, S.M. Nanoemulsions: Formation, Structure, and Physical Properties. *J. Phys. Condens. Matter.* 2006, 18, R635–R666.
Maynard, A. D.; Aitken, R. J.; Butz, T.; Colvin, V.; Donaldson, K.; Oberdörster, G.; Philbert, M. A.; Ryan, J.; Seaton, A.; Stone, V.; Tinkle, S. S.; Tran, L.; Walker, N. J.; Warheit, D. B. Safe Handling of Nanotechnology. *Nature* 2006, 444, 267.
McClements, D. Edible Nanoemulsions: Fabrication, Properties, and Functional Performance. *Soft Matter*. 2011, 7, 2297–2316.
McClements, D. J. Design of Nano-Laminated Coatings to Control Bioavailability of Lipophilic Food Components. *J. Food Sci.* 2010, 75, R30–R42.
McClements, D. J. Nanoscale Nutrient Delivery Systems for Food Applications: Improving Bioactive Dispersibility, Stability, and Bioavailability. *J. Food Sci.* 2015a, 80, N1602–N1611.
McClements, D. J. Reduced-Fat Foods: The Complex Science of Developing Diet-Based Strategies for Tackling Overweight and Obesity. *Adv. Nutr.* 2015b, 6, 338S–352S.
McClements, D. J. Theoretical Prediction of Emulsion Color. *Adv. Colloid Interface Sci.* 2002, 97, 63–89.
McClements, D. J.; Rao, J. Food-Grade Nanoemulsions: Formulation, Fabrication, Properties, Performance, Biological Fate, and Potential Toxicity. *Crit. Rev. Food Sci. Nutr.* 2011, 51, 285–330.
McClements, D. J.; Xiao, H. Potential Biological Fate of Ingested Nanoemulsions: Influence of Particle Characteristics. *Food Funct.* 2012, 3, 202–220.
Moghassemi, S.; Hadjizadeh, A. Nano-Niosomes as Nanoscale Drug Delivery Systems: An Illustrated Review. *J. Control. Release.* 2014, 185, 22–36.

Muxika, A.; Etxabide, A.; Uranga, J.; Guerrero, P.; de la Caba, K. Chitosan as a Bioactive Polymer: Processing, Properties and Applications. *Int. J. Biol. Macromol.*2017, 105, 1358–1368.

National Research Council. *Toxicity Testing in the 21st Century: A Vision and a Strategy*; National Academies Press, Washington, 2007.

Neethirajan, S.; Jayas, D.S. Nanotechnology for the Food and Bioprocessing Industries. *Food Bioproc. Tech.*2011, 4, 39–47.

Nogueira, C. M.; de Azevedo, W. M.; Dagli, M. L. Z.; Toma, S. H.; Leite, A. Z. de A.; Lordello, M. L.; Nishitokukado, I.; Ortiz-Agostinho, C. L.; Duarte, M. I. S.; Ferreira, M. A.; Sipahi, A. M. Titanium Dioxide Induced Inflammation in the Small Intestine. *World J. Gastroenterol.* 2012, 18, 4729–4735.

Oh, Y. A.; Oh, Y. J.; Song, A. Y.; Won, J. S.; Song, K. B.; Min, S. C. Comparison of Effectiveness of Edible Coatings Using Emulsions Containing Lemongrass Oil of Different Size Droplets on Grape Berry Safety and Preservation. *LWT - Food Sci. Technol.* 2017, 75, 742–750.

Osheba, A.; Sorour, M.; Abdou. Effect of Chitosan Nanoparticles as Active Coating on Chemical Quality and Oil Uptake of Fish Fingers. *J. Agr. Env. Sci.* 2013, 2, 01–14.

Otoni, C. G.; Pontes, S. F. O.; Medeiros, E. A. A.; Soares, N. de F. F. Edible Films from Methylcellulose and Nanoemulsions of Clove Bud (*Syzygium aromaticum*) and Oregano (*Origanum vulgare*) Essential Oils as Shelf Life Extenders for Sliced Bread. *J. Agric. Food Chem.*, 2014, 62, 5214–5219.

Oun, A. A.; Rhim, J. W. Preparation and Characterization of Sodium Carboxymethyl Cellulose/Cotton Linter Cellulose Nanofibril Composite Films. *Carbohydr. Polym.* 2015, 127, 101–109.

Park, E.-J.; Bae, E.; Yi, J.; Kim, Y.; Choi, K.; Lee, S. H.; Yoon, J.; Lee, B. C.; Park, K. Repeated-Dose Toxicity and Inflammatory Responses in Mice by Oral Administration of Silver Nanoparticles. *Environ. Toxicol. Pharmacol.* 2010, 30, 162–168.

Patil, U. S.; Adireddy, S.; Jaiswal, A.; Mandava, S.; Lee, B. R.; Chrisey, D. B. In Vitro/In Vivo Toxicity Evaluation and Quantification of Iron Oxide Nanoparticles. *Int. J. Mol. Sci.* 2015, 16, 24417–24450.

Patwa, R.; Kumar, A.; Katiyar, V. Effect of Silk Nano-Disc Dispersion on Mechanical, Thermal, and Barrier Properties of Poly (Lactic Acid) Based Bionanocomposites. *J. Appl. Polym. Sci.* 2018, 135, 46671.

Pereda, M.; Dufresne, A.; Aranguren, M. I.; Marcovich, N. E. Polyelectrolyte Films Based on Chitosan/Olive Oil and Reinforced with Cellulose Nanocrystals. *Carbohydrate Polymers* 2014, 101, 1018–1026.

Pereira Jr, V. A.; de Arruda, I. N. Q.; Stefani, R. Active Chitosan/PVA Films with Anthocyanins from *Brassica oleraceae* (Red Cabbage) as Time–Temperature Indicators for Application in Intelligent Food Packaging. *Food Hydrocolloids* 2015, 43, 180–188.

Peters, R.; Kramer, E.; Oomen, A. G.; Herrera Rivera, Z. E.; Oegema, G.; Tromp, P. C.; Fokkink, R.; Rietveld, A.; Marvin, H. J. P.; Weigel, S.; Peijnenburg, A. A. C. M.; Bouwmeester, H. Presence of Nano-Sized Silica during In Vitro Digestion of Foods Containing Silica as a Food Additive. *ACS Nano*. 2012, 6, 2441–2451.

Pietroiusti, A.; Magrini, A.; Campagnolo, L. New Frontiers in Nanotoxicology: Gut Microbiota/Microbiome-Mediated Effects of Engineered Nanomaterials. *Toxicol. Appl. Pharmacol.* 2016, 299, 90–95.

Pilon, L.; Spricigo, P. C.; Miranda, M.; de Moura, M. R.; Assis, O. B. G.; Mattoso, L. H. C.; Ferreira, M. D. Chitosan Nanoparticle Coatings Reduce Microbial Growth on Fresh-Cut Apples While Not Affecting Quality Attributes. *Int. J. Food Sci. Technol.* 2015, 50, 440–448.

Piorkowski, D. T.; McClements, D. J. Beverage Emulsions: Recent Developments in Formulation, Production, and Applications. *Food Hydrocoll.* 2014, 42, 5–41.

Pirsa, S.; Shamusi, T.; Kia, E. M. Smart Films Based on Bacterial Cellulose Nanofibers Modified by Conductive Polypyrrole and Zinc Oxide Nanoparticles. *J. Appl. Poly. Sci.* 2018, 135, 46617.

Pourjavaher, S.; Almasi, H.; Meshkini, S.; Pirsa, S.; Parandi, E. Development of a Colorimetric pH Indicator Based on Bacterial Cellulose Nanofibers and Red Cabbage (*Brassica oleraceae*) Extract. *Carb. Poly.* 2017, 156, 193–201.

Powell, J. J.; Faria, N.; Thomas-McKay, E.; Pele, L. C. Origin and Fate of Dietary Nanoparticles and Microparticles in the Gastrointestinal Tract. *J. Autoimmun.* 2010, 34, J226–J233.

Pulit-Prociak, J.; Stokłosa, K.; Banach, M. Nanosilver Products and Toxicity. *Environ. Chem. Lett.* 2015, 13, 59–68.

Ramezani, Z.; Zarei, M.; Raminnejad, N. Comparing the Effectiveness of Chitosan and Nanochitosan Coatings on the Quality of Refrigerated Silver Carp Fillets. *Food Control* 2015, 51, 43–48.

Rao, J.; McClements, D. J. Impact of Lemon Oil Composition on Formation and Stability of Model Food and Beverage Emulsions. *Food Chem.* 2012, 134, 749–757.

Raspopov, R. V.; Trushina, É.; Gmoshinskiĭ, I. V.; Khotimchenko, S. A. Bioavailability of Nanoparticles of Ferric Oxide When Used in Nutrition. Experimental Results in Rats. *Vopr. Pitan.* 2011, 80, 25–30.

Realini, C. E.; Marcos, B. Active and Intelligent Packaging Systems for a Modern Society. *Meat Sci.* 2014, 98, 404–419.

Sagalowicz, L.; Leser, M.E. Delivery Systems for Liquid Food Products. *Curr. Opin. Colloid Interface Sci.* 2010, 15, 61–72.

Sani, M. A.; Ehsani, A.; Hashemi, M. Whey Protein Isolate/Cellulose Nanofibre/TiO2 Nanoparticle/Rosemary Essential Oil Nanocomposite Film: Its Effect on Microbial and Sensory Quality of Lamb Meat and Growth of Common Foodborne Pathogenic Bacteria During Refrigeration. *Int. J. Food Microbiol.* 2017, 251, 8–14.

Sekhon, B. S. Food Nanotechnology – an Overview. *Nanotechnol. Sci. Appl.* 2010, *3*, 1–15.

Seo, M. H.; Chang, Y. H.; Lee, S.; Kwak, H. S. The Physicochemical and Sensory Properties of Milk Supplemented with Ascorbic Acid-soluble Nano-chitosan During Storage. *Int. J. Dairy Technol.* 2011, 64, 57–63.

Serpa, A.; Velásquez-Cock, J.; Gañán, P.; Castro, C.; Vélez, L.; Zuluaga, R. Vegetable Nanocellulose in Food Science: A Review. *Food Hydrocolloids*, 2016, 57, 178–186.

Shah, S. W. A.; Qaisar, M.; Jahangir, M.; Abbasi, K. S.; Khan, S. U.; Ali, N.; Liaquat, M. Influence of CMC- and Guar Gum-Based Silver Nanoparticle Coatings Combined with Low Temperature on Major Aroma Volatile Components and the Sensory Quality of Kinnow (Citrus Reticulata). *Int. J. Food Sci. Technol.* 2016, 51, 2345–2352.

Shak, K. P. Y.; Pang, Y. L.; Mah, S. K. Nanocellulose: Recent Advances and Its Prospects in Environmental Remediation. *Beilstein J. Nanotechnol.* 2018, 9, 2479–2498.

Sharma, V. K.; Siskova, K. M.; Zboril, R.; Gardea-Torresdey, J. L. Organic-Coated Silver Nanoparticles in Biological and Environmental Conditions: Fate, Stability, and Toxicity. *Adv. Colloid. Interface Sci.* 2014, 204, 15–34.

Sharoba, Prof. A.; Khalaf, H. H.; El-Tanahi, H. H.; Morsy, M. Stability of Antimicrobial Activity of Pullulan Edible Films Incorporated with Nanoparticles and Essential Oils and Their Impact on Turkey Deli Meat Quality. *J. Food Dairy Sci. Mansoura Univ.* 2013, 4, 557–573.

Shin, G. H.; Kim, J. T.; Park, H. J. Recent Developments in Nanoformulations of Lipophilic Functional Foods. *Trends Food Sci. Technol.* 2015, 46, 144–157.

Sirelkhatim, A.; Mahmud, S.; Seeni, A.; Kaus, N. H. M.; Ann, L. C.; Bakhori, S. K. M.; Hasan, H.; Mohamad, D. Review on Zinc Oxide Nanoparticles: Antibacterial Activity and Toxicity Mechanism. *Nano-Micro Lett.* 2015, 7, 219–242.

Smolander, M.; Hurme, E.; Latva-Kala, K.; Luoma, T.; Alakomi, H. L.; Ahvenainen, R. Myoglobin-based Indicators for the Evaluation of Freshness of Unmarinated Broiler Cuts. *Inn. Food Sci. Emerg. Technol.* 2002, 3, 279–288.

Sohail, M.; Sun, D. W.; Zhu, Z. Recent Developments in Intelligent Packaging for Enhancing Food Quality and Safety. *Crit. Rev. Food Sci. Nutr.* 2018, 58, 2650–2662.

Song, Z.-M.; Chen, N.; Liu, J.-H.; Tang, H.; Deng, X.; Xi, W.-S.; Han, K.; Cao, A.; Liu, Y.; Wang, H. Biological Effect of Food Additive Titanium Dioxide Nanoparticles on Intestine: An in Vitro Study. *J. Appl. Toxicol.* 2015, 35, 1169–1178.

Sozer, N.; Kokini, J.L. Nanotechnology and Its Applications in the Food Sector. *Trends Biotechnol.* 2009, 27, 82–89.

Tada-Oikawa, S.; Ichihara, G.; Fukatsu, H.; Shimanuki, Y.; Tanaka, N.; Watanabe, E.; Suzuki, Y.; Murakami, M.; Izuoka, K.; Chang, J.; Wu, W.; Yamada, Y.; Ichihara, S. Titanium Dioxide Particle Type and Concentration Influence the Inflammatory Response in Caco-2 Cells. *Int. J. Mol. Sci.* 2016, 17, 576.

Tang, H.; Xiong, H.; Tang, S.; Zou, P. A Starch-Based Biodegradable Film Modified by Nano Silicon Dioxide. *J. Appl. Polym. Sci.* 2009, 113, 34–40.

te Kulve, H.; Konrad, K.; Alvial Palavicino, C.; Walhout, B. Context Matters: Promises and Concerns Regarding Nanotechnologies for Water and Food Applications. *Nanoethics* 2013, 7, 17–27.

Teng, Z.; Luo, Y.; Wang, Q. Nanoparticles Synthesized from Soy Protein: Preparation, Characterization, and Application for Nutraceutical Encapsulation. *J. Agric. Food Chem.* 2012, 60, 2712–2720.

Tian, X.; Yan, D.; Lu, Q.; Jiang, X. Cationic Surface Modification of Nanocrystalline Cellulose as Reinforcements for Preparation of the Chitosan-Based Nanocomposite Films. *Cellulose* 2017, 24, 163–174.

Toker, R. D.; Kayaman-Apohan, N.; Kahraman, M. V. UV-Curable Nano-Silver Containing Polyurethane Based Organic–Inorganic Hybrid Coatings. *Prog. Org. Coat.* 2013, 76, 1243–1250.

Üner, M.; Yener, G. Importance of Solid Lipid Nanoparticles (SLN) in Various Administration Routes and Future Perspectives. *Int. J. Nanomedicine.* 2007, 2, 289–300.

Vu, C. H. T.; Won, K. Leaching-resistant Carrageenan-based Colorimetric Oxygen Indicator Films for Intelligent Food Packaging. *J. Agric. Food Chem.* 2014, 62, 7263–7267.

Vu, C. H. T.; Won, K. Novel Water-resistant UV-activated Oxygen Indicator for Intelligent Food Packaging. *Food Chem.* 2013, 140, 52–56.

Wan, Z.-L.; Guo, J.; Yang, X.-Q. Plant Protein-Based Delivery Systems for Bioactive Ingredients in Foods. *Food Funct.* 2015, 6, 2876–2889.

Wang, W.; Liu, Y.; Jia, H.; Liu, Y.; Zhang, H.; He, Z.; Ni, Y. Effects of Cellulose Nanofibers Filling and Palmitic Acid Emulsions Coating on the Physical Properties of Fish Gelatin Films. *Food Biophys.* 2017, 12, 23–32.

Wang, Y.; Liu, L.; Zhou, J.; Ruan, X.; Lin, J.; Fu, L. Effect of Chitosan Nanoparticle Coatings on the Quality Changes of Postharvest Whiteleg Shrimp, Litopenaeus vannamei, During Storage at 4 C. *Food Bioproc. Technol.* 2015, 8, 907–915.

Wang, Y.; Yuan, L.; Yao, C.; Ding, L.; Li, C.; Fang, J.; Sui, K.; Liu, Y.; Wu, M. A Combined Toxicity Study of Zinc Oxide Nanoparticles and Vitamin C in Food Additives. *Nanoscale* 2014, 6, 15333–15342.

Warheit, D. B.; Brown, S. C.; Donner, E. M. Acute and Subchronic Oral Toxicity Studies in Rats with Nanoscale and Pigment Grade Titanium Dioxide Particles. *Food Chem. Toxicol.* 2015, 84, 208–224.

Watson, S.; Beydoun, D.; Scott, J.; Amal, R. Preparation of Nanosized Crystalline TiO_2 Particles at Low Temperature for Photocatalysis. *J. Nanopart. Res.* 2004, 6, 193–207.

Weir, A.; Westerhoff, P.; Fabricius, L.; Hristovski, K.; von Goetz, N. Titanium Dioxide Nanoparticles in Food and Personal Care Products. *Environ. Sci. Technol.* 2012, 46, 2242–2250.

Woranuch, S.; Yoksan, R. Eugenol-loaded Chitosan Nanoparticles: II. Application in Bio-based Plastics for Active Packaging. *Carb. Polym.* 2013, 96, 586–592.

Wu, C.; Wang, L.; Hu, Y.; Chen, S.; Liu, D.; Ye, X. Edible Coating from Citrus Essential Oil-loaded Nanoemulsions: Physicochemical Characterization and Preservation Performance. *RSC Adv.* 2016, 6, 20892–20900.

Wu, H.; Yin, J.-J.; Wamer, W. G.; Zeng, M.; Lo, Y. M. Reactive Oxygen Species-Related Activities of Nano-Iron Metal and Nano-Iron Oxides. *J. Food Drug Anal.* 2014, 22, 86–94.

Yada, R. Y.; Buck, N.; Canady, R.; DeMerlis, C.; Duncan, T.; Janer, G.; Juneja, L.; Lin, M.; McClements, D. J.; Noonan, G.; Oxley, J.; Sabliov, C.; Tsytsikova, L.; Vázquez-Campos, S.; Yourick, J.; Zhong, Q.; Thurmond, S. Engineered Nanoscale Food Ingredients: Evaluation of Current Knowledge on Material Characteristics Relevant to Uptake from the Gastrointestinal Tract. *Compr. Rev. Food Sci. F.* 2014, 13, 730–744.

Yang, J.; Han, S.; Zheng, H.; Dong, H.; Liu, J. Preparation and Application of Micro/Nanoparticles Based on Natural Polysaccharides. *Carbohydr. Polym.* 2015, 123, 53–66.

Yang, W.; Owczarek, J. S.; Fortunati, E.; Kozanecki, M.; Mazzaglia, A.; Balestra, G. M.; … Puglia, D. Antioxidant and Antibacterial Lignin Nanoparticles in Polyvinyl Alcohol/Chitosan Films for Active Packaging. *Ind. Crops Prod.* 2016, 94, 800–811.

Yang, Y.; Doudrick, K.; Bi, X.; Hristovski, K.; Herckes, P.; Westerhoff, P.; Kaegi, R. Characterization of Food-Grade Titanium Dioxide: The Presence of Nanosized Particles. *Environ. Sci. Technol.* 2014, 48, 6391–6400

Yang, Y.; Marshall-Breton, C.; Leser, M. E.; Sher, A. A.; McClements, D. J. Fabrication of Ultrafine Edible Emulsions: Comparison of High-Energy and Low-Energy Homogenization Methods. *Food Hydrocoll.* 2012, 29, 398–406.

Youssef, A. M.; Abdel-Aziz, M. S. Preparation of Polystyrene Nanocomposites Based on Silver Nanoparticles Using Marine Bacterium for Packaging. *Poly-Plast Technol.* 2013, 52, 607–613.

Yu, H.-Y.; Zhang, H.; Song, M.-L.; Zhou, Y.; Yao, J.; Ni, Q.-Q. From Cellulose Nanospheres, Nanorods to Nanofibers: Various Aspect Ratio Induced Nucleation/Reinforcing Effects on Polylactic Acid for Robust-Barrier Food Packaging. *ACS Appl. Mater. Interfaces.* 2017, 9, 43920–43938.

Zambrano-Zaragoza, M.; Mercado-Silva, E.; Ramirez-Zamorano, P.; Cornejo-Villegas, M.; Gutiérrez-Cortez, E.; Quintanar, D. Use of Solid Lipid Nanoparticles (SLNs) in Edible Coatings to Increase Guava (*Psidiumguajava*L.) Shelf-Life. *Food Res. Int.* 2013, 51, 946–953.

Zarei, M.; Ramezani, Z.; Ein-Tavasoly, S.; Chadorbaf, M. Coating Effects of Orange and Pomegranate Peel Extracts Combined with Chitosan Nanoparticles on the Quality of Refrigerated Silver Carp Fillets. *J. Food Proc. Pres.* 2015, 39, 2180–2187.

Zimmermann, M. B.; Hilty, F. M. Nanocompounds of Iron and Zinc: Their Potential in Nutrition. *Nanoscale.* 2011, 3, 2390–2398.

CHAPTER 7

Campylobacteriosis: Emerging Foodborne Zoonosis

AROCKIASAMY ARUN PRINCE MILTON[1*], GOVINDARAJAN BHUVANA PRIYA[2], MADESH ANGAPPAN[1], KASANCHI M. MOMIN[1], SANDEEP GHATAK[1], and PORTEEN KANNAN[3]

[1]*Division of Animal Health, ICAR Research Complex for NEH Region, Umiam, Meghalaya 793103, India*

[2]*College of Agriculture, Central Agricultural University (Imphal), Kyrdemkulai, Meghalaya 793104, India*

[3]*Department of Veterinary Public Health and Epidemiology, Madras Veterinary College, Chennai, Tamil Nadu 600007, India*

[*]*Corresponding author. E-mail: vetmilton@gmail.com*

ABSTRACT

Among a plethora of foodborne pathogens, *Campylobacter* is well recognized as a leading cause of bacterial food-borne diarrheal disease worldwide. Campylobacteriosis is one of the most commonly reported zoonotic diseases with an estimate of 400–500 million diarrheal cases/year in humans. Although campylobacteriosis is the group description of diseases caused by the members of genus *Campylobacter*, enteritis caused by *C. jejuni* and *C. coli* is the only form of main public health significance. Now, it is well established that poultry, chiefly fresh and frozen broiler meat, is the most important reservoir of *Campylobacter* spp. The clinical spectrum of human campylobacteriosis ranges from a noninflammatory, nonbloody, watery diarrhea to severe inflammatory diarrhea with fever, abdominal pain, and the infection frequently leads to chronic sequelae, such as Guillain–Barré syndrome, Miller Fisher syndrome, and reactive arthritis. The foremost focus of efforts to diminish human campylobacteriosis remains on poultry, especially chicken as it is the widely consumed, relatively cheap meat, and is the main source of the disease. Infected broiler birds carry excessive *Campylobacter* load in their intestinal tract. Therefore, farm-level intervention to prevent or reduce the *Campylobacter* colonization is crucial in any control policy. Besides hygienic procedures at the farm, control interventions during post-harvest processing can reduce the *Campylobacter* numbers on the retail or finished product.

7.1 INTRODUCTION

One of the leading causes of diarrheal disease in human beings that aroused out of obscurity is *Campylobacter* group of organisms, namely *Campylobacter jejuni* and *Campylobacter coli* which was encountered by Butzler et al. (1973) at Belgium and by Skirrow (1977) at England (Butzler and Skirrow, 1979; Karmali and Fleming, 1979). *C. jejuni* is the most commonly isolated species in human gastroenteritis cases whereas *C. coli* contributes only 3–5% (Karmali et al., 1983). *C. jejuni* are also isolated from wild birds, captive birds and animals, poultry, domestic animals, and primates (Luechtefeld et al., 1982; Milton et al., 2017). In the United States, an estimate states that 2.4 million foodborne illnesses are caused by *C. jejuni* annually out of 5.2 million foodborne infections contributed by other bacteria (CDC, 2005; Mead et al., 1999). In Europe, cases of campylobacteriosis recorded and estimated by the European Food Safety Authority (EFSA) annually were about 9 million and it contributes EUR 2.4 billion loss to public health systems (Manzano et al., 2015). Campylobacteriosis is one of the most commonly reported zoonotic diseases with an estimate of 400–500 million Campylobacteriosis cases/year in humans (Ruiz-Palacios, 2007). In 2014, in Europe, The European Centre for Disease Prevention and Control and EFSA had graded campylobacteriosis to be one of the predominant zoonotic diseases with 214,268 confirmed cases (EET, 2012). The symptoms of campylobacteriosis in humans may vary from mild self-limiting to severe inflammatory diarrhea which sometimes may progress to develop auto-immune diseases such as Miller Fisher syndrome and Guillain–Barrè syndrome.

7.2 HISTORY

In 1886, Theodore Escherich observed and reported for the first time about a spiral-shaped nonculturable bacteria (Vandamme et al., 2010). In 1906, two veterinarians from Great Britain specifically identified the presence of peculiar organism (*Campylobacter*) in large numbers in the uterine mucosa of a pregnant sheep (Zilbauer et al., 2008). Following this in 1913, these microorganisms were isolated from aborted cattle fetuses by McFadyean and Stockman. Smith and Orcutt (1927) named a set of bacteria isolated from cattle diarrhoeic feces as *Vibrio jejuni.* Later in 1944, Doyle named a different vibrio isolated from feces of pigs with diarrhea as *Vibrio coli* (Vandamme et al., 2010). In 1963, Sebald and Veron first proposed the genus *Campylobacter* considering certain distinguishing characteristics from true vibrio. Those characteristics were nonfermentative metabolism, microaerophilic growth requirements, and low DNA base composition (On, 2001). Afterwards, a study conducted by Butzler et al. (1973) elevated the attention in *Campylobacter* taking into account their high incidences in human diarrhea cases (On, 2001). Although campylobacters are generally recognized as a human pathogen since 1980, they have been known to cause disease in animals since 1909. Presently, the family *Campylobacteraceae* encompasses two genera, *Campylobacter* and *Arcobacter* (Vandamme, 2000). Genera *Helicobacter* and *Wolinella* of the family *Helicobacteraceae* are the two genetically related closest genera (Vandamme, 2000).

7.3 THE BACTERIA

The name *Campylobacter* was originally derived from "Kampulos," a Greek word which means "Baxtron" rod and bent/curved. Campylobacters are small, Gram-negative bacteria measuring 0.2–0.8 μm × 0.5–5 μm, slender rods, spirally curved (S), and when clustered appear as gull-winged structure (V). Most of the members of this genus are motile by a single polar flagellum which is unsheathed and exhibit a corkscrew-like motility fashion, the exception being *C. showae* that possess peritrichous flagella and *C. gracilis* that is nonmotile (Debruyne et al., 2005). Campylobacters are microaerophilic organisms requiring 85% N_2, 10% CO_2, and 3–5% O_2 for its optimum growth (Park, 2002). Campylobacters are also thermophilic because of their ability to grow at a temperature up to 42 °C (optimum 41.5 °C) but not at 30 °C as they do not possess any cold shock proteins. Levin (2007) termed them as "thermotolerant" instead of thermophilic, as they cannot grow at a temperature of more than 55 °C. This thermotolerant nature of campylobacters decreases the ability of them to grow and multiply outside the host system and in food matrices at processing/storage (Park, 2002). They also form viable but nonculturable cells (VBNC) under conditions that do not favor their growth (Portner et al., 2007). All the campylobacters of the genus oxidize and ferment carbohydrates except *C. gracilis* whose energy source is either amino acids or the intermediates of the TCA cycle (Vandamme, 2000). *C. jejuni* reduces nitrate, hydrolyzes hippurate, and indoxyl acetate. The campylobacters possess a smaller genome with a size of just 1.6–1.7 Mbp, rich in AT with a GC ratio of about 30%. The G+C content of this pathogen is lower than other bacteria ranging from 29 to 38 mol%. Majority strains of campylobacters are resistant to cephalothin and fluoroquinolone group of antibiotics. Currently, there are 28 species and 9 subspecies under the genus *Campylobacter* (Silva et al., 2011; Kaakoush et al., 2015; Caceres et al., 2017).

7.4 SOURCE AND TRANSMISSION

Now, it is well established that poultry, chiefly fresh and frozen broiler meat, is the most important reservoir of *Campylobacter* spp. *C. jejuni* and *C. coli*, associated with *Campylobacter* enteritis are mainly the commensal organisms of the intestinal tract of birds and other animals, occasionally as enteric pathogens in calves, lambs, and puppies (Skirrow, 1990) with *C. coli* mainly in pigs. Animals serve to be the primary reservoir host of *Campylobacter* infection and man serves to be only transient hosts. Manure from farm animals may contaminate water sources, fulfilling the transmission cycle between natural reservoirs. Being zoonotic, transmission to human beings is either by direct or indirect means. Veterinarians, farmers, butchers, slaughterhouse workers, poultry processors, and sometimes the general public are at higher risk of infection due to their close contact with animals or their products and infected pets. This direct, close contact transmission from the above-mentioned sources though well-documented cause only minor infections. Indirect transmission via milk, meat, and water is most important and is the potential source of infection to human beings (Kaakoush et al., 2015). *Campylobacter* count per bird in an uneviscerated chicken is up to 2.4×10^7 (Hood et al., 1988), and the broiler chickens those sold in shops are mostly contaminated with campylobacters. Freshly processed birds might have counts upto1.5×10^6 campylobacters per bird. Damaging effect during the process of freezing and thawing favors the survivability of a small number of organisms on frozen birds (Stern et al., 1985). *C. jejuni* easily colonizes the poultry intestine with a minimum of 35 organisms in 1-day-old chicks and in about 4 weeks most of the birds get colonized that are maintained in commercial operations (Humphrey et al., 1993; Kapperud et al., 1993). Farm personnel (Humphrey*et* et al., 1993; Kapperud et al., 1993; Kazwala et al., 1990), drinking water which is not chlorinated (Pearson et al., 1993) and beetles in the poultry milieu (Jacobs-Reitsma et al., 1995) serve as reservoirs for *Campylobacter* infection. Campylobacters cannot survive in a dry environment, which states that poultry feeds are an improbable source of infection. Consumption of undercooked poultry meat and handling of live birds are identified as some of the risk factors associated with sporadic campylobacteriosis (Kaakoush et al., 2015). Consumption of grilled pork or sausage, intake of untreated milk or water or the milk containers pecked by birds, contact with pets such as dogs, cats or its younger ones with diarrhea and traveling away may account for a lesser percentage of risk factors accounting for sporadic campylobacteriosis (Silva et al., 2011; Kaakoush et al., 2015). Human–human transmission is very less common but the serotypes affecting human is reported in cattle, poultry, and vice versa is reported, signifying that *C. jejuni* is transmitted to humans mainly by animal and animal products (Nielsen et al., 1997).

7.5 VIRULENCE FACTORS AND PATHOGENESIS

As campylobacters do not share their pathogenesis with any other bacterial pathogens, the specific disease-causing mechanisms have not yet been explicated for *Campylobacter* spp. The virulence factors for campylobacters include flagellar mediated motility, adherent ability to gut mucosa, invasiveness, and production of toxins (Dastia et al., 2010; Silva et al., 2011). The survival of bacteria in the GI tract under different environmental conditions is mainly due to its flagella. The flagella also enhance the motility and colonization of the pathogen in the small intestine where the mucosa is highly viscous. The cytolethal distending toxins (CDT) are one of the defined important virulence factors produced by *Campylobacter* spp. and are well characterized than the CDTs produced by other Gram-negative bacteria (Asakura et al., 2008; Silva et al., 2011). CDT is a holotoxin made up of three subunits encoded by three genes (*cdtA, cdtB,* and *cdtC*). CDT's performs their action by arresting the G2/M phase of the eukaryotic cell cycle, restricting them to enter the mitotic phase, and subsequently causing cell death (Zilbauer et al., 2008). In spite of its inadequate knowledge of virulence mechanisms, in general, the pathogen has to ability to resist gastric acids and bile salts that favor them in reaching the intestinal tract. Campylobacter utilizes flagella to colonize the small intestine and moves to its target site that is the colon and invades the mucosa resulting in cellular inflammation which, in turn, produces cytotoxins that abrupt the absorption capacity of the intestine. Despite all this, the host's immune system and the virulent strain of the pathogen determine the severity of the disease (Zilbauer et al., 2008; Silva et al., 2011).

7.6 GLOBAL INCIDENCE

Evidence is sufficient to suggest that there has been a constant rise in the global incidence of *Campylobacter* infection in the past decade. The numbers of campylobacteriosis cases have increased in Europe, North America, and Australia. While data on the incidence of campylobacteriosis in Asia, Africa, and the Middle East are patchy, several reports are available in the public domain indicating the public health importance of the organism and suggest that this infection is endemic in these regions. Due to underreporting, the reported cases of campylobacteriosis in humans are likely to correspond to only the tip of the iceberg. In developing countries, *Campylobacter* infection is generally limited to children, with illness ratio declining with age. This is suggestive of the development of protective immunity owing to exposure in early life (Rao et al., 2001). This is the reason for high asymptomatic infections in these regions, which could contribute to the rapid transmission of infection due to the excretion of pathogens without manifesting any symptoms (Havelaar et al., 2009). Similarly, asymptomatic excretion is also reported in developed nations through a number of studies (de Wit et al., 2001). In England and Wales, UK, 143 campylobacteriosis outbreaks were documented between 1992 and 2009. Of these 143, 114 were attributed to food or water contamination, 2 due to animal contacts, and 22 were due to unknown transmission modes (Little et al., 2010). Between 1999 and 2008, in the United States, based on CDC records, 4936 outbreaks of *Campylobacter* were reported (Batz et al., 2012). According to the US National Outbreak Reporting System, 1550 *Campylobacter* infections and 52 hospitalizations were reported out of 56 and 13 confirmed and suspected outbreaks, respectively, in 1 year (2009–2010) (Hall et al., 2013). A recent detailed evaluation of the epidemiology of *Campylobacter* infection in 27 member states of the European Union indicated an incidence range of 29.9–13500 per 1 lakh population, with the highest incidence in Bulgaria and lowest in Finland and Sweden (Havelaar et al., 2013).

Very limited epidemiological data on campylobacteriosis is available for Asia and the Middle East. Between 2005 and 2006, investigation of the causative agent of gastroenteritis in three hospitals of China revealed that 4.84% of patients were PCR positive for *C. jejuni* out of 3061 patients with diarrhea and the highest prevalence being reported in patients aged lesser than 7 (Huang et al., 2009). In another hospital in Beijing, China from 2005 to 2009, 14.9% of patients with gastroenteritis were detected to be positive for *Campylobacter* species with 127 and 5 patients with *C. jejuni* and *C. coli*, respectively (Chen et al., 2011). Between 2005 and 2006, in the Miyagi Prefecture, Tohoku region of Japan, the numbers of acute gastroenteritis events associated with *Campylobacter, Vibrio parahaemolyticus*, and *Salmonella* infections were estimated to be 1512, 100, and 209 per 1 lakh population per year, respectively. This suggests that *Campylobacter* is accountable for most of the bacterial gastroenteritis episodes in this region (Kubota et al., 2011). Recent bacterial culture-based hospital data from Kolkata, India documented that, 7% out of 3186 patients with gastroenteritis were positive for *Campylobacter* species, with the highest isolates being identified as *C. jejuni* (70%) (Mukherjee et al., 2013). In Israel, the annual incidence of *Campylobacter* infection per 100,000 population has increased from 31.04 cases in 1999 to 90.09 cases in 2010 with the highest incidence of 356.12 cases in children under 2 years of age (Weinberger et al., 2013). Limited data from African countries have indicated that campylobacteriosis is most prevalent in children (Kaakoush et al., 2015). In Blantyre, Malawi of Africa, a 10-year study (1997–2007) report states that 21% of 1941 children hospitalized for diarrhea were detected real-time PCR positive for *C. jejuni* and *C. coli* with 85% accounting to *C. jejuni* (Mason et al., 2013).

7.7 HUMAN CAMPYLOBACTERIOSIS

Although campylobacteriosis is the group description of diseases caused by the members of genus *Campylobacter*, enteritis caused by *C. jejuni* and *C. coli* is the only form of main public health significance (Nachamkin and Blaser, 2000). It has been reported that 50–70% of human campylobacteriosis cases have been accounted for the consumption of poultry meat and its products (Allos, 2001). The postharvest broiler meat handling, processing, and consumption may attribute for 20%–30% of human campylobacteriosis cases, whereas 50%–80% cases may be due to the whole chicken reservoir (EFSA, 2010). The infective dose in children, immunocompromised, and elderly individuals is 500–800 cfu of *C. jejuni* (Black et al., 1988). Earlier, this disease was only considered as a mild illness but a high rate of individuals seeking medical care in recent days reflects the severity of this infection. Recent surveys reveal that 1 in 7 patients in the United Kingdom and 1 in 4 patients in the Netherlands visit a general medical practitioner and roughly 1% of these cases are hospitalized (Tam et al., 2012; Havelaar et al., 2012). The clinical spectrum of human campylobacteriosis ranges from a noninflammatory, nonbloody, watery diarrhea to severe inflammatory diarrhea with fever and abdominal pain mimicking acute peritonitis. The incubation period is generally 2–5 days, sometimes extending up to 10 days. *Campylobacter* enteritis is severe with fever, bloody stool, and severe abdominal pain in developed countries compared to developing countries (Taylor, 1992). The incubation period is generally 2–5 days, sometimes extending up to 10 days. *Campylobacter* enteritis is severe with fever, bloody stool, and severe abdominal pain in developed countries compared to developing countries (Taylor, 1992). In developing countries, the infection is characterized by fever, watery stool, abdominal pain, vomiting, and the presence of leukocytes in the stool. The patients are also dehydrated, malnourished, and underweight (Rao et al., 2001; Coker et al., 2002). Bacteraemia is noticed in less than 1% *Campylobacter* enteritis cases and happens most often in immune-compromised patients (Butzler, 2004). *C. jejuni* also causes extraintestinal infections thus leading to long-term complications like meningitis, pancreatitis, septicemia, abortion, cholecystitis, Guillain-Barré syndrome (GBS), reactive arthritis, Miller Fisher syndrome (SMF), etc. (Butzler, 2004; Hannu et al., 2005). Rarely young patients especially teenagers develop acute appendicitis leading to peritonitis and in majority cases, inflammation of ileum and jejunum with mesenteric adenitis arise. GBS, generally characterized by acute flaccid paralysis is a rare autoimmune disorder of the peripheral nervous system affecting the myelin and axons of nerves. Globally, nearly 25%–40% cases of GBS suffer from *C. jejuni* infection 1–3 weeks preceding the illness (Mishu and Blaser, 1993; Willison et al., 2016). Molecular mimicry between lipooligosaccharide on the *C. jejuni* cell envelope and ganglioside epitopes on the nerves of a human produces a cross-reactive immune response followed by nerve damage. The patient endures from muscle weakness and paralysis (Perera et al., 2007). *C. coli* are also attributed to Guillain–Barre′ syndrome (Eberle and Kiess, 2012).

7.8 DETECTION METHODS

7.8.1 CULTURE METHODS

International Organization for Standardization (ISO) methods are available for isolation and confirmation of *C. jejuni* and *C. coli* (ISO 10272: 2017). For isolation of *Campylobacter* from fecal or intestinal samples, enrichment is not required; samples can be directly plated on to the selective agar. For isolation from food samples, lots of enrichment broths are available (Corry et al., 1995). Generally, enrichment broths comprise of a basal medium, such as nutrient broth, or *Brucella* broth supplemented with antibiotics (Corry et al., 1995). Although maximum enrichment broths were supplemented with sheep blood or lysed horse blood, it has been found that it is not required to isolate *Campylobacter* from poultry meat (Liu et al., 2009). Plating media for selective isolation of *Campylobacter* may be grouped into blood-based and charcoal-based. Routine blood-based agar media are Skirrow agar, Preston agar, Butzler agar, and campy-cefex agar. Some charcoal-based media are CCDA (charcoal cefoperazone deoxycholate agar), CAT agar (cefoperazone, amphotericin, and teicoplanin), and Karmali agar or CSM (charcoal-selective medium). Worldwide, CCDA agar is the most commonly used solid media for the isolation of *Campylobacter* (Bolton and Coates, 1983). On blood-based media like Skirrow agar, the characteristic *Campylobacter* colonies are round, slightly pink, smooth, shiny, and convex with a regular edge. On charcoal containing agars, typical colonies are flat, greyish, moistened, spreading, and may have a metallic sheen (OIE Terrestrial Manual, 2017). The majority of the selective plate media are supplemented with antimicrobials, such as vancomycin and cefoperazone for inhibiting other enteric bacteria. Presently, the combination of 4 mg of amphotericin B and 33 mg of cefoperazone per liter of the medium seems to be best to isolate Campylobacter from broiler chicken (Williams and Oyarzabal, 2012). Suspected colonies are confirmed using biochemical tests such as catalase, oxidase, hippurate, and indoxyl acetate. Hippurate hydrolysis test is the only biochemical test that can distinguish *C. jejuni* (positive test) and *C. coli* (negative test) (Gharst et al., 2013). *Campylobacter* spp. has been conventionally isolated under microaerophilic conditions. Several methods have been developed and routinely deployed for creating a microaerophilic environment. One of the simple methods in vogue is the use of simple sachets (like Oxoid BR56, Oxoid, UK) that produce carbon dioxide from citric acid and sodium bicarbonate or hydrogen from sodium borohydride with palladium as catalyst (Sails et al., 1998). These sachets are used in a sealed anaerobic jar or container. Another method is the evacuation–replacement system, in which a pump is used to evacuate the air in the jar and replaces with air containing a microaerobic mix (Gharst et al., 2013). Various oxygen-quenchers, like lysed horse blood, alkaline hematin, defibrinated animal blood, iron salts, ferrous sulfate, charcoal, norepinephrine, sodium pyruvate, sodium metabisulfite, have been used in liquid and solid media for the isolation of *Campylobacter* spp.

7.8.2 SERODIAGNOSIS

Early researchers have concentrated on the discovery of conserved *Campylobacter* antigens, such as flagellin, lipopolysaccharides, and other protein antigens, for developing immunodiagnostic assays (Nachamkin and Hart, 1986; Lamoureux et al., 1997; Brooks et al., 1998; Kawatsu et al., 2008). Subsequent to the development of immune assays to detect *Campylobacter*, many commercial assays have hit the market like ImmunoCARD STAT! CAMPY kit, ProSpectT *Campylobacter* EIA kit, Premier CAMPY EIA kit (Granato et al., 2010), VIDAS® 30 system (Reis et al., 2018). A study has compared three commercial enzyme immune assay (EIA) kits with conventional culture method and real-time PCR and produced results in favor of commercial EIA kits with specificity and sensitivity above 98% as compared to 94.1% for the microbiological culturing technique. However, they have also advocated that cross-reactivity is the major disadvantage with commercial EIA kits leading to false-positive results (Granato et al., 2010). More recently Masdor et al. (2017) have developed a surface plasma resonance sensor method with an improved limit of *C. jejuni* detection. Similarly, a nano-based (cotton swab) immunological assay developed by Alamer et al. (2018) was proven to be more user friendly and potentially practical adapting to various poultry specific uses. In this assay, cotton swabs were immersed into various cocktails of colored nanobead conjugated *C. jejuni* specific monoclonal antibody. The color intensity change was captured by a smartphone and quantitated using the National Institute of Health ImageJ computer program. This assay has achieved a limit of detection of 10 CFU per ml without any cross-reactivity.

7.8.3 MOLECULAR DETECTION METHODS

A range of polymerase chain reaction (PCR)-based detection assays have been developed and routinely deployed for the identification of *Campylobacter* (On, 1996; Vandamme, 2000). A study has evaluated the specificity of eleven PCR-based assays for the identification of *C. jejuni* and *C. coli* (On and Jordan, 2003). Many studies have developed PCR assays for the detection of *C. jejuni* and *C. coli* using different target genes including 16S rRNA gene (Marshall et al., 1999), 23S rRNA gene, (Hurtado et al., 1997), *glyA* gene (Al Rashid et al., 2000), *gyrA* gene (Menard et al., 2005), *fla* gene (Comi et al., 1996), *lpxA* gene (Klena et al., 2004), or a multiplex of the 16S rRNA, *hipO* (hippuricase), and aspartokinase genes (Linton et al., 1997). In addition, two commercial real-time PCR assays are available in the market: IqCheck™ *Campylobacter* (Bio-Rad, Hercules, CA) and BAX® System for *C. jejuni/coli/lari* (DuPont, Qualicon, Wilmington, DE). A rapid method to distinguish *C. jejuni* and *C. coli* strains is a duplex qPCR, targeting *mapA* gene of *C. Jejuni* and *CeuE* gene of *C. coli* (Best et al., 2003). Yang et al. (2003) have also developed a qPCR assay targeting *VS1* of *C. jejuni* for quantification in milk, poultry, water, and environmental samples. Another qPCR method frequently used to identify and distinguish between *C. jejuni, coli*, and *lari* is developed by Mayr et al. (2010). In recent years, rapid and simple loop-mediated isothermal amplification (LAMP)-based methods have been developed (Dong et al., 2014; Sabike et al., 2016). A commercial LAMP-based kit is also available in the market (Loopamp®, Eiken Chemical, Co., Ltd. (Tokyo, Japan)). In the latest study (Romero and Cook, 2018), a simple and rapid LAMP assay to detect *Campylobacter* was developed successfully evaluated for its applicability in chicken and turkey abattoir. This method is totally equipment free and does not require any culture or DNA-extraction step and generates results in 1 h. A key disadvantage of the nucleic acid-based method is that they cannot discriminate live against dead cells as they detect or quantify all DNA present that is capable of annealing to the oligonucleotides. To alleviate this important issue, a pre-enrichment step may be included to recover only live cells (Kralik and Ricchi, 2017).

7.9 TREATMENT AND ANTIBIOTIC RESISTANCE

Campylobacter enteritis is usually self-limiting and do not necessitate antibiotic coverage. Without antibiotic treatment, *Campylobacter* remains shedding in feces for 2–7 weeks postinfection. Standard recommendations advise antibiotic treatment to patients with comorbidities or severe infections characterized by bloody stool, fever, severe abdominal cramping, or patients having severe diarrhea (>7 days) or having >8 bowel movements per day. Patients whose immune system is compromised (like HIV) should receive antibiotic therapy (Butzler, 2004; Ternhag et al., 2007). Supportive treatments like fluid and electrolyte replacement are generally recommended therapies. Intravenous fluids are crucial for expanding volume in severely dehydrated individuals otherwise oral rehydration is sufficient (Blaser, 1990). When antibiotic treatment is suggested, erythromycin is the drug of option, given its low cost, efficacy, and low toxicity yet other antibiotics, particularly the fluoroquinolones, tetracycline, and newer generation macrolides, like azithromycin, are also indicated (Butzler, 2004). Another alternative drug of choice for systemic infection of *Campylobacter* spp. is gentamicin (Aarestrup and Engberg, 2001). Globally, antibiotic resistance in *Campylobacter* is emerging and has already been documented by many researchers and WHO has also recognized as an important public health problem (Takkinen et al., 2003; McDermott et al., 2005; Moore et al., 2006). The growing resistance to erythromycin, fluoroquinolones, and tetracycline might compromise the treatment efficacy (Aarestrup and Engberg, 2001; Gibreel and Taylor, 2006; Alfredson and Korolik, 2007). The authorization and use of fluoroquinolones in poultry in the USA and Europe triggered the rise of resistance to fluoroquinolone drugs in *Campylobacter* spp. isolated from human patients and livestock (Takkinen et al., 2003; Smith and Fratamico, 2010). In the USA, the prevalence of fluoroquinolone-resistant *C. jejuni* was 0% in 1990, growing to 13% in 1997, and further to 18% in 1999, subsequent to the approval of fluoroquinolones use in poultry farming in 1995 (Butzler, 2004). A higher prevalence of multidrug-resistant *Campylobacter* has been demonstrated in animal isolates than human isolates (EFSA, 2009). The prevalence of such resistant strains is very little in countries where antibiotics usage in poultry production is uncommon (Norstrom et al., 2007).

7.10 PREVENTION AND CONTROL

The foremost focus of efforts to diminish human campylobacteriosis remains on poultry, especially chicken as it is the widely consumed, relatively cheap meat, and is the main source of the disease. Infected broiler birds carry excessive *Campylobacter* load in their intestinal tract. Therefore, farm-level intervention to prevent or reduce the *Campylobacter* colonization is crucial in any control policy. Besides hygienic procedures at the farm, control interventions during postharvest processing can reduce the *Campylobacter* numbers on the retail or finished product.

7.10.1 ON-FARM PREVENTIVE AND CONTROL MEASURES

Every new broiler bird production cycle begins with *Campylobacter* free flock. In the all-in-all-out broiler production system, poultry sheds are properly cleaned and disinfected before the arrival of the new batch to prevent the carry-over contamination from the previous batch of birds. However, birds become colonized with *Campylobacter* (Newell et al., 2011). Studies based on experiments indicate that the ingestion of mere 40 organisms is sufficient for colonization. Once the first chicken has been colonized, then it sheds up to 10^7cfu bacteria per gram of feces. Then it spreads to the entire flock and all the birds become colonized within a very few days. Thus, the main prerequisite to maintain a *Campylobacter*-negative flock is to prevent the colonization of the first bird (Cawthraw et al., 1996). Three important strategies have been suggested to control *Campylobacter* on the poultry at farm-level (1) reducing environmental exposure (strict biosecurity measures), (2) increasing host resistance to reduce gut carriage (vaccination, competitive exclusion), (3) employing antimicrobial alternatives to lessen or eliminate *Campylobacter* from previously colonized birds (Lin, 2009). Prevention and control measures to reduce or eliminate *Campylobacter* colonization in birds can be achieved by various methods. Those include hygienic and biosecurity measures (Ellis-Iversen et al., 2009), water treatment (Chaveerach et al., 2004), vaccination (Solis de los Santos et al., 2010; Skanseng et al., 2010; van Gerwe et al., 2010), passive immunization (Buckley et al., 2010; Layton et al., 2010), application of prebiotics, and probiotics (Schoeni and Wong, 1994; Tsubokura et al., 1997; Willis and Reid, 2008).

The use of antibiotics in food animals is no longer a suitable approach as this has given rise to antibiotic resistance and thereby limited the efficacy of antibiotics in human practice. Some potential was found in the use of probiotics (lactic acid bacteria strains) and could be studied further (Hariharan et al., 2004). The application of the competitive exclusion principle (Nurmi concept) has not been found promising in controlling *Campylobacter* (Mead, 2002). However, administration of bacteriophages specific to *C. jejuni* has successfully reduced *C. jejuni* from 0.5 to 5 log cfu/g of caecal contents in 5 days post-application (Carrillo et al., 2005). In another study, a reduction of 2 log cfu per gram of feces was achieved after administering a 3-phage lytic cocktail to infected chickens (Carvalho et al., 2010). Developing a suitable and effective cocktail bacteriophage treatment by minimizing the development of phage resistance seems a feasible means of reducing the infection levels in infected flocks, although it is difficult to eliminate the organism. Feeding medium-chain fatty acids is another option to reduce the *Campylobacter* colonization of broiler gut (Van Gerwe et al., 2010; Solis de los Santos et al., 2010). Studies concerning the use of essential oils to reduce colonization are in progress (Hermans et al., 2010). To date, there is no single effective vaccine available to control Campylobacter infections in poultry. An effective vaccine should prevent *Campylobacter* colonization or reduce the numbers (> 2 log units) in chickens. De Zoete et al. (2007) have systematically reviewed all the vaccines so far developed against *Campylobacter* in chickens.

7.10.2 POSTHARVEST OR PROCESSING PLANT CONTROL MEASURES

In the slaughterhouse, segregation of *Campylobacter*-positive batch from negative batch and slaughtering of the positive batch has proved to be a fruitful method of reducing the contamination spread. By this method, certified campylobacter free chicken has been produced in Denmark. A 4-h gel-based PCR technique was used for the rapid testing of flocks for segregation. Proper cleaning after slaughtering positive batch was essential to the certification (Krause et al., 2006). Postharvest control measures for controlling colonization of *Campylobacter* in poultry can be achieved by various methods. These include multistage scalds tanks (Hinton et al., 2004), chilling (Boysen and Rosenquist, 2009), irradiation (Collins et al., 1996; Cox and Pavic, 2010), freezing (Stern and Kotula, 1982), and chemical decontamination (Cox and Pavic, 2010).

Cogan et al. (1999) demonstrated that washing with hypochlorite and hot water has significantly reduced the contamination. Spraying of carcasses or dipping with hypochlorite or citric acid or lactic acid has been usually applied but their effectiveness is limited 1–1.5 log units (Ellebroek et al., 2007). Chemical decontamination of carcasses is completely allowed in the US but specific approval is required in EU and, currently, no such chemical decontaminant is approved for use on chicken carcasses in EU. In the US, chemicals used are organic acids, acidified sodium chlorite, trisodium phosphate, and quaternary ammonium compounds. Physical decontamination methods like ultraviolet, ultrasound, etc., have been specifically used on chicken carcasses to reduce *Campylobacter*, but they can attain only maximal reductions of 1–2 log units. Although irradiation methods are highly effective, poor consumer acceptability and difficulty in implementing under high throughput conditions are the major shortcomings. Freezing of chicken carcasses for 3 weeks has been found promising in reducing *Campylobacter* risks although not entirely eliminated (Sandberg et al., 2006). Freezing of carcasses from *Campylobacter*-positive chicken flocks can reduce the bacterial concentration by 2–3 log units and this method has been successfully used in Iceland to reduce human campylobacteriosis cases (Stern et al., 2003). Nevertheless, given the consumer preference of fresh meat and logistic viewpoints, freezing would be difficult to implement, particularly in countries with a high prevalence of *Campylobacter* in poultry (Havelaar et al., 2007).

The reduction of *Campylobacter* numbers in chicken carcasses can lead to an analogous reduction in human *Campylobacter* infections, which is a significant food safety issue. However, there is no reliable, effective, and practical intervention method available to prevent or reduce *Campylobacter* in poultry to date. Except for biosecurity measures, all other measures are still under the process of development. More research is needed on the application of promising intervention measures like bacteriocin treatment and vaccination.

KEYWORDS

- **Campylobacteriosis**
- ***C. jejuni***
- ***C. coli***
- **food-borne**
- **poultry**

REFERENCES

Aarestrup, F. M.; Engberg, J. Antimicrobial Resistance of Thermophilic *Campylobacter*. *Vet. Res.* 2001, 32, 311–321.

Al Rashid, S. T.; Dakuna, I.; Louie, H.; Ng, D.; Vandamme, P.; Johnson, W.; Chan, V. L. Identification of *Campylobacter jejuni, C. coli, C. lari, C. upsaliensis, Arcobacter butzleri*, and *A. butzleri*-like Species Based on the *glyA* Gene. *J. Clin. Microbiol.* 2000, 38, 1488–1494.

Alamer, S.; Eissa, S.; Chinnappan, R.; Zourob, M. A Rapid Colorimetric Immunoassay for the Detection of Pathogenic Bacteria on Poultry Processing Plants Using Cotton Swabs and Nanobeads. *Microchim. Acta.* 2018, 185, 1–10.

Alfredson, D. A.; Korolik, V. Antibiotic Resistance and Resistance Mechanismsin *Campylobacter jejuni* and *Campylobacter coli*. *FEMS Microbiol. Lett.* 2007, 277, 123–132.

Allos, B. M. *Campylobacter jejuni* Infections: Update on Emerging Issues and Trends. *Clin. Infect. Dis*. 2001, 32, 1201–1206.

Asakura, M.; Samosornsuk, W.; Hinenoya, A.; Misawa, N.; Nishimura, K.; Matsuhisa, A.; Yamasaki, S. Development of a Cytolethal Distending Toxin (*cdt*) Gene-based Species-specific Multiplex PCR Assay for the Detection and Identification of *Campylobacter jejuni*, *Campylobacter coli* and *Campylobacter fetus*. *FEMS Immunol. Med. Microbiol.* 2008, 52, 260–266.

Batz, M. B.; Hoffmann, S.; Morris, J. G. Jr. Ranking the Disease Burden of 14 Pathogens in Food Sources in the United States Using Attribution Data from Outbreak Investigations and Expert Elicitation. *J. Food Prot.* 2012, 75, 1278 –1291.

Best, E. L.; Powel, E. J.; Swift, C., Kathleen, A. G.; Frost, J. A. Applicability of a Rapid Duplex Real-time PCR Assay for Speciation of *Campylobacter jejuni* and *Campylobacter coli* Directly from Culture Plates. *FEMS Microbiol.* 2003, 229, 237–241.

Black, R. E.; Levine, M. M.; Clements, M. L.; Hughes T. P.; Blaser, M. J. Experimental *Campylobacter jejuni* Infection in Humans. *J. Infect. Dis.* 1988, 157, 472–479.

Blaser, M. J. *Campylobacter* species. In: Principles and Practice of Infectious Diseases. Mandell GL, Douglas RG, Bennett JE, editors. 3rd ed. New York: Churchill Livingstone, 1990, 194, 1649–1658.

Bolton, F. J.; Coates, D. Development of a Blood-free *Campylobacter* Medium: Screening Tests on Basal Media and Supplements, and the Ability of Selected Supplement to Facilitate Aerotolerance. *J. Appl. Bacteriol.* 1983, 54, 115–125.

Boysen, L.; Rosenquist, H. Reduction of Thermotolerant *Campylobacter species* on Broiler Carcases Following Physical Decontamination at Slaughter. *J. Food. Protec.* 2009, 72, 497–502.

Brooks, B.; Mihowich, W. J. G.; Blais, B. W.; Yamazaki, H. Specificity of Monoclonal Antibodies to *Campylobacter jejuni* Lipopolysaccharide Antigens. *Immunol. Investig.* 1998, 27, 257–265.

Buckley, A. M.; wang, J. H.; Hudson, D. L.; Grant, A. J.; Jones, M. A.; Maskell, D. J.; Stevens, M. P. Evaluation of Live-attenuated Salmonella Vaccines Expressing Campylobacter Antigens for Control of *C. jejuni* in Poultry. *Vaccine.* 2010, 28, 1094–1105.

Butzler, J. P. Campylobacter, from Obscurity to Celebrity. *Clin. Microbiol. Infect.* 2004, 10, 868–876.

Butzler, J. P.; Dekeyser, P.; Detrain, M.; Dehaen, F. Related Vibrios in Stools. *J. Pediatrics.* 1973, 82, 493–495.

Butzler, J. P.; Skirrow, M. B. *Campylobacter* Enteritis. *Clin. Gastroenterol.* 1979, 8, 737–765.

Caceres, A.; Munoz, I.; Iraola, G.; Diaz-Viraque, F.; Collado, L. *Campylobacter ornithocola* sp. nov., a Novel Member of the *Campylobacter lari* Group Isolated from Wild Bird Faecal Samples. *Int. J. Syst. Evol. Microbiol.* 2017, 67, 1643–1649.

Carrillo, C. L.; Atterbury, R. J.; El-Shibiny, A.; Connerton, P. L.; Scott, A.; Connerton, I. F. Bacteriophage Therapy to Reduce *Campylobacter jejuni* Colonization of Broiler Chickens. *Appl. Environ. Microbiol.* 2005, 71, 6554–6563.

Carvalho, C. M.; Gannon, B. W.; Halfhide, D. E.; Santos, S. B.; Hayes, C. M.; Roe, J. M.; Azeredo, J. The in Vivo Efficacy of Two Administration Routes of a Phage Cocktail to Reduce Numbers of *Campylobacter coli* and *Campylobacter jejuni* in Chickens. *BMC Microbiol.* 2010, 10, 232. doi:10.1186/1471-2180-10-232.

Cawthraw, S. A.; Wassenaar, T. M.; Ayling, R.; Newell, D. G. Increased Colonization Potential of *Campylobacter jejuni* Strain 81116 After Passage Through Chickens and its Implication on the Rate of Transmission Within Flocks. *Epidemiol. Infect.* 1996, 117, 213–215

CDC, Campylobacter infections. Atlanta, GA: Department of Health and Human Services, Centers for Disease Control, Division of Bacterial and Mycotic Diseases; 2005. Available at: http://www.cdc.gov/ncidod/dbmd/diseaseinfo/campylobacter_g.htm.

Chaveerach, P.; Keuzenkamp, D. A.; Urlings, H. A.; Lipman, L. J.; Van Knapen, F. In Vitro Study on the Effect of Organic Acids on *Campylobacter jejuni/coli* Populations in Mixtures of Water and Feed. *Poult. Sci.* 2002, 81, 621–628.

Chen, J.; Sun, X. T, Zeng, Z.; Yu, Y. Y. *Campylobacter* Enteritis in Adult Patients with Acute Diarrhea from 2005 to 2009 in Beijing, China. *Chin. Med. J (Engl).* 2011, 124, 1508 –1512.

Cogan, T. A.; Bloomfield, S. F.; Humphrey, T. J. The Effectiveness of Hygiene Procedures for Prevention of Cross-contamination from Chicken Carcasses in the Domestic Kitchen. *Lett. Appl. Microbiol.* 1999, 29, 354–358.

Coker, A. O.; Isokpehi, R. D.; Thomas, B. N.; Amisu, K. O.; Obi, C. L. Human Campylobacteriosis in Developing Countries. *Emerg. Infect. Dis.* 2002, 8, 237–244

Collins, C. I.; Murano, E. A. Wesley, I. V. Survival of *Arcobacter butzleri* and *Campylobacter jejuni* After Irradiation Treatment in Vacuum-packaged Ground Pork. *J. Food Protect.* 1996, 59, 1164–1166.

Comi, G.; Pipan, C.; Botta, G.; Cocolin, L.; Cantoni, C.; Manzano, M. A. Combined Polymerase Chain Reaction and Restriction Endonuclease Enzyme Assay for Discriminating Between *Campylobacter coli* and *Campylobacter jejuni*. *FEMS Immunol. Med. Microbiol.* 1996, 16, 45–49.

Corry, J. E. L.; Post, D. E.; Colin, P.; Laisney, M. J. Culture Media for Isolation of Campylobacters. *Int. J. Food. Microbiol.* 1995, 26, 43–76.

Cox, J.M.; Pavic, A. Advances in Enteropathogen Control in Poultry Production. *J. Appl. Microbiol.* 2010, 108, 745–755.

Dastia, J. I.; Tareena, A. M.; Lugerta, R.; Zautnera,A. E.; Groß, U. *Campylobacterjejuni*: a Brief Overview on Pathogenicity-associated Factors and Disease-mediating Mechanisms. *Int. J. Med. Microbiol.* 2010, 300, 205–211.

de Wit, M. A.; Koopmans, M. P.; Kortbeek, L. M.; Wannet, W. J.; Vinje, J.; van Leusden, F.; Bartelds, A. I.; van Duynhoven, Y. T. Sensor, a Population-based Cohort Study on Gastroenteritis in the Netherlands: Incidence and Etiology. *Am. J. Epidemiol.* 2001, 154, 666 –674.

De Zoete, M. R.; Van Putten, J. P. M.; Wagenaar, J. A. Vaccination of Chickens Against *Campylobacter*. *Vaccine* 2007, 25, 5548–5557.

Debruyne, L.; Gevers, D.; Vandamme, P. "Taxonomy of the family Campylobacteraceae," in Campylobacter, 3rd Ed; I. Nachamkin and M. J. Blaser, Eds; ASM: Washington, DC, 2005, 3–27.

Dong, D.; Zou, D.; Liu, H.; Yang, Z.; Huang, S.; Liu, N.; He, X.; Liu, W.; Huang, L. Rapid Detection of *Pseudomonas aeruginosa* Targeting the *tox*A Gene in Intensive Care Unit Patients from Beijing, China. *Front. Microbiol.* 2015, 6, 1100.

Eberle, K. N.; Kiess, A. S. Phenotypic and Genotypic Methods for Typing *Campylobacter jejuni* and *Campylobacter coli* in Poultry. *Poult. Sci.* 2012, 91, 255–264.

EFSA. The Community Summary Report on Trends and Sources of Zoonoses and Zoonotic. Agents in the European Union in 2007. *EFSA. J.* 2009, 223, 223–440.

Ellebroek, L.; Lienau, J. A.; Alter, T.; Schlichting, D. Effectiveness of Different Chemical Decontamination Methods on the *Campylobacter* Load of Poultry Carcasses. *Fleischwirtschaft.* 2007, 87, 224–227.

Ellis-Iversen, J.; Jorgensen, F.; Bull, S.; Powell, L.; Cook, A. J.; Humphrey, T. J. Risk Factors for *Campylobacter* Colonisation During Rearing of Broiler Flocks in Great Britain. *Prev. Vet. Med.* 2009, 89, 178–184.

European Food Safety Authority (EFSA). Scientific Opinion on Quantification of the Risk Posed by Broiler Meat to Human Campylobacteriosis in the EU. *EFSA J.* 2010, 8, 1437–1526.

Eurosurveillance Editorial Team (EET). The European Union Summary Report on Trends and Sources of Zoonoses, Zoonotic Agents and Food-borne Outbreaks in 2010. *Euro Surveill.* 2012, 17.

Gharst, G.; Oyarzabal, O. A.; Hussain, S. K. Review of Current Methodologies to Isolate and Identify *Campylobacter* spp. from Foods. *J. Microbiol. Methods.* 2013, 95, 84–92.

Gibreel, A.; Taylor, D. E. Macrolide Resistance in *Campylobacter jejuni* and *Campylobacter coli. J. Antimicrob. Chemother.* 2006, 58, 243–255.

Granato, P. A.; Chen, L.; Holiday, I.; Rawling, R. A.; Novak-Weekley, S. M.; Quinlan, T.; Musser, K. A. Comparison of Premier CAMPY Enzyme Immunoassay (EIA), ProSpecT Campylobacter EIA, and ImmunoCard STAT! CAMPY Tests with Culture for Laboratory Diagnosis of *Campylobacter* Enteric Infections. *J. Clin. Microbiol.* 2010, 48, 4022–4027.

Hall, A. J.; Wikswo, M. E.; Manikonda, K.; Roberts, V. A.; Yoder, J. S; Gould, L. H. Acute Gastroenteritis Surveillance Through the National Outbreak Reporting System, United States. *Emerg. Infect. Dis.* 2013, 19, 1305–1309.

Hannu, T.; Mattila, L.; Rautelin, H.; Siitonen, A.; Leirisalo-Repo, M. Three Cases of Cardiac Complications Associated with *Campylobacter jejuni* Infection and Review of the Literature. *Eur. J. Clin. Microbiol. Infect. Dis.* 2005, 24, 619–622.

Hariharan, H.; Murphy, G. A.; Kempf, I. *Campylobacter jejuni*: Public Health Hazards and Potential Control Methods in Poultry: A Review. *Vet. Med.* 2004, 49, 441–446.

Havelaar, A. H.; Haagsma, J. A.; Mangen, M. J. J.; Kemmeren, J. M.; Verhoef, L. P.; Vijgen, S. M.; Wilson, M.; Friesema, I. H.; Kortbeek, L. M.; van Duynhoven, Y. T.; van Pelt, W. Disease Burden of Foodborne Pathogens in The Netherlands, 2009. *Int. J. Food. Microbiol.* 2012, 156, 231–238.

Havelaar, A. H.; van Pelt, W.; Ang, C. W.; Wagenaar, J. A.; van Putten, J. P; Gross, U.; Newell, D. G. Immunity to *Campylobacter*: Its Role in risk Assessment and Epidemiology. *Crit. Rev. Microbiol.* 2009, 35, 1–22.

Havelaar, A.H.; Ivarsson, S.; Lofdahl, M.; Nauta, M. J. Estimating the True Incidence of Campylobacteriosis and Salmonellosis in the European Union, 2009. *Epidemiol. Infect.* 2013, 141, 293–302.

Hermans, D.; Martel, A.; Van Deun K.; Verlinden, M.; Van Immerseel, F.; Garmyn, A.; Messens, W.; Heyndrickx, M.; Haesebrouck, F.; Pasmans, F. Intestinal Mucus Protects

Campylobacter jejuni in the Ceca of Colonized Broiler Chickens Against The Bactericidal Effects of Medium-chain Fatty Acids. *Poult. Sci.* 2010, 89, 1144–1155.

Hinton, A. J. R.; Cason, J. A.; Hume, M. E.; Ingram, K. Use of MIDI-fatty Acid Methyl Ester Analysis to Monitor the Transmission of *Campylobacter* During Commercial Poultry Processing. *J. Food. Prote.* 2004, 67, 1610–1616.

Hood, A. M.; Pearson, A. D.; Shahamat, M. The Extent of Surface Contamination of Retailed Chickens with *Campylobacter jejuni* Serogroups. *Epidemiol. Inf.* 1988, 100, 17–25.

Huang, J.; Yang, G.; Meng, W.; Wu, L.; Zhu, A.; Jiao, X. 2010. An Electrochemical Impedimetric Immunosensor for Label-free Detection of *Campylobacter jejuni* in Diarrhea Patients' Stool Based on O-carboxymethyl Chitosan Surface Modified Fe3O4 Nanoparticles. *Biosens. Bioelectron.* 2010, 25, 1204–1211.

Humphrey, T. J.; Henley, A.; Lanning, D. G. The Colonization of Broiler Chickens with *Campylobacter jejuni*; Some Epidemiologic Investigations. *Epidemiol. Infect.* 1993, 110, 601–607.

Hurtado, A.; Owen, R. J. A Molecular Scheme Based on 23S rRNA Gene Polymorphisms for Rapid Identification of *Campylobacter* and *Arcobacter* Species. *J. Clin. Microbiol.* 1997, 35, 2401–2404.

ISO. 2017. ISO 10272-1:2017. Microbiology of the Food Chain—Horizontal Method for Detection and Enumeration of *Campylobacter* spp. Available at https://www.iso.org/standard/63225.html. Accessed on 28 May, 2019.

Jacobs-Reitsma, W. F.; van de Giessen, A. W.; Bolder, N. M.; Mulder, R. W. A. W. Epidemiology of *Campylobacter* spp. at Two Dutch Broiler Farms. Epidemiol. Infect. 1995, 114, 413–421.

Kaakoush, N. O.; Castano-Rodríguez, N.; Mitchell, H. M; Man, S.M. Global Epidemiology of *Campylobacter* Infection. *Clin. Microbiol. Rev.* 2015, 28, 687–720.

Kapperud, G.; Skjerve, E.; Vik, L.; Hauge, K.; Lysaker, A.; Aalmen, I.; Ostroff, S.M.; Potter, M. Epidemiological Investigation of Risk Factors for *Campylobacter* Colonization in Norwegian Broiler Flocks. *Epidemiol. Infect.* 1993, 111, 45–55.

Karmali, M. A.; Fleming, P. C. *Campylobacter* enteritis. *Can. Med. Assoc. J.* 1979, 120, 1525-1532.

Karmali, M. A.; Penner, J. L.; Fleming, P. C.; Williams, A.; Hennessy, J. N. The serotype and biotype distribution of clinical isolates of *Campylobacter jejuni* and *Campylobacter coli* over a three-year period. *J. Infect. Dis.* 1983, 147, 243–246. http://dx.doi.org/10.1093/infdis/147.2.243.

Kawatsu, K.; Kumeda, Y.; Taguchi, M.; Yamazaki-Matsune, W.; Kanki, M.; Inoue, K. Development and Evaluation of Immunochromatographic Assay for Simple and Rapid Detection of *Campylobacter jejuni* and *Campylobacter coli* in Human Stool Specimens. *J. Clin. Microbiol.* 2008, 46, 1226–1231.

Kazwala, R. R.; Collins, J. D.; Hannan, J.; Crinion, R. A. P.; O'Mahony, H. Factors Responsible for the Introduction and Spread of *Campylobacter jejuni* Infection in Commercial Poultry Production. *Vet. Rec.* 1990, 126, 305–306.

Klena, J. D.; Parker, C. T.; Knibb, K.; Ibbitt, J. C.; Devane, P. M. L.; Horn, S. T.; Miller W. G.; Konkel, M. E. Differentiation of *Campylobacter coli, Campylobacter jejuni, Campylobacter lari*, and *Campylobacter upsaliensis* by a Multiplex PCR Developed from the Nucleotide Sequence of the Lipid A Gene *lpxA*. *J. Clin. Microbiol.* 2004, 42, 5549–5557.

Kralik, P.; Ricchi, M. A Basic Guide to Real Time PCR in Microbial Diagnostics: Definitions, Parameters, and Everything. *Front. Microbiol.* 2017, 8, 1–9.

Krause, M.; Josefsen, M. H.; Lund, M.; Jacobsen, N. R.; Brorsen, L.; Moos, M.; Stockmarr, A.; Hoorfar, J. Comparative, Collaborative, and On-site Validation of a TaqMan PCR Method as a Tool for Certified Production of Fresh, *Campylobacter*- Free Chickens. *Appl. Environ. Microbiol.* 2006, 72, 5463–5468.

Kubota, K.; Kasuga, F.; Iwasaki, E.; Inagaki, S.; Sakurai, Y.; Komatsu, M.; Toyofuku, H.; Angulo, F. J., Scallan, E.; Morikawa, K. Estimating the Burden of Acute Gastroenteritis and Foodborne Illness Caused by *Campylobacter, Salmonella*, and *Vibrio parahaemolyticus* by Using Population Based Telephone Survey Data, Miyagi Prefecture, Japan, 2005 to 2006. *J. Food Prot.* 2011, 74, 1592–1598.

Lamoureux, M.; Mackay, A.; Messier, S.; Fliss, I.; Blais, B. W.; Holley, R. A.; Simard, R. E. Detection of *Campylobacter jejuni* in Food and Poultry Viscera Using Immunomagentic Separation and Micro Titre Hybridization. *J. Appl. Microbiol.* 1997, 83, 641–651.

Layton, S. L.; Morgan, M. J.; Cole, K.; Kwon, Y. M.; Donoghue, D. J.; Hargis, B. M.; Pumford, N. R. Evaluation of Salmonella-vectored Campylobacter Peptide Epitopes for Reduction of *Campylobacter jejuni* in Broiler Chickens. *Clin. Vaccine. Immunol.* 2010, 18, 449–454.

Levin, R. E. *Campylobacter jejuni*: A Review of its Characteristics, Pathogenicity, Ecology, Distribution, Subspecies Characterization and Molecular Methods of Detection. *Food. Biotechnol.* 2007, 21, 271–347.

Lin, J. Novel Approaches for *Campylobacter* Control in Poultry. *Foodborne Pathog. Dis.* 2009, 6, 755–765.

Linton, D.; Lawson, A. J.; Owen, R. J.; Stanley, J. PCR Detection, Identification to Species Level, and Fingerprinting of *Campylobacter jejuni* and *Campylobacter coli* Direct from Diarrheic Samples. *J. Clin. Microbiol.* 1997, 35, 2568–2572.

Little, C. L.; Gormley, F. J.; Rawal, N.; Richardson, J. F. A Recipe for Disaster: Outbreaks of Campylobacteriosis Associated with Poultry Liver Pate in England and Wales. *Epidemiol. Infect.* 2010, 138, 1691–1694.

Liu, L.; Hussain, S. K.; Miller, R. S.; Oyarzabal, O. A. Research Note: Efficacy of miniVIDAS for the Detection of *Campylobacter* spp. From Retail Broiler Meat Enriched in Bolton Broth with or Without the Supplementation of Blood. *J. Food Prot.* 2009, 72, 2428–2432.

Luechtefeld, N. W.; Wang, W. L. L. Animal Reservoirs of *Campylobacter jejuni*, in *Campylobacter*: Epidemiology, Pathogenesis, and Biochemistry. Newell, D. G., Ed.; MTP Press: Lancaster, U.K. 1982, 249.

Manzano, M.; Cecchini, F.; Fontanot, M.; Iacumin, L.; Comi, G.; Melpignano, P. OLED-based DNA Biochip for *Campylobacter* spp. Detection in Poultry Meat Samples. *Biosens. Bioelectron.* 2015, 66, 271–276.

Marshall, S. M.; Melito, P. L.; Woodward, D. L.; Johnson, W. M.; Rodgers, F. G.; Mulvey, M. R. Rapid Identification of *Campylobacter, Arcobacter*, and *Helicobacter* Isolates by PCR Restriction Fragment Length Polymorphism Analysis of the *16S rRNA* Gene. *J. Clin. Microbiol.* 1999, 37, 4158–4160.

Masdor, N. A.; Altintas, Z.; Tothill, I. E. Surface Plasmon Resonance Immunosensor for the Detection of *Campylobacter jejuni. Chem. Aust.* 2017, 5, 1–15.

Mason, J.; Iturriza-Gomara, M.; O'Brien, S. J.; Ngwira, B. M.; Dove, W., Maiden, M. C.; Cunliffe, N. A. *Campylobacter* Infection in Children in Malawi is Common and is Frequently Associated with Enteric Virus Coinfections. *PLoS One*. 2013, 8, e59663.

Mayr, A. M.; Lick, S.; Bauer, J.; Thärigen, D.; Busch, U.; Huber, I. Rapid Detection and Differentiation of *Campylobacter jejuni, Campylobacter coli,* and *Campylobacter lari* in Food, Using Multiplex Real-time PCR. *J. Food Prot.* 2010, 73, 241–250.

McDermott, P. F.; Bodeis-Jones, S. M.; Fritsche, T. R.; Jones, R. N.; Walker, R. D.; The Campylobacter Susceptibility Testing Group. Broth Microdilution Susceptibility Testing of *Campylobacter jejuni* and the Determination of Quality Control Ranges for Fourteen Antimicrobial Agents. *J. Clin. Microbiol.* 2005, 43, 6136–6138.

Mead, G. C. Factors Affecting Intestinal Colonization of Poultry by *Campylobacter* and Role of Microflora in Control. *Worlds Poult. Sci. J.* 2002, 58, 169–178.

Mead, P. S.; Slutsker, L.; Dietz, V.; McCaig, L. F.; Bresee, J. S.; Shapiro, C.; Griffin, P. M.; Tauxe, R. V. Food-related Illness and Death in the United States. *Emerg. Infect. Dis.* 1999, 5, 607–625

Ménard, A.; Dachet, F.; Prouzet-Mauleon, V.; Oleastro, M.; Mégraud, F. Development of a Real-time Fluorescence Resonance Energy Transfer PCR to Identify the Main Pathogenic *Campylobacter* spp. *Clin. Microbiol. Infect.* 2005, 11, 281–287.

Mishu, B.; Blaser, M. J. Role of Infection due to *Campylobacter jejuni* in the Initiation of Guillain-Barre syndrome. *Clin. Infect. Dis.* 1993, 17, 104–108.

Milton AAP, Agarwal RK, Priya GB et al. Prevalence of Campylobacter jejuni and Campylobacter coli in captive wildlife species in India. *Iranian J. Vet. Res.* 2017, 18, 177–182.

Moore, J.; Barton, M.; Blair, I.; Corcoran, D.; Dooley, J.; Fanning, S.; Kempf, I.; Lastovica, A.; Lowery, C.; and Seal, B. The Epidemiology of Antibiotic Resistance in *Campylobacter* spp. *Microbes. Infect.* 2006, 8, 1955–1966.

Mukherjee, P.; Ramamurthy, T.; Bhattacharya, M. K.; Rajendran, K.; Mukhopadhyay, A. K. *Campylobacter jejuni* in Hospitalized Patients with diarrhea, Kolkata, India. *Emerg. Infect Dis.* 2013, 19, 1155–1156.

Nachamkin, I.; Blaser, M. J. *Campylobacter*, 2nd ed; American Society for Microbiology: Washington; 2000.

Nachamkin, I.; Hart, A. M. Common and Specific Epitopes of *Campylobacter* Flagellin Recognized by Monoclonal Antibodies. *Appl. Environ. Microbiol.* 1986, 53, 438–440.

Newell, D. G.; Elvers, K. T.; Dopfer, D.; Hansson, I.; Jones, P.; James, S. Biosecurity Based Interventions and Strategies to Reduce *Campylobacter spp.* on Poultry Farms. *Appl. Environ. Microbiol.* 2011, 77, 8605–8614. Epub 2011/10/11.

Nielsen, E. M.; Engberg, J.; Madsen, M. Distribution of Serotypes of *Campylobacter jejuni* and *C. coli* from Danish Patients, Poultry, Cattle, and Swine. *FEMS Immunol.* Med. *Microbiol.* 1997. 19, 47–56.

Norström, M.; Johnsen, G.; Hofshagen, M.; Tharaldsen, H.; Kruse, H. Antimicrobial Resistance in *Campylobacter jejuni* from Broilers and Broiler House Environments in Norway. *J. Food Prot.* 2007, 70, 736–737.

OIE Terrestrial Manual. Chapter 2.9.3. Infection with *Campylobacter jejuni* and *C. coli.* World Organization for Animal Health (OIE), 2017.

On, S. L. W. Identification Methods for *Campylobacters, Helicobacters*, and Related Organisms. *Clin. Microbiol. Rev.* 1996, 9, 405–422.

On, S. L. W. Taxonomy of *Campylobacter, Arcobacter, Helicobacter* and Related Bacteria: Current Status, Future Prospects and Immediate Concerns. *J. Appl. Microbiol.* 2001, 90, 1S–15S.

On, S. L. W.; Jordan, P. J. Evaluation of 11 PCR Assays for Species-level Identification of *Campylobacter jejuni* and *Campylobacter coli. J. Clin. Microbiol.* 2003, 41, 330–336.

Park, S. F. The physiology of *Campylobacter* Species and its Relevance to Their Role as Food Borne Pathogens. *Int. J. Food. Microbiol.* 2002, 74, 177–188.

Pearson, A. D.; Greenwood, M.; Healing, T. D.; Rollins, D.; Shahamat, M.; Donaldson, J.; Colwell, R. R. Colonization of Broiler Chickens by Waterborne *Campylobacter jejuni. Appl. Environ. Microbiol.* 1993, 59, 987–996.
Perera, V. N.; Nachamkin, I.; Ung, H.; Patterson, J. H.; McConville, M. J.; Coloe, P.J.; Fry, B. N. Molecular Mimicry in *Campylobacter jejuni*: Role of the Lipo-oligosaccharide Core Oligosaccharide in Inducing Antiganglioside Antibodies. *FEMS. Immunol. Med. Microbiol.* 2007, 50, 27–36.
Portner, D. C.; Leuschner, R. G. K.; Murray, B. S. Optimising the Viability During Storage of Freeze-dried Cell Preparations of *Campylobacter jejuni. Cryobiology* 2007, 54, 265–270.
Rao, M. R.; Naficy, A. B.; Savarino, S. J.; Abu-Elyazeed, R.; Wierzba, T. F.; Peruski, L. F.; Abdel-Messih, I.; Frenck, R.; Clemens, J. D. Pathogenicity and Convalescent Excretion of *Campylobacter* in Rural Egyptian Children. *Am. J. Epidemiol.* 2001, 154, 166–173.
Reis, L. P.; Meneze, L. D. M.; Lima, G. K.; de Souza Santos, E. L.; Dorneles, E. M. S.; de Assis, D. C. S.; Lage, A. P.; Cancado, S. D.; de Figueiredo T. C. Detection of *Campylobacter* spp. in Chilled and Frozen Broiler Carcasses Comparing Immunoassay, PCR and Real Time PCR Methods. *Ciênc. Rural, Santa Maria.* 2018, 48, e20161034.
Romero, M. R.; Cook, N. A Rapid LAMP-Based Method for Screening Poultry Samples for *Campylobacter* Without Enrichment. *Front. Microbiol.* 2018, 9, 2401.
Ruiz-Palacios, G. M. The Health Burden of *Campylobacter* Infection and the Impact of Antimicrobial Resistance: Playing Chicken. *Clin. Infect. Dis.* 2007, 44, 701–703.
Sabike, I. I.; Uemura, R.; Kirino, Y.; Mekata, H.; Sekiguchi, S.; Okabayashi, T.; Goto, Y.; Yamazaki, W. Use of Direct LAMP Screening of Broiler Fecal Samples for *Campylobacter jejuni* and *Campylobacter coli* in the Positive Flock Identification Strategy. *Front. Microbiol.* 2016, 7, 1582. doi: 10.3389/fmicb.2016.01582.
Sails, A. D.; Wareing, D. R. A.; Bolton, F. J. Evaluation of Three Microaerobic Systems for the Growth and Recovery of *Campylobacter* spp. In Lastovica, A.J., Newell, D.G., Lastovica, E.E., Eds.; *Campylobacter, Helicobacter and Related Organisms*. The Rustica Press: Pinelands, South Africa, 1998, 39–42.
Sandberg, M.; Nygård, K.; Meldal, H.; Valle, P. S.; Kruse, H.; Skjerve, E. Incidence Trend and Risk Factors for *Campylobacter* Infections in Humans in Norway. BMC Public Health 2006, 6, 179. doi:10.1186/1471- 2458-6-179.
Schoeni, J. L.; Wong, A. C. Inhibition of *Campylobacter jejuni* Colonisation in Chicks by Defined Competitive Exclusion Bacteria. *Appl. Environ. Microbiol.* 1994, 60, 1191–1197.
Silva, J.; Leite, D., Fernandes, M.; Mena, C., Gibbs, P. A.; Teixeira, P. Campylobacter spp. as a Foodborne Pathogen: A Review. *Front. Microbiol.* 2011, 2, 200.
Skanseng, B.; Kaldhusdal, M.; Moen, B.; Gjevre, A. G.; Johannessen, G. S.; Sekelja, M.; Trosvik, P.; Rudi, K. Prevention of Intestinal *Campylobacter jejuni* Colonization in Broilers by Combinations of In-feed Organic Acids. *J. Appl. Microbiol.* 2010, 109, 1265–1273.
Skirrow MB. Campylobacter Enteritis: A "New" Disease. *Br. Med. J.* 1977, 2(6078), 9–11.
Skirrow, M. B. *Campylobacter* and *Helicobacter* Infections of Man and Animals. In T. Parker and L. H. Collier, Eds.; *Topley and Wilson's Principles of Bacteriology, Virology and Immunity*; Edward Arnold: London, 1990, 8, 529–542.
Smith, J. L.; Fratamico, P. M. Fluoroquinoloneresistance in *Campylobacter*. *J. Food. Prot.* 2010, 73, 1141–1152.
Solis de los Santos, F. S.; Hume, M.; Venkitanarayanan, K.; Donoghue, A. M.; Hanning, I.; Slavik, M.F.; Aguiar, V.F.; Metcalf, J. H.; Reyes-Herrera, I.; Blore, P. J.; Donoghue, D. J. Caprylic Acid Reduces Enteric *Campylobacter* colonization in Market-aged Broiler

Chickens but does not Appear to Alter Cecal Microbial Populations. *J. Food. Prot.* 2010, 73, 251–257.

Stern, N. J.; Hiett, K. L.; Alfredsson, G. A.; Kristinsson, K. G.; Reiersen, J.; Hardardottir, H.; Briem, H.; Gunnersson, E.; Georgsson, F.; Lowman, R.; Berndtson, E.; Lammerding, A. M.; Paoli, G. M.; Musgrove, M. T. *Campylobacter* spp. in Icelandic Poultry Operations and Human Disease. *Epi. Infect.* 2003, 130, 23–32.

Stern, N. J.; Kotula, A. W. Survival of *Campylobacter jejuni* Inoculated into Ground Beef. *Appl. Environ. Microbiol.* 1982, 44, 1150.

Stern, N. J.; Rothenberg. P. J.; Stone, J. M. Enumeration and Reduction of *Campylobacter jejuni* in Poultry and Red Meats. *J. Food. Protect.* 1985, 48, 606–610.

Takkinen, J.; Ammon, A.; Robstad, O.; Breuer, T.; *Campylobacter* Working Group. European Survey on *Campylobacter* Surveillance and Diagnosis 2001. *EuroSurveill.* 2003, 8, 207–213.

Tam, C. C.; Rodrigues, L. C.; Viviani, L.; Dodds, J. P.; Evans, M. R.; Hunter PR, et al. Longitudinal Study of Infectious Intestinal Disease in the UK (IID2 study): Incidence in the Community and Presenting to General Practice. *Gut* 2012, 61, 69–77. pmid:21708822.

Taylor, D. N. *Campylobacter* Infections in Developing Countries. In Nachamkin, I., Blaser, M. J., Tompkins, L. S., Eds; *Campylobacter jejuni: Current Status and Future Trends*; Washington: Clin. Microbiol. Rev. 1992, 20–30.

Ternhag, A.; Asikainen, T.; Giesecke, J.; Ekdahl, K. A Meta-analysis on the Effects of Antibiotic Treatment on Duration of Symptoms Caused by Infection with Campylobacter Species. *Clin. Infect. Dis.* 2007, 44, 696–700.

Tsubokura, K.; Berndtson, E.; Bogstedt, A.; Kaijser, B.; Kim, M.; Ozeki, M.; Hammarstrom, L. Oral Administration of Antibodies as Prophylaxis and Therapy in *Campylobacter jejuni*-infected Chickens. *Clin. Exp. Immunol.* 1997, 108: 451–455.

Van Gerwe, T.; Bouma, A.; Klinkenberg, D.; Wagenaar, J. A.; Jacobs-Reitsma, W. F.; Stegeman, A. Medium Chain Fatty Acid Feed Supplementation Reduces the Probability of *Campylobacter jejuni* Colonization in Broilers. *Vet. Microbiol.* 2010, 143, 314–318.

Vandamme, P. Taxonomy of the Family Campylobacteraceae. In *Campylobacter*; I. Namchamkin, M. J. Blaser, Eds.; ASM: Washington, DC; 2000, 3–27.

Vandamme, P.; Debruyne, L.; De Brandt, E.; Falsen, E. Reclassification of *Bacteroides ureolyticus* as *Campylobacter ureolyticus* comb. nov., and Emended Description of the Genus Campylobacter. *Int. J. Syst. Evol. Microbiol.* 2010, 60, 2016–2022.

Williams, A.; Oyarzabal, O. A. Prevalence of *Campylobacter* spp. in skinless, Boneless Retail Broiler Meat from 2005 Through 2011 in Alabama, USA. *BMC Microbiol.* 2012, 12, 184.

Willis, W. L.; Reid, L. Investigating the Effects of Dietary Probiotic Feeding Regimens on Broiler Chicken Production and *Campylobacter jejuni* Presence. *Poult. Sci.* 2008, 87, 606–611.

Willison, H. J.; Jacobs, B. C.; Van Doorn, P. A. Guillain-Barré Syndrome. *Lancet.* 2016, 388 (10045), 717–727.

Yang, C.; Jiang, Y.; Huang, K.; Zhu, C.; Yin, Y. Application of Real-time PCR for Quantitative Detection of *Campylobacter jejuni* in Poultry, Milk and Environmental Water. *FEMS Immunol. Med. Microbiol.* 2003, 38, 265–271.

Zilbauer, M.; Dorrell, N.; Wren, B. W.; Bajaj-Elliott, M. *Campylobacter jejuni* Mediated Disease Pathogenesis: An Update. *Trans. R. Soc. Trop. Med. Hyg.* 2008, 120, 123–129.

CHAPTER 8

Enzymatic Modification of Ferulic Acid Content in Arabinoxylans from Maize Distillers Grains: Effect on Gels Rheology

JORGE A. MÁRQUEZ-ESCALANTE and ELIZABETH CARVAJAL-MILLAN

Laboratory of Biopolymers, Research Center for Food and Development (CIAD, AC), Carretera Gustavo Enrique Astiazarán Rosas No. 46, Col. La Victoria, Hermosillo 83304, Sonora, Mexico

[*]Corresponding author.
E-mail: marquez1jorge@gmail.com; ecarvajal@ciad.mx

ABSTRACT

Maize distillers grain (MDG) is a by-product of distillery plants produced in large amounts worldwide. MDG is used in livestock feed, however, it is rich in arabinoxylans (AX) which could add value and favor the sustainable development of distillery plants. The aim of this work was to modify the ferulic acid (FA) content in AX from MDG by using a ferulate esterase and to investigate the rheological characteristics of the gels formed. FA content in AX and ferulate esterase treated AX (FAX) was 5.0 and 3.7 μg/mg polysaccharide, respectively. AX and FAX formed gels at 2% (w/v) with elasticity (G') and viscosity (G'') values of 195 and 112 Pa and 0.7 and 0.5 Pa, respectively. The gelation time was lower in AX (13 min) than in FAX (16 min). The mechanical spectrum of AX and FAX gels exhibited G' higher than G'' and G' independent of frequency whereas G'' was dependent on frequency. These results confirm the importance of FA content in the rheological characteristics of the AX-based gels. Enzymatic modification of FA content in AX from MDG could offer an interesting opportunity for the reliable design of tailored gels which could add value to this maize by-product.

8.1 INTRODUCTION

Maize is the largest source for bioethanol as a transportation fuel in areas such as the United States, European Union, and China (Cheng and Timilsina, 2011). The dry milling of maize produces several co-products, with bran representing the main fraction, which normally ends up in maize distillers grains (MDG). The MDG potential utilization for the production of high added-value product increased and has been used as a potential source for the extraction of ferulated arabinoxylans (AX) (Mendez-Encinas et al., 2018). AX are formed by a linear chain of β-1,4 xyloses, which may or may not be substituted by α-1,3 and/or α-1,2 arabinose. Consequently, there may be three different forms of substitution: unsubstituted, mono-substituted, and di-substituted. The arabinose to xylose ratio (A/X) or degree of substitution is referred to the number of arabinose units attached to the xylose backbone. Besides the arabinose units, other minor substituents are glucuronic acid and galactose (Vinkx and Delcour, 1996). The substitution degree as well as the distribution of side chains are important factors in the physical–chemical properties of the AX (Izydorczyk and Biliaderis, 1995). AX can contain ferulic acid (FA) in their structure, in such a way that some arabinose residues are ester-linked on O-5 to FA (Smith and Hartley, 1983). The FA in AX allows these polysaccharides to gel in the presence of certain oxidizing agents. AX undergoing oxidative gelation was first reported by Durham (Durham, 1925) and Fausch et al. (1963) demonstrated that the FA esterified to AX is involved in this oxidative gelation. The gelling mechanism of AX may involve chemical ($FeCl_3$) or enzymatic (laccase and peroxidase/H_2O_2) oxidizing agents, which induce the formation of phenoxy free radical. The coupling of these phenoxy free radicals results in the formation of dimers (di-FA) and trimers (tri-FA) of FA, which allows the chains of polysaccharides to cross-link forming an aqueous gel (Geissman and Neukom, 1973; Figueroa-Espinoza and Rouau, 1998). Laccases (E.C.1.10.3.2) catalyze the oxidation of phenolic compounds using molecular oxygen as the final electron acceptor (Selinheimo et al., 2007). The AX gels are neutral and have no odor or color. Due to their covalent nature, these gels have interesting characteristics such as high water absorption capacity and stability to pH, temperature, and ionic changes. These characteristics and their meso- and macroporous structure make the AX gels highly suitable for the formulation of colon-targeted drug delivery systems (Carvajal-Millan et al., 2005). The AX gels present some advantages when compared to other polysaccharides gels. AX gels are quickly formed, their cross-linking is much stronger and stable to heat and pH, and do not exhibit syneresis after prolonged storage (Izydorczyk and Biliaderis, 1995). The FA content has been considered of great importance in the AX gelation since a good correlation between the elasticity of the AX gels formed and the initial FA content has been reported (Carvajal-Millan et al., 2005). The use of ferulate esterase to reduce FA content in AX extracted from ragi and wheat flour has been previously studied (Latha and Muralikrishna, 2009); however, those authors only reported a decrease in the relative viscosity of FAX under peroxidase exposure which was related to the partial removal of FA, without a rheological investigation of the gels. The aim of this work was to study the modification of FA content in gelling AX from MDG by using a ferulate esterase and to evaluate the rheological characteristics of the gels formed.

8.2 MATERIALS AND METHODS

8.2.1 MATERIALS

The MDG) was provided by a feed industry in Northern Mexico. Laccase (EC 1.10.3.2) from *Trametes versicolor* and other chemical products were purchased from Sigma Chemical Co. (St Louis, MO, USA). Ferulate esterase (E.C. 3.1.1.73) was a kind gift from Biocatalysts Limited (Cardiff, UK).

8.*2.2 **METHODS***

8.2.2.1 ARABINOXYLANS FROM MDG

The AX were extracted as reported by Mendez-Encinas et al. (2018) (Vinkx and Delcour, 1996). AX presented an arabinose to xylose (A/X) ratio of 1.1, a FA content of 5.0 μg/mg AX, and a molecular weight of 200 kDa.

8.2.2.2 DEFERULATION OF AX

Ferulate esterase activity was determined by reverse-phase high-performance liquid chromatography (RP-HPLC) using ethyl ferulate as a substrate. An assay mix (1 mL) containing enzyme extract (3×10^{-3} mg/mL), ethyl ferulate (5 mM), and MOPS (100 mM, pH 6) was incubated in darkness and gentle stirring at 40 °C for 60 min. The reaction was stopped by adding 200 μL of glacial acetic acid. FA released in solution was recovered by adding ethyl acetate (3 mL, 2×) and centrifuged at 6000 rpm (3820×*g*), 25 °C for 5 min. The supernatant was concentrated by evaporation at 40 °C under nitrogen in the dark. The FA extract obtained was suspended in methanol/water/acetic acid (40/59/1) solution, filtered (0.45 μm), and injected onto an Alliance HPLC System (e2695 Separation Module; Waters Co., Milford, MA, USA) at 25 °C. A Supelcosil LC-18-BD column (250 × 4.6 mm, Supelco Inc., Bellefonte, PA, USA) was used with an isocratic solution of methanol/water/acetic acid (40/59/1) at 0.6 ml/min. The FA was detected by UV absorbance at 280 nm in a 2998 Photodiode Array Detector (Waters Co., Milford). The assay was carried out in triplicate. One unit (U) of activity was defined as the amount of enzyme releasing 1 μmol of FA per minute at pH 6 and a temperature of 40 °C. For AX deferulation, an AX solution (2% w/v) was prepared in MOPS (100 mM, pH 6.0) and stirred at 4 °C for 12 h. The AX solution was equilibrated at 40 °C for 5 min, and the deferulation reaction was performed by the addition of 721 μU of ferulate esterase/mg AX at 40 °C in the dark over 24 h with gentle stirring. The de-esterification was stopped by adding glacial acetic acid at a ratio of 1:5 with respect to the assay mixture. The reaction mixture was precipitated with 75% ethanol at 4 °C for 12 h. The precipitate was recovered and dried by solvent exchange (80% (v/v) ethanol, absolute ethanol, and acetone) to give FAX.

8.2.2.3 FOURIER TRANSFORM INFRARED SPECTROSCOPY (FTIR)

FT-IR spectra of dry AX and FAX powder were recorded on a Nicolet FT-IR spectrophotometer (Nicolet Instrument Corp. Madison, WI, USA) as described previously (Martínez-López et al., 2013). Samples were pressed into KBr pellets (2 mg sample/200 mg KBr). A blank KBr disk was used as a background. Spectra were recorded in absorbance mode from 4000 to 400 cm^{-1}.

8.2.2.4 FA CONTENT

After saponification, FA content was determined by RP-HPLC (Vansteenkiste et al., 2004). Two mL of 2N NaOH was added to 50 mg of AX in a nitrogen-enriched atmosphere. The samples were incubated under agitation at 50 rpm, 35 °C for 2 h in darkness (KS 3000 ic control, IKA, Wilmington,

NC). Subsequently, 100 μL of 3,4,5-trimethoxy-trans-cinnamic acid (TMCA) and 5 mL of 4N hydrochloric acid were added to both samples. pH was adjusted to 2. Phenolics were extracted twice with 5 mL of diethyl ether and then evaporated to dryness at 40 °C under a gentle stream of nitrogen (Dri-Block DB-3A, Techne, UK). The phenolic extract was recovered in 1 mL of methanol/water (50/50), filtered (0.45 μm nylon membrane), and quantified using an HPLC system (Waters Alliance e2695, Waters Co.). A Supelcosil LC-18-BD column (250 × 4.6 mm) (Supelco, Inc.) and a photodiode array detector (Waters 2998, Waters Co.) were used. Detection was followed by UV absorbance at 320 nm.

8.2.2.5 GELATION OF AX AND FAX

The gelation of AX and FAX was carried out according to Carvajal-Millan et al. (2005). Laccase (1.675 nkat/mg sample) was added to 2% (w/v) AX and FAX solutions in 0.05 M citrate-phosphate buffer (pH 5.5).

8.2.2.6 RHEOLOGY

The gelation kinetics and rheological properties of AX and FAX with and without insulin were monitored using small amplitude oscillatory shear (Carvajal-Millan et al., 2005). Rheological measurements were carried out using a Discovery HR-2 rheometer (TA Instruments, New Castle, DE, USA) at a frequency of 0.25 Hz and 5% strain for 1 h. The mechanical spectrum of the gel was obtained by frequency sweep from 0.1 to 10 Hz at 5% strain. All measurements were carried out at 25 °C. The rheological analysis was performed in duplicate.

8.3 RESULTS AND DISCUSSION

8.3.1 ENZYMATIC DE-ESTERIFICATION OF AX

The initial FA content in AX was 5.0 μg/mg polysaccharide which diminished to 3.7 μg/mg polysaccharide when the sample was treated with ferulate esterase, indicating that 26% of the FA esterified to AX chains was removed. In the present study, AX and FAX presented an average FA residue content of 5.2 and 3.8 FA per 1000 xylose units, respectively.

Previous studies have reported 2-6 FA per 1000 xylose residues in AX from wheat, triticale, rye, and Barley (Carvajal-Millan et al., 2005). In a previous study, Carvajal-Millan et al. (2005) reported the preparation of a range of wheat flour AX samples with FA content varying from 0.1 to 2.3 μg/mg AX corresponding to a FA residues content from 0.1 to 2.5 per 1000 xyloses, respectively, by using chemical de-esterification. Those authors found that in AX presenting less than 0.5 FA residues per 1000 xyloses, gelation is no longer possible. However, in the present study, it was not possible to decrease the FA content in AX beyond 3.7 μg/mg polysaccharide even at longer ferulate esterase incubation time. AX used in the present study present a higher A/X ratio (1.10) in relation to that reported by Carvajal-Millan et al. (2005) in AX from the wheat endosperm (A/X=0.59) which may difficult the FA de-esterification from arabinoses attached to the xylose backbone.

8.3.2 FOURIER-TRANSFORM INFRARED SPECTROSCOPY

Vibrational spectral signatures of AX and FAX were collected by FTIR and are presented in Figure 8.1. AX and FAX registered bands in the spectral region of 1200–850 cm^{-1} that have been reported to be related to this polysaccharide with a main band centered at 1035 cm^{-1} that could be assigned to C–OH bending and small shoulders at 1158 and 897 cm^{-1} that have been associated with the antisymmetric C-O-C stretching mode of the glycosidic link and β(1→4) linkages. The amide I band (1640 cm^{-1}), mainly assigned to polypeptide carbonyl stretching (Cyran and Saulnier, 2005), was also detected suggesting the presence of residual protein in the samples as reported by Mendez-Encinas et al. (2018) (Vinkx and Delcour, 1996). The bands observed at 3413 and 2854 cm^{-1} have been previously attributed to the stretching of the OH groups and CH_2 groups, respectively (Séné et al., 1994). An absorbance band was registered at 1720 cm^{-1} which could be related to esterification with aromatic esters such as FA (Martínez-López et al., 2013).

Lignins, proteins, and phenolic acids, such as FA, have specific absorption bands between 1700 and 1500 cm^{-1}. The region of 3500–1800 cm^{-1} is the fingerprint region of polysaccharides related to AX with two bands (3400 cm^{-1} corresponding to the stretching of the OH groups and 2900 cm^{-1} corresponding to the CH_2 groups) (Kacuráková et al., 1999).

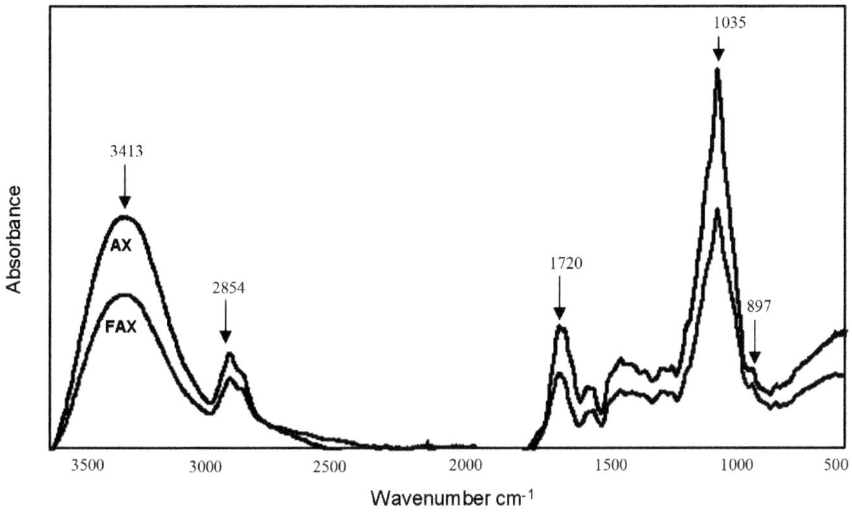

FIGURE 8.1 FTIR spectra of AX and FAX.

8.3.3 GELATION OF AX AND FAX

The gelation kinetics of AX and FAX solutions are shown in Figure 8.2. A similar behavior was observed for both solutions. Initially, G' (elastic modulus) and G'' (viscous modulus) gradually increased their value (where $G' < G''$), followed by a rapid increase until nearly constant values. The G' values for AX and FAX gels were 195 Pa and 112 Pa, respectively. In addition, the gelation time ($G' = G''$) was lower in AX (13 min) than in FAX (16 min). The G' value decrease and the higher gelation time registered in FAX in relation to AX are related to the partial removal of FA in the molecule after ferulate esterase treatment.

These results were similar to those reported for gels obtained from wheat endosperm AX, which were de-esterified by a chemical process (Carvajal-Millan et al., 2005) confirming the fundamental role of FA in the elastic properties of AX gels.

The mechanical spectra of AX and FAX gels after 1 h of gelation is presented in Figure 8.3. Whatever the ferulation degree of AX used, a typical gel behavior was found with the G' independent of frequency and higher than the G'' (Doublier and Cuvelier, 1996). G' values decreased from 195 to 112 Pa when the initial FA content of AX was from 5.0 to 3.7 μg/mg AX. Nevertheless, G'' values were almost superposed indicating that the viscous contribution to the gel structure was similar. From G' and G'' values in the AX and FAX gels at 1 Hz (frequency used during polysaccharide gelation), tan delta (G''/G') values were calculated. It was found that tan delta values increased from 0.0036 to 0.0045 when the AX initial FA content was reduced from 5.0 to 3.7 μg/mg AX. This augmentation in tan delta value indicates an increase in the polymer chain flexibility in the gel (Doublier and Cuvelier, 1996). These mechanical spectra are similar to others reported for wheat and corn AX gels (Mendez-Encinas et al., 2018; Carvajal-Millan et al., 2005; Martínez-López et al., 2013).

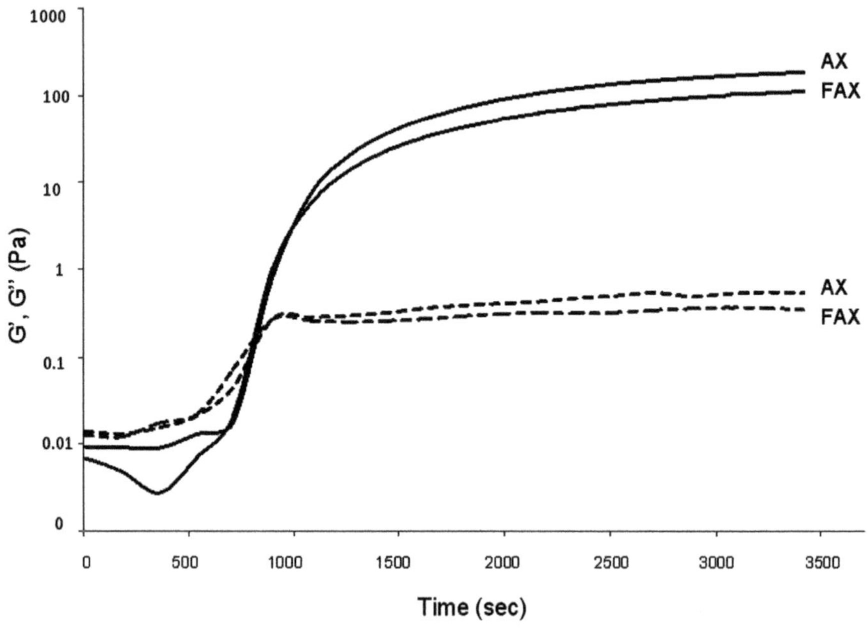

FIGURE 8.2 Monitoring the storage (G' continuous line) and loss (G'' discontinuous line) modulus of AX and FAX solutions at 2% (w/v) during gelation by laccase. Rheological tests at 25 °C, 1% Hz and 5% strain.

The results discussed above indicate that by tailoring the FA content of the AX, gels with different rheological characteristics can be obtained. These differences in the polysaccharide network formed could induce changes in the functional properties of the material generated.

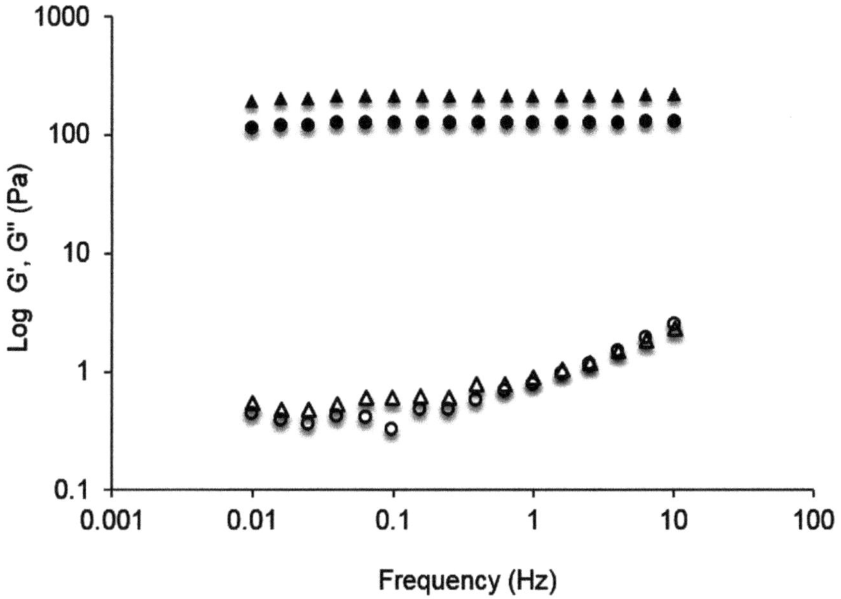

FIGURE 8.3 Mechanical spectra of AX and FAX gels at 2% (w/v) after 1 h of laccase exposure. Triangle: AX; circle: FAX; *G'*: black symbol; *G''*: white symbol. Rheological tests were performed at 25 °C and 5% strain.

8.4 CONCLUSIONS

The AX from MDG can be enzymatically modified in FA content allowing a reduction in the initial FA content of the polysaccharide of 26% after FAX. The elasticity of the gels formed with FAX decrease by 43% in relation to that registered in AX gels, ratifying the fundamental role of FA in the gelation capability and this polysaccharide and the rheological characteristics of the gels formed. Partial de-esterification of FA in AX from MDG via ferulate esterase could be a useful tool to design AX gels with specific rheological characteristics.

FUNDING

This research was funded by Fondo Institucional CONACyT—Problemas Nacionales 2015, Mexico (Grant 2015-01-568 to E. Carvajal-Millan).

KEYWORDS

- **polysaccharide**
- **ferulate esterase**
- **cross-linking**
- **tailoring**
- **rheology**

REFERENCES

Carvajal-Millan, E.; Guilbert, S.; Morel, M. H.; Micard, V. Storage Stability of Arabinoxylans Gels. *Carbohydr Polym* 2005, 59, 181–188.

Carvajal-Millan, E.; Landillon, V.; Morel, M. H.; Rouau, X.; Doublier, J. L.; Micard, V. Arabinoxylan Gels: Impact of the Feruloylation Degree on Their Structure and Properties. *Biomacromolecules* 2005, 6, 309–317.

Cheng, J. J.; Timilsina, G.R. Status and Barriers of Advanced Biofuel Technologies: A Review. *Renew Energ* 2011, 36, 3541–3549.

Cyran, M. R.; Saulnier, L. Cell Wall Fractions Isolated from Outer Layers of Rye Grain by Sequential Treatment with α-Amylase and Proteinase: Structural Investigation of Polymers in Two Ryes with Contrasting Bread Making Quality. *J Agric Food Chem* 2005, 53, 9213–9224.

Doublier, J. L.; Cuvelier, G. Gums and Hydrocolloids: Functional Aspects. In Carbohydrates in Food, Editor Eliasson, A.C. Marcel Dekker, New York, USA, 1996, 283–318.

Durham, R. K. Effect of Hydrogen Peroxide on Relative Viscosity Measurements of Wheat and Flour Suspensions. *Cereal Chem* 1925, 2, 297–305.

Fausch, H.; Kunding, W.; Neukom, H. Ferulic acid as a component of glycoprotein from wheat flour. *Nature* 1963,199, 287.

Figueroa-Espinoza, M. C.; Rouau, X. Oxidative Cross-linking of Pentosans Bye a Fungal Laccase and a Horseradish Peroxidase: Mechanism of Linkage Between Feruloylated Arabinoxylans. *Cereal Chem* 1998, 75, 259–265.

Geissman, T.; Neukom, H. On the Composition of the Water-soluble Wheat Flour Pensosanes and Their Oxidative Gelation. *Lebenson Wiss Technol* 1973, 6, 59–62.

Izydorczyk, M. S.; Biliaderis, C. G. Cereal Arabinoxylans: Advances in Structure and Physicochemical Properties. *Carbohydr Polym* 1995, 28, 33–48.

Kacuráková, M.; Wellner, N.; Ebringerova, A.; Hromadkova, Z.; Wilson, R. H.; Belton, P. S. Characterisation of Xylan-type Polysaccharides and Associated Cell Wall Components by FT-IR and FT-Raman Spectroscopies. *Food Hydrocolloid* 1999, 13, 35–41.

Latha, G. M.; Muralikrishna, G. Effect of Finger Millet (Eleusine coracana, Indaf-15) Malt Esterases on the Functional Characteristics of Non-starch Polysaccharides. *Food Hydrocolloid* 2009, *23*, 1007–1014. https://doi.org/10.1016/j.foodhyd.2008.07.02114.

Martínez-López, A. L.; Carvajal-Millan, E.; Rascón-Chu, A.; Márquez-Escalante, J.; Martínez-Robinson, K. Gels of Ferulated Arabinoxylans Extracted from Nixtamalized and

Non-nixtamalized Maize Bran: Rheological and Structural Characteristics. CyTA 2013, 11, 22–28.

Mendez-Encinas, M. A.; Carvajal-Millan, E.; Yadhav, M.; Lopez-Franco, L.; Rascon-Chu, A.; Lizardi-Mendoza, J.; Brown-Bojorquez, F.; Silva-Campa E.; Pedroza-Montero, M. Partial Removal of Protein Associated with Arabinoxylans: Impact on the Viscoelasticity, Cross-linking Content and Microstructure of the Gels Formed. *J. Appl Polym Sci* 2019, 136, 47300–47320.

Selinheimo E.; Autio K.; Kruus K.; Buchert J. Elucidating the Mechanism of Laccase and Tyrosinase in Wheat Bread Making. *J Agric Food Chem* 2007, 25, 6357–65.

Séné, C. F. B.; McCann, M. C.; Wilson, R. H.; Grinter, R. Fourier-Transform Raman and Fourier-Transform Infrared Spectroscopy. An investigation of Five Higher Plant Cell Walls and Their Components. *Plant Physiol.* 1994, 106, 1623–1631.

Smith, M. M.; Hartley, R. D. Occurrence and Nature of Ferulic Acid Substitution of Cell-wall Polysaccharides in Graminaceous Plants. *Carbohydr Res* 1983, 118, 65–80.

Vansteenkiste, E.; Babot, C.; Rouau, X.; Micard, V. Oxidative Gelation of Feruloylated Arabinoxylan as Affected by Protein. Influence on Protein Enzymatic Hydrolysis. *Food Hydrocolloid* 2004, 18, 557–564.

Vinkx, C. J. A.; Delcour, J. A. Rye (Secale cereale L.) Arabinoxylan. A Critical Review. *J Cereal Sci* 1996, 24, 1–14.

CHAPTER 9

Stability and Quality of Fruit Juices Incorporated with Probiotic *Lactobacilli*

DIPANKAR KALITA[1*], MANABENDRA MANDAL[2], and CHARU LATA MAHANTA[1]

[1]*Department of Food Engineering and Technology, School of Engineering, Tezpur University, Assam, India*

[2]*Department of Molecular Biology and Biotechnology, School of Sciences, Tezpur University, Assam, India*

[*]*Corresponding author. E-mail: dkalita@tezu.ernet.in*

ABSTRACT

In the present study, the probiotic characteristics of the three different strains of *Lactobacilli* and their survivability in freshly prepared juices from four different fruits The potential isolates of the *Lactobacilli* and a suitable carrier juice were identified based on the viability of the bacteria, pH, total soluble sugars (TSS), titratable acidity, and color of the probiotic fruit juices. The viability of the bacterial cells and quality parameters of the probiotic juices were also studied in refrigerated storage condition. The effect of the addition of probiotics on the phytochemical and antioxidant properties of the selected fruit juices was also studied. The probiotic juices were analyzed for total phenolic content, total flavonoid content ,and antioxidant properties like DPPH radical scavenging activity, ferric reducing antioxidant property. Reverse-phase high-performance liquid chromatography of the probiotic juices were also carried out to study the changes in the profile of polyphenols, organic acids, and sugars during storage. Mineral constituents of the juice were also determined by atomic absorption spectroscopy.

9.1 INTRODUCTION

Foods are consumed with the aim to satisfy hunger and provide the necessary nutrients for humans, promote a state of physical and mental well-being, improve health, prevent and/or reducing diet-related diseases. Consumers are increasingly becoming aware of the relationship between food and health which has led to an explosion of interest in "healthy foods"; this phenomenon could be partly attributed to the increasing cost of healthcare, the steady increase in life expectancy, the increase in population with diet-related chronic diseases, and the desire of older people for an improved quality of their later years (Granato et al., 2010).

Nowadays, healthy foods mean "functional foods and functional foods are those containing or prepared with bioactive compounds, such as dietary fiber, oligosaccharides, and active "friendly" bacteria that promote the equilibrium of intestinal bacterial strains. Probiotics belong to an emerging generation of active ingredients, which include prebiotics, phytonutrients, and lipids (Jankovic et al., 2010). The label "functional food" was introduced in 1980 in Japan, which was the first country that stated a specific regulatory approval process for functional foods, known as Foods for Specified Health Use (FOSHU) (Perricone et al., 2014). On the other hand, in Europe, the interest in functional foods started in the 1990s, when the European Commission created a commission called Functional Food Science in Europe to explore the concept of functional foods through a science-based approach (Corbo et al., 2014). The market of functional foods is characterized by an increasing trend, and some researchers reported that probiotic foods represent around 60%–70% of functional foods (Tripathi and Giri, 2014).

The word probiotic comes from the Greek word "προ-βίος" that means "for life"; thus, probiotics are live microorganisms (mainly bacteria but also yeasts) that confer a beneficial effect on the host if administered in proper amounts (Perricone et al., 2014). Dairy fermented products have been traditionally considered as the best carriers for probiotics; but, nowadays, up to 70% of the world population is affected by lactose-intolerance. Furthermore, the use of milk-based products may be also limited by allergies, cholesterol diseases, dyslipidemia, and vegetarianism; therefore, several raw materials have been extensively investigated to determine if they are suitable substrates to produce novel nondairy functional foods (Perricone et al., 2014).

Recently, beverages based on fruits, vegetables, cereals, and soybeans have been proposed as new products containing probiotic strains; particularly, fruit juices have been reported as a novel and appropriate medium for probiotic for their content of essential nutrients. Moreover, they are usually referred to as healthy foods, designed for young and old people (Luckow et al., 2006). Many authors reported on the effects of juices on health; for example, Sutton (2007) demonstrated that aqueous extracts of kiwifruit and avocado had very low cytotoxicity and high anti-inflammatory activity in a Crohn's gene-specific assay. No aqueous extracts of kiwifruit, blueberry and avocado had similarly high anti-inflammatory activity, with slightly higher cytotoxicity than the aqueous extracts. Fenech et al. (2005) studied the effect of the intake of nine micronutrients (vitamin E, calcium, folate, retinol, nicotinic acid, β-carotene, riboflavin, pantothenic acid, and biotin) on genome damage and repair; these compounds can be easily found in juices. Furthermore, fruit juices have shown negative effects on some pathogenic microorganisms, improving the growth of probiotics because berries, such as blueberry, blackberry, and raspberry, possess antimicrobial effects toward many pathogens (Ranadheera et al., 2014).

Therefore, juice fortification with probiotic microorganisms is a challenge and a frontier goal, as juices could combine nutritional effects with the added value of a health benefit from a probiotic. Maintaining the viability (the recent trend is to have one billion viable cells per portion—100 g of product) and the activity of probiotics in foods till the end of shelf-life are two important criteria to be fulfilled in juices, where low pH represents a drawback. The most commonly used probiotic bacterial genera are *Lactobacillus* and *Bifidobacterium*, while *Saccharomyces cerevisiae* var. *boulardii* is the yeast used and these serve as probiotics both in dairy and nondairy functional foods (Ranadheera et al., 2010). Several strains of *L. plantarum, L. rhamonsus, L. acidophilus,* and *L. casei* can grow in fruit matrices due to their tolerance to acidic environments (Peres et al., 2012).

9.2 MATERIALS

The fruit samples viz. litchi (*Litchi chinensis* Sonn.) and pineapple *(Ananas comosus* L. Merr) were procured from the local fruit market, Tezpur, Assam during the season. The fruits and probiotic strains were selected from previous experiments on the basis of their suitability to develop health-promoting functional fruit drinks. Chemicals used in the study were of analytical grade purchased from Sigma-Aldrich, Merck, and Himedia. All the standards were purchased from Sigma-Aldrich. *Lactobacillus* isolates, *Lactobacillus plantarum* MTCC2621 (LP), and *L. rhamnosus* MTCC1480 (LR) were obtained from Microbial Type Culture Collection and Gene Bank (MTCC) (IMTECH, CSIR, Chandigarh, India).

9.2.1 INOCULUM PREPARATION

The freeze-dried cultures of LP and LR were activated in sterile glycerol (50% v/v). The glycerol stock culture was stored at frozen (−20 °C) in sterile screw-cap tubes. The identity of all the probiotic bacteria was confirmed using biochemical methods (Lorca and de Valdez, 2001). The probiotic organisms were grown individually by inoculating into 10 mL sterile de Man Rogosa and Sharp (MRS) broth (Himedia Laboratories Pvt. Ltd, Mumbai, India) and incubated at 37 °C for 2 days under aerobic condition. The cells were harvested by centrifuging (Sigma, Germany) at 1500×*g* for 15 min at 4 °C. Before inoculation into fruit juices, the harvested cells were washed twice with sterile saline water (0.85% w/v NaCl) to remove any residual MRS.

9.2.2 FRUIT JUICE PREPARATION

The fresh and ripe fruit samples were washed and the juice was extracted using a household juicer (Philips). The juice was strained through a muslin cloth and pasteurized at 90 °C for 1 min with constant stirring.

9.2.3 INOCULATION OF SUBSTRATES

Pasteurized juice (100 mL) was taken into sterile Erlenmeyer flasks. Each flask containing 100 mL juice was inoculated with 1% culture with *Lactobacillus plantarum* MTCC2621 and *L. rhamnosus* MTCC1480 under aseptic conditions and labeled as LP and LR, respectively. The control had no culture. The flasks were then incubated at 37 °C. After 12 h of fermentation at 37 °C, the flasks were kept at refrigerated condition (4 ± 1 °C) for 6 weeks. At an interval of 7 days, 10 mL of juice was taken out from each flask and used for further analysis.

9.2.4 ENUMERATION OF BACTERIA

The enumeration of free probiotic cells was performed using the method described by Yoon et al. (Lorca and de Valdez, 2001) and expressed in CFU/mL (colony-forming unit). Enumeration of the probiotic bacteria in fruit juice was performed using the same formula presented on weekly basis over a period of 6 weeks, using MRS agar after incubation at 37 °C for 24 h under aerobic conditions.

9.2.5 BIOCHEMICAL ANALYSIS OF THE PROBIOTIC FRUIT JUICE DURING STORAGE

At an interval of 7 days, 10 mL of juice was taken out from each flask and tested for biochemical parameters viz. pH, titratable acidity and total soluble sugar, and color change. The pH of each juice sample was measured in a pH meter (Eutech, Germany) after proper calibration. Titratable acidity expressed as g lactic acid/100 g was determined by titration against 0.1N sodium hydroxide using phenolphthalein as an endpoint indicator. The total soluble sugar was estimated in terms of °Brix by handheld refractometer (Erma, Japan). All experiments were performed in triplicate to determine the mean and standard error of the mean.

9.2.6 COLOR ANALYSIS

The color of the fermented litchi juice was determined by a Hunter ColorLab UltraScan-Vis colorimeter, UK. The colorimeter was calibrated and measurements were made through a 0.375-inch port/viewing area. The reflectance instruments determined three color parameters: lightness (L), redness (a), and yellowness (b). Numerical values of L (light/dark), a (a+ redness/green a–) and b (b + yellowness/blueness b–) were converted into ΔE (total color difference) which was calculated using the following equation.

9.2.7 SENSORY EVALUATION

Sensory attributes of color, flavor, taste, odor, mouth feel, after taste, and overall acceptability of probiotic juices were evaluated by a trained panel of 10 members. A 9-point Hedonic scale reading (1–4 dislike extremely to dislike slightly, 5-neither like nor a dislike, 6–9 like slightly to like extremely) was performed on day 28 of product storage.

9.2.8 DETERMINATION OF TOTAL PHENOLIC CONTENT

Total phenolic content in the sample extracts was assessed using the Folin–Ciocalteau assay (Van de Guchte et al., 2002) with slight modification. For the analysis, 20 μL each of filtered juice, gallic acid standard or blank were taken in separate test tubes and to each 1.58 mL of distilled water was added, followed by 100 μL of Folin–Ciocalteau reagent, mixed well and within 8 min, 300 μL of sodium carbonate was added. The samples were vortexed immediately and the tubes were incubated in the dark for 30 min at 40 °C. The absorbance was then measured at 765 nm in a UV–Vis spectrophotometer (Cecil, Aquarius7400). The results were expressed in mg GAE/100 mL.

9.2.9 DETERMINATION OF TOTAL FLAVONOID CONTENT

The flavonoid content was determined by aluminum trichloride method. (Mitsuoka, 1992). Briefly, 0.5 mL of the filtered juice was mixed with 1.5 mL of 95% ethanol, 0.1mL of 10% aluminum trichloride, 0.1 mL of 1 M potassium acetate, and 2.8 mL of deionized water. After incubation at room temperature for 40 min, the reaction mixture absorbance was measured at 415 nm against deionized water blank in a UV–Vis spectrophotometer (Cecil, Aquarius 7400). Results were expressed as quercetin equivalent (mg QE/100 mL) of the sample.

9.2.10 DETERMINATION OF FERRIC REDUCING ANTIOXIDANT PROPERTY (FRAP)

FRAP activity of the samples was measured by the method of Benzie and Strain (Tannock et al., 2000). Briefly, a 40 μL aliquot of properly diluted sample extract was mixed with 3 mL of FRAP solution. The reaction mixture was incubated at 37 °C for 4 min and the absorbance was determined at 593 nm in a UV–Vis spectrophotometer (Cecil, Aquarius 7400) against a blank that was prepared using distilled water. FRAP solution was prewarmed at 37 °C and prepared freshly by mixing 2.5 mL of a 10 mM 2,4,6-TPTZ [2,4,6-tri(2-pyridyl)-1,3,5-triazine] solution in 40 mM hydrochloric acid with 2.5 mL of 20mM ferric chloride and 25 mL of 0.3M acetate buffer (pH 3.6). A calibration curve was prepared, using an aqueous solution of ferrous sulfate (1–10 mM). FRAP values were expressed as μM Fe^{2+} equivalent per 100 mL of sample.

9.2.11 DETERMINATION OF DPPH RADICAL SCAVENGING ACTIVITY

Radical scavenging activity of the sample extracts was measured by determining the inhibition rate of DPPH (2,2-diphenyl-1-picrylhydrazyl) radical

(Teixeira et al., 1997). Precisely, 100 μL of extracts was added to 1.4 mL DPPH radical methanolic solution (10^{-4} M). The absorbance at 517 nm was measured at 30 min against blank (100 μL methanol in 1.4 mL of DPPH radical solution) using a UV–Vis Spectrophotometer (Cecil Aquarius 7400). The results were expressed in terms of radical scavenging activity.

Radical scavenging activity (%) = [(Ao–As)/ Ao] × 100

where Ao is absorbance of control blank and As is the absorbance of sample extract.

9.2.12 QUANTIFICATION OF POLYPHENOLS BY HIGH-PERFORMANCE LIQUID CHROMATOGRAPHY

High-performance liquid chromatography (HPLC) was performed to analyze the major phenolic compounds in the juice. The separation module consisted of a Waters HPLC (Waters) equipped with a C18 Symmetry 300™ C_{18} (5 μm, 4.6 × 250 mm) column with a binary pump (Waters, 1525) and a UV–Vis detector (Waters, 2489). The samples were eluted with a gradient system consisting of solvent A [acidified ultrapure water (0.1% acetic acid, pH 3.2)] and solvent B (methanol), used as a mobile phase. The flow rate was maintained at 0.8mL/min and wavelengths used for UV–Vis detector were 254 and 325 nm. The temperature of the column was maintained at 25 °C and the injection volume was 20 μL. The gradient system started at 80% solvent A (0–8 min), 65% A (9–12 min), 45% A (13–16 min), 30% A (17–20 min), 20% A (21–30 min), 10% of A (31–34 min) and then washing of the column with 65% A (35–39 min) and lastly 80% A (40–45 min) was followed. The juice samples were centrifuged at 15000×*g* using a Sigma—18K centrifuge (Sigma, Germany) for 15 min. The juice supernatant was then filtered through a Whatman membrane filter (0.2 μm) before injection into HPLC. The ethanolic extract was evaporated under vacuum in a rotary vacuum evaporator (Roteva, Medica Equipments) and then redissolved in 1 mL methanol. A sample volume of 20 μL was used. The standards (Sigma-Aldrich) used for comparison and identification were (±) catechin, quercetin, gallic acid, coumaric acid, caffeic acid, syringic acid, ferulic acid, chlorogenic acid, kaempferol, and rutin hydrate. The peaks of the phenolic compounds were monitored and the concentration of phenolic compounds was determined using the external calibration curve of standard compound at 0.62, 1.25, 2.5, 5, 10, and 20 mg/L.

2.13 HPLC ANALYSIS OF THE ORGANIC ACIDS

For the detection of the organic acids in probiotic fruit juice, 5 mL aliquots of the juice was taken on a weekly basis and frozen in 50 mL Falcon tubes (Tarsons, India). The samples were diluted with 70 µL of 15.5 N nitric acid and 4930 µL of 0.009% sulfuric acid and then mixed gently. The samples were then centrifuged at 14,000×*g* for 10 min (Sigma, Germany). The supernatant was removed using a L sterile syringe and then filtered into HPLC vials using a Whatman 0.2 µm PTFE membrane filter. The analysis of organic acid concentration was performed using the method described by Ong et al. (Sheehan et al., 2007). The HPLC apparatus used consisted of a Waters 1525 binary pump and 2489 UV/Vis detector and Breeze software. The column was a Waters Symmetry™ C18 5 µm 4.6 × 300 mm and was heated to 65 °C when used. The mobile phase consisted of 0.009 H_2SO_4 which was filtered through Whatman 0.45 µm membrane filter and degassed with bath sonicator (JSGW, India). The mobile phase was set at a flow rate of 0.6 mL/min. A sample volume of 20 µL was used for both standards and samples, and detection was achieved at 220 nm. The organic acid standard kit was obtained from Sigma Aldrich® Inc. All samples were run for 40 min in gradient mode and all analyses were carried out in triplicate. All instrument control, analysis, and data processing were performed via Waters® Breeze® Chromatography Data Software.

9.2.14 MINERAL ANALYSIS

The samples were digested in Digestion System using a mixture of nitric acid and sulfuric acid in 1:1 ratio and the aqueous solutions were injected to Atomic Absorption Spectrophotometer AAS (Thermo, iCE 3500) and analyzed in reference to the calibration of 3 standard concentrations made from certified single element AA standards (Sigma Aldrich®).

9.3 RESULTS AND DISCUSSION

9.3.1 ENUMERATION OF BACTERIA

In order for probiotics to stay alive in the unfavorable conditions of the gastrointestinal tract and reach the intestine in adequate numbers, they must be present at a concentration of at least 10^7 CFU/mL in the product at the end of shelf-life; this, in the case of fruit juices, corresponds to approximately 10^9 CFU per portion (Sheehan et al., 2007).

The bacterial counts of the three strains in probiotic litchi and pineapple juices kept under refrigerated condition are given in Table 9.1. The enumeration was done for 6 weeks against control (pasteurized juice). There was no bacterial colony in the control juice for up to 6 weeks. All three strains showed good viability up to 4 weeks in all four juices under refrigerated condition. *L. plantarum* (Lp) and *L. rhamnosus* (Lr) maintained the required CFU/mL up to 6 weeks in cold storage. At 0 weeks the bacterial count of Lp and Lr were 9.5 and 9.1 log CFU/mL in litchi juice which reduced to 8.1 and 8.0 log CFU/mL after 6 weeks, whereas the cell count of La was found below 10 CFU/mL after 4 weeks in litchi juice. The reduction of the bacterial count of La may be due to the presence of secondary metabolites and low pH of the fruit juice. Lp and Lr showed better viability up to 6 weeks compared to La in litchi and pineapple juices.

The population of probiotic Lp, Lr, and La in orange and guava juices is given in Table 9.2. The bacterial count of Lp and Lr was 9.1 and 8.7 log CFU/mL in orange juice at 0 weeks which was well maintained up to 6 weeks of storage. On the other hand, La had survived up to 4 weeks only. A similar trend was also found in probiotic pineapple juice. The results obtained in this study were in agreement with the results of the study by Sheehan et al. (2007), which showed that probiotic *Lactobacillus* and *Bifidobacterium* strains survived better in pineapple juice than in cranberry juice.

The cell counts of Lp and Lr in guava juice were 8.8 and 8.5 log CFU/mL at 0 weeks. The number of bacteria slowly reduced and reached 6.2 and 5.9 log CFU/mL after 5 weeks of storage. The count of *L. acidophilus* was found to be 8.4 log CFU/mL at 0 weeks and after that, there was a sharp decline of the viability that reached below 10 CFU/mL after 4 weeks. Studies were carried out to explore the suitability of juices such as tomato, beetroot, and cabbage juices as raw materials for the production of probiotic drinks using *L. plantarum*, *L. acidophilus*, and *L. casei* as probiotic bacteria cultures (King et al., 2007; Yoon et al., 2004). From the above results, it was observed that Lp and Lr grew well as compared to La in all four fruit juices and maintained required cell counts up to 6 weeks (Pereira et al., 2011). Results also revealed that litchi and pineapple were more suitable as carriers of the probiotic bacteria than orange and guava. Similar results have been reported by previous researchers, (Nualkaekul and Charalampopoulos, 2011) who observed that cells survived well in litchi, blackcurrant, and pineapple juices, which can be attributed to the high pH of these juices (Saarela et al.,

TABLE 9.1 Enumeration of Three Strains of *Lactobacillus* (log CFU/mL) in Probiotic Litchi and Pineapple Juices During Storage at Refrigerated Condition (4 ± 1 °C)

Week	Log CFU/mL							
	Litchi				Pineapple			
	Control	Lp	Lr	La	Control	Lp	Lr	La
0	ND	9.5 ± 0.3^{f}	9.1 ± 0.4^{d}	8.9 ± 0.6^{d}	ND	9.5 ± 0.3^{e}	9.1 ± 0.4^{e}	8.4 ± 0.4^{d}
1	ND	9.1 ± 0.5^{e}	9.0 ± 0.2^{d}	7.0 ± 0.7^{c}	ND	9.3 ± 0.7^{d}	9.2 ± 0.7^{f}	7.1 ± 0.2^{c}
2	ND	8.8 ± 0.6^{d}	8.7 ± 0.2^{c}	6.2 ± 0.2^{b}	ND	8.7 ± 0.5^{c}	8.8 ± 0.2^{d}	6.8 ± 0.7^{b}
3	ND	8.5 ± 0.5^{c}	8.5 ± 0.1^{b}	3.4 ± 0.2^{a}	ND	8.5 ± 0.4^{b}	8.7 ± 0.6^{d}	3.2 ± 0.4^{a}
4	ND	8.3 ± 0.7^{b}	8.3 ± 0.3^{a}	<10 CFU/mL	ND	8.3 ± 0.2^{b}	8.2 ± 0.2^{c}	<10 CFU/mL
5	ND	8.2 ± 0.4^{a}	8.1 ± 0.6^{a}	<10 CFU/mL	ND	8.3 ± 0.4^{b}	7.8 ± 0.4^{b}	<10 CFU/mL
6	ND	8.1 ± 0.1^{a}	8.0 ± 0.4^{a}	<10 CFU/mL	ND	8.2 ± 0.3^{a}	7.5 ± 0.4^{a}	<10 CFU/mL

Results are mean ± S.D of triplicates. Same letter within a row means no significant difference at $p \leq 0.05$ by DMRT.
ND: Not detectable, Lp: L. plantarum, Lr: L. rhamnosus, and La: L. acidophilus.

TABLE 9.2 Enumeration of Three Strains of *Lactobacillus* ($\log_{10}$ CFU/mL) in Probiotic Orange and Guava Juices During Storage at Refrigerated Condition (4 ± 1 °C)

Week	Log CFU/mL							
	Orange				Guava			
	Control	Lp	Lr	La	Control	Lp	Lr	La
0	ND	9.1 ± 0.7^{e}	8.7 ± 0.3^{e}	8.5 ± 0.4^{d}	ND	8.8 ± 0.3^{f}	8.5 ± 0.6^{e}	8.4 ± 0.7^{e}
1	ND	8.7 ± 0.4^{d}	8.6 ± 0.4^{d}	8.3 ± 0.2^{c}	ND	8.5 ± 0.2^{e}	8.3 ± 0.8^{e}	8.0 ± 0.9^{d}
2	ND	8.5 ± 0.7^{c}	8.5 ± 0.6^{c}	7.2 ± 0.2^{b}	ND	8.1 ± 0.6^{d}	7.9 ± 0.2^{d}	7.6 ± 0.5^{c}
3	ND	8.4 ± 0.6^{c}	8.3 ± 0.8^{c}	7.0 ± 0.1^{b}	ND	7.5 ± 0.7^{c}	7.5 ± 0.9^{c}	5.1 ± 0.8^{b}
4	ND	8.0 ± 0.5^{b}	8.2 ± 0.4^{c}	6.7 ± 0.2^{a}	ND	7.2 ± 0.4^{c}	6.2 ± 0.4^{b}	3.2 ± 0.4^{a}
5	ND	8.0 ± 0.5^{b}	7.7 ± 0.3^{b}	<10 CFU/mL	ND	6.2 ± 0.1^{b}	5.9 ± 0.2^{a}	<10 CFU/mL
6	ND	8.0 ± 0.4^{a}	7.4 ± 0.2^{a}	<10 CFU/mL	ND	5.5 ± 0.4^{a}	<10 CFU/mL	<10 CFU/mL

Results are mean ± S.D of triplicates. Same letter within a row means no significant difference at $p \leq 0.05$ by DMRT.

ND: Not detectable, Lp: *L. plantarum*, Lr: *L. rhamnosus* and La: *L. acidophilus*.

2006). Both *L. plantarum* and *L. rhamnosus* were capable of surviving the low pH and high acidic conditions in fermented cabbage juice during cold storage at 4 °C and also grew well in cabbage juice at 30 °C with a viable cell count of 10 × 10^8 CFU/mL after 48 h of fermentation at 30 °C (Yoon et al., 2006). Besides this, several other factors that could affect the survival and growth of the probiotics, accumulation of metabolic end products such as lactic acid and other organic acids, diacetyl, acetaldehyde, acetoin, , would reduce cell viability in the product (Shah and Jelen, 1990).

9.3.2 BIOCHEMICAL ANALYSIS OF THE PROBIOTIC FRUIT JUICE DURING STORAGE

9.3.2.1 CHANGE IN pH

The low pH of fruit juice compared to the fairly neutral pH of milk (6.6–6.7) is likely the most important determinant for the poor probiotic viability in this food matrix. *Bifidobacteria* are generally sensitive to pH values below 4.6 (Boylston et al., 2004). However, *Lactobacillus* strains are clearly more acid-resistant than the strains of other *Bifidobacterium* species (Mättö et al., 2004). It is known that fruit juices can inhibit the growth of lactic acid bacteria (strain-specific effect) and that the inhibition is mainly due to low of pH of the juices (Vinderola et al., 2002).

The changes in pH due to the addition of probiotics were studied in four fruit juices for 6 weeks (Tables 9.3 and 9.4). It is interesting to point out that there were slight changes in pH in probiotic litchi and pineapple juices when fortified with Lp and Lr. The cells survived well in litchi and pineapple juices; one reason for this was probably the high pH of these juices, which was around pH 4.6–4.7. But the change of pH was observed when these juices were fortified with La. The pH of litchi juice changed gradually from 4.60 to 3.93 and in the case of pineapple, pH changed from 4.75 to 3.73 after 6 weeks of probiotication (Table 9.3). There was a significant change of pH in probiotic orange and guava juices when these juices were fermented with Lp and Lr. Moreover, the change of pH was also observed when fortified with La in both the juices during storage at refrigerated condition (Table 9.4). Though all three strains were able to withstand pH of 4.5 and 3.3, only Lp (*L. plantarum*) and Lr (*L. rhamnosus*) were able to survive at pH 3.66 for up to 6 weeks. Therefore, these two cultures were finally selected for fortification of different fruit juices. Similar results were also found for *L.*

TABLE 9.3 Change in pH of the Probiotic Litchi and Pineapple Juices During Storage Refrigerated Condition (4 ± 1 °C)

Week	pH							
	Litchi				Pineapple			
	Control	Lp	Lr	La	Control	Lp	Lr	La
0	4.65 ± 0.09^{c}	4.63 ± 0.04^{d}	4.66 ± 0.11^{d}	4.60 ± 0.15^{f}	4.68 ± 0.02^{b}	4.70 ± 0.05^{f}	4.76 ± 0.10^{f}	4.75 ± 0.05^{g}
1	4.61 ± 0.04^{c}	4.54 ± 0.01^{c}	4.45 ± 0.15^{c}	4.46 ± 0.09^{e}	4.65 ± 0.02^{b}	4.54 ± 0.21^{e}	4.65 ± 0.17^{e}	4.66 ± 0.09^{f}
2	4.59 ± 0.02^{b}	4.47 ± 0.03^{c}	4.31 ± 0.09^{b}	4.42 ± 0.05^{e}	4.61 ± 0.03^{b}	4.49 ± 013^{e}	4.63 ± 0.12^{e}	4.52 ± 0.05^{e}
3	4.58 ± 0.01^{b}	4.40 ± 0.12^{c}	4.27 ± 0.10^{b}	4.28 ± 0.12^{d}	4.58 ± 0.02^{a}	4.41 ± 0.12^{d}	4.55 ± 0.14^{d}	4.28 ± 0.12^{d}
4	4.55 ± 0.03^{b}	4.33 ± 0.11^{b}	4.25 ± 0.07^{b}	4.18 ± 0.08^{c}	4.54 ± 0.04^{a}	4.33 ± 0.07^{c}	4.41 ± 0.04^{c}	4.05 ± 0.08^{c}
5	4.52 ± 0.02^{a}	4.22 ± 0.04^{a}	4.22 ± 0.08^{b}	4.09 ± 0.11^{b}	4.53 ± 0.01^{a}	4.20 ± 0.02^{b}	4.05 ± 0.06^{b}	3.88 ± 0.11^{b}
6	4.51 ± 0.01^{a}	4.18 ± 0.13^{a}	4.15 ± 0.06^{a}	3.93 ± 0.05^{a}	4.50 ± 0.02^{a}	4.08 ± 0.15^{a}	3.86 ± 0.11^{a}	3.73 ± 0.05^{a}

Results are mean ± S.D of triplicates. Same letter within a row means no significant difference at $p \leq 0.05$ by DMRT.
Lp: *L. plantarum*, Lr: *L. rhamnosus* and La: *L. acidophilus*.

TABLE 9.4 Change in pH of the Probiotic Orange and Guava Juices During at Storage Refrigerated Condition (4 ± 1 °C)

Week	pH							
	Orange				Guava			
	Control	Lp	Lr	La	Control	Lp	Lr	La
0	4.19 ± 0.06^{b}	4.63 ± 0.04^{d}	4.66 ± 0.11^{d}	4.60 ± 0.15^{g}	5.32 ± 0.06^{b}	5.22 ± 0.06^{b}	5.28 ± 0.03^{d}	5.27 ± 0.04^{e}
1	4.18 ± 0.03^{b}	4.54 ± 0.01^{d}	4.25 ± 0.15^{c}	4.16 ± 0.09^{f}	5.29 ± 0.03^{b}	5.18 ± 0.03^{b}	5.16 ± 0.02^{c}	5.06 ± 0.16^{d}
2	4.17 ± 0.01^{b}	4.37 ± 0.03^{c}	4.11 ± 0.09^{b}	4.02 ± 0.05^{e}	5.25 ± 0.01^{b}	5.17 ± 0.08^{b}	5.18 ± 0.01^{c}	4.92 ± 0.11^{c}
3	4.13 ± 0.02^{a}	4.10 ± 0.02^{b}	4.07 ± 0.10^{b}	3.88 ± 0.12^{d}	5.23 ± 0.03^{a}	5.13 ± 0.03^{a}	5.15 ± 0.02^{c}	4.88 ± 0.07^{c}
4	4.11 ± 0.01^{a}	4.03 ± 0.01^{b}	4.00 ± 0.07^{b}	3.65 ± 0.08^{c}	5.20 ± 0.02^{a}	5.10 ± 0.03^{a}	5.11 ± 0.02^{c}	4.65 ± 0.10^{b}
5	4.09 ± 0.02^{a}	4.02 ± 0.04^{b}	3.92 ± 0.05^{b}	3.58 ± 0.11^{b}	5.20 ± 0.01^{a}	5.07 ± 0.03^{a}	4.80 ± 0.14^{b}	4.58 ± 0.06^{b}
6	4.08 ± 0.01^{a}	3.88 ± 0.03^{a}	3.66 ± 0.06^{a}	3.33 ± 0.05^{a}	5.18 ± 0.01^{a}	5.06 ± 0.04^{a}	4.66 ± 0.12^{a}	4.33 ± 0.13^{a}

Results are mean ± S.D of triplicates. Same letter within a row means no significant difference at $p \leq 0.05$ by DMRT.
Lp: *L. plantarum*, Lr: *L. rhamnosus* and La: *L. acidophilus*.

plantarum of by other researchers when fruit juices were fortified with LAB (Vinderola et al., 2002; Do Espírito Santo et al., 2011).

9.3.2.2 CHANGE IN TOTAL SOLUBLE SUGARS (TSS) AND TITRABLE ACIDITY (TA)

Figure 9.1 illustrates the effect of probiotication using three species of lactic acid bacteria in all four fruit juices. There is a declining trend in TSS content in all the probiotic juices due to the addition of probiotics. On the other hand, the TA in all the probiotic juices increased with time of storage. At the initial stage (0 weeks) the TSS of the juice was around 17, 12, 11, and 18 °Brix, respectively, for litchi, pineapple, orange, and guava when fortified with Lp. The TSS of these juices reduced to 11, 7, 5, and 13 °Brix after 6 weeks of

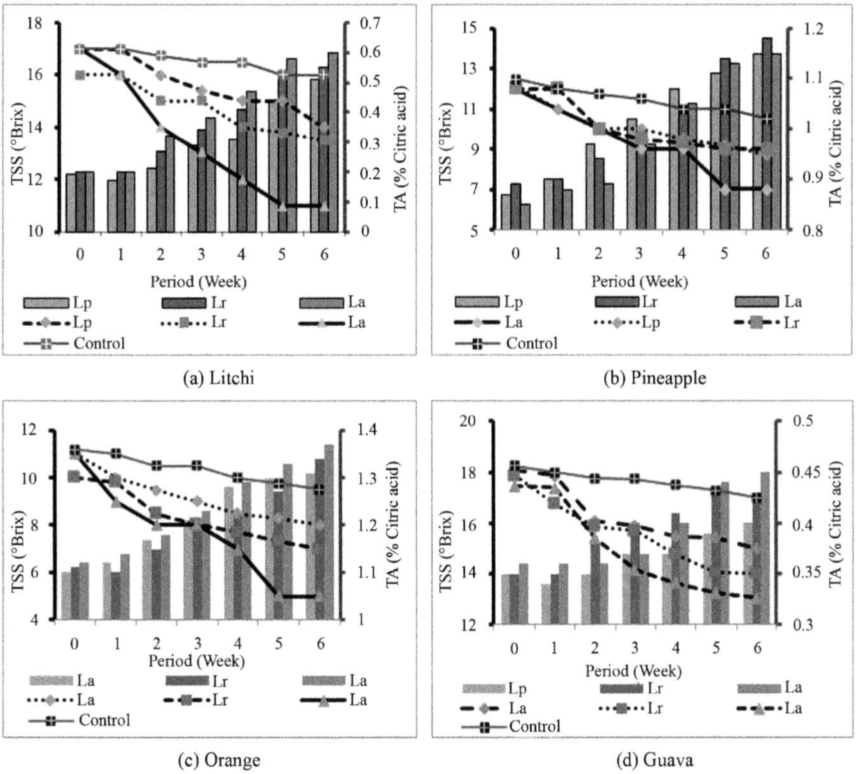

FIGURE 9.1 Changes in total soluble sugars and titrable acidity of probiotic juices: (a) litchi, (b) pineapple, (c) orange, and (d) guava during storage at refrigerated condition (4 ± 1 °C).

cold storage. A similar trend also found with the other two strains of *Lactobacillus*. The highest fall was observed in case of La which sharply declined in all probiotic juices after 1 week and reached as low as °Brix after 6 weeks. TA of the probiotic juices increased from 0.2 to 0.5, 0.9 to 1.15, 1.11 to 1.31, and 0.34 to 0.40 % in litchi, pineapple, orange, and guava, respectively, when inoculated with Lp. However, the increase in TA was more prominent in the case of juices fortified with La. The decrease in acidity was concurrent with the decrease in the sugar content of the fruit juices (Nagpal et al., 2012). These results indicated that all three strains of *Lactobacillus* were not only able to survive but also utilized and fermented the fruit sugars for their cell synthesis and metabolism. Yoon et al. (2004) also observed a decrease in sugar and pH and an increase in acidity when tomato juice was inoculated and incubated with *L. plantarum, L. acidophilus*, and *L. casei.*

In the present investigation, the two strains, La (*L. plantarum*) and Lr (*L. rhamnosus*) were observed to not only survive but also utilize the fruit juices for their growth and synthesis of the secondary metabolites, as indicated by the decrease in sugar and pH, and increase in acidity. However, La (*L. acidophilus*) was found to consume sugars at a faster rate than the other two species. The results of reduced TSS were well supported by the similar trend in increased acidity in the juices inoculated with La (*L. plantarum*) and Lr (*L. rhamnosus*) (Nagpal et al., 2012).

The images of litchi and orange juices at 0 day and after 6 weeks of probiotication are presented in Figure 9.2.

FIGURE 9.2 Probiotic juice with *Lactocacillus plantarum* at 0 day and after 6 weeks at refrigerated condition (4 ± 1°C).

9.3.2.3 COLOR ANALYSIS

The color values of the juice samples and their changes upon the addition of probiotics are presented in Table 9.5. In litchi juice, the *L* values decreased when fortified with Lp, whereas an increase was observed in Lr and La. Similarly, the *a* and *b* values also significantly decreased in most of the cases with some exceptions. This revealed that the juice became darker due to the addition of Lp. When the litchi juice was fortified with Lr and La, increase in brightness was observed due to increase in the *b* values.

TABLE 9.5 Change in Color of the Probiotic Fruit Juices During Storage at 4 ± 1°C

Parameters	0 Day	After 6 weeks		
		L. plantarum	*L. rhamnosus*	*L. acidophilus*
Litchi				
L	31.86 ± 0.09^{b}	29.41 ± 0.14^{a}	36.51 ± 0.07^{c}	38.99 ± 0.18^{d}
A	-1.13 ± 0.04^{b}	-1.03 ± 0.05^{d}	-1.28 ± 0.01^{a}	-1.09 ± 0.03^{c}
B	-1.95 ± 0.07^{a}	-1.38 ± 0.07^{b}	0.71 ± 0.008^{c}	-0.01 ± 0.005^{d}
△E	–	4.66 ± 0.03^{a}	5.38 ± 0.03^{b}	7.72 ± 0.09^{c}
Pineapple				
L	24.59 ± 0.16^{a}	23.27 ± 0.10^{a}	22.03 ± 0.04^{a}	27.34 ± 0.12^{a}
A	-1.31 ± 0.06^{a}	-0.18 ± 0.09^{d}	-0.51 ± 0.09^{c}	-0.87 ± 0.03^{b}
B	2.35 ± 0.05^{d}	0.92 ± 0.02^{a}	1.26 ± 0.05^{b}	2.14 ± 0.07^{c}
△E	–	2.45 ± 0.05^{a}	3.32 ± 0.01^{b}	3.52 ± 0.01^{c}
Orange				
L	18.28 ± 0.14^{a}	20.57 ± 0.06^{b}	22.19 ± 0.0^{c}	23.65 ± 0.04^{c}
A	1.38 ± 0.03^{d}	0.02 ± 0.006^{a}	0.18 ± 0.02^{c}	0.09 ± 0.01^{b}
B	1.02 ± 0.01^{a}	1.70 ± 0.02^{c}	1.55 ± 0.14^{b}	1.74 ± 0.02^{d}
△E	–	2.06 ± 0.01^{a}	4.05 ± 0.03^{b}	5.53 ± 0.01^{c}
Guava				
L	23.86 ± 0.21^{c}	22.36 ± 0.04^{d}	20.59 ± 0.08^{a}	21.84 ± 0.09^{b}
A	0.02 ± 0.001^{b}	-0.03 ± 0.006^{a}	-0.19 ± 0.04^{c}	-0.35 ± 0.02^{d}
B	0.62 ± 0.02^{b}	0.29 ± 0.03^{a}	1.45 ± 0.01^{c}	1.77 ± 0.07^{d}
△E	–	0.34 ± 0.06^{a}	1.41 ± 0.07^{b}	2.32 ± 0.05^{c}

N Results are mean ± S.D of triplicates. Same letter within a row means no significant difference at $p \leq 0.05$ by DMRT.

In the case of pineapple, no significant change was observed in "*L*" value but the redness increased in all the juices irrespective of the probiotic strains inoculated. When the orange juice was inoculated with probiotics, the brightness of the juice was increased due to an increase of the *b* values whereas, in the case of guava, a decrease in lightness was observed along with an increase in *b* values. The total color change ΔE (reference value was 0 days of storage) increased in the fermented fruit juices during storage. This increase might be attributed to the increase in the yellow intensity in the fermented samples. The human eye cannot perceive small color variations and therefore, instrumental color measurement is helpful. $\triangle E$ values measure the overall color change (Pereira et al., 2013). It was observed that the fermented fruit juices did not show a perceptive color change during storage.

9.3.2.4 SENSORY EVALUATION

The overall acceptability of any fruit juice is mainly influenced by the quality of the product, in which the most important attribute is the product's taste, followed by nutritional value, odor, and price (Krasaekoopt and Kitsawad, 2010). This implied that most consumers buy fruit juice due to taste rather than other qualities. The attitude of the panelists on probiotic fruit juices containing probiotic bacteria are shown in Figures 9.3–9.5.

The attribute of the overall acceptability of litchi juice was scored between 7.2 and 8.0 across the strain used. In other juices, the overall acceptability was scored in the acceptable range of 6.5–7.5, which remained almost the same up to 4 weeks of storage. After 6 weeks of storage, all probiotic fruit juices were found to have lower acceptability as compared to 4 weeks which might be due to changes in taste and mouthfeel. Similarly, sensory parameters of appearance, color, taste, mouthfeel, and overall acceptability were found to be lower for 6 weeks stored juices than for 4 weeks. The results of the sensory analysis showed that changes in color values of the probiotic litchi juices during 6 weeks of storage were in good agreement with the overall acceptability of the juices.

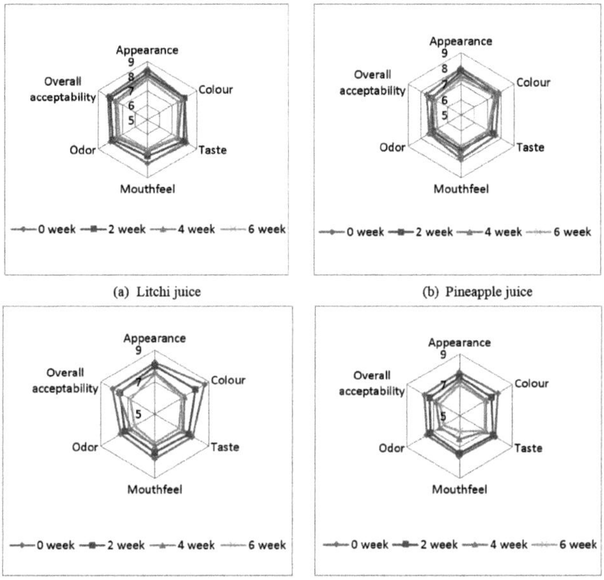

(a) Litchi juice (b) Pineapple juice

(c) Orange juice (d) Guava juice

FIGURE 9.3 Sensory score for probiotic fruit juices juice inoculated with *L. plantarum* during 6 weeks of storage.

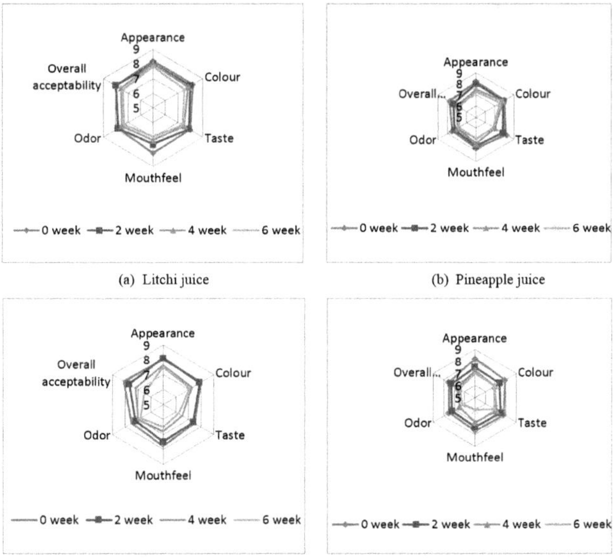

(a) Litchi juice (b) Pineapple juice

(c) Orange juice (d) Guava juice

FIGURE 9.4 Sensory score for probiotic fruit juices juice inoculated with *L. rhamnosus* during 6 weeks of storage.

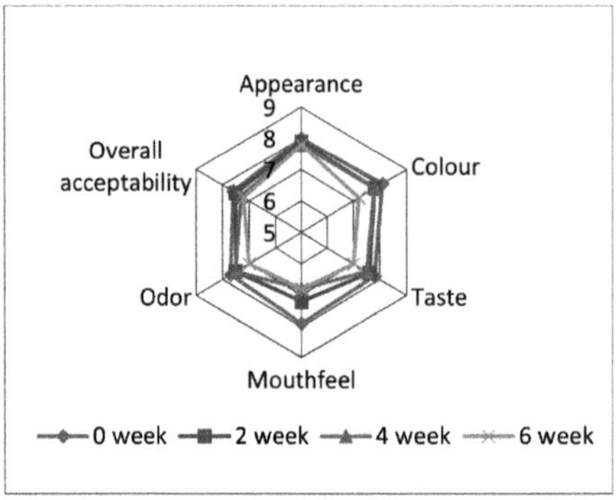

(a) Litchi juice

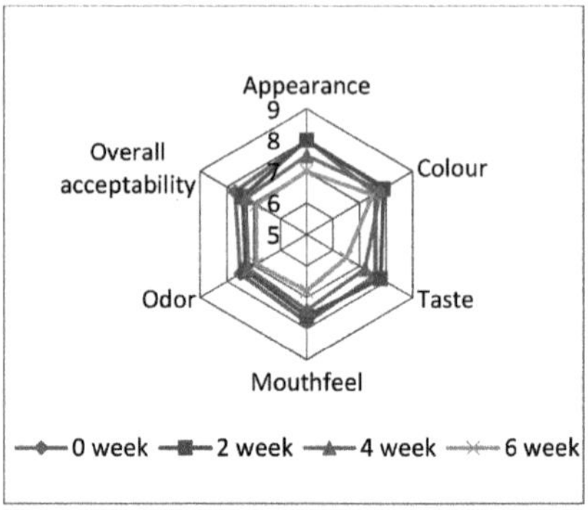

(b) Pineapple juice

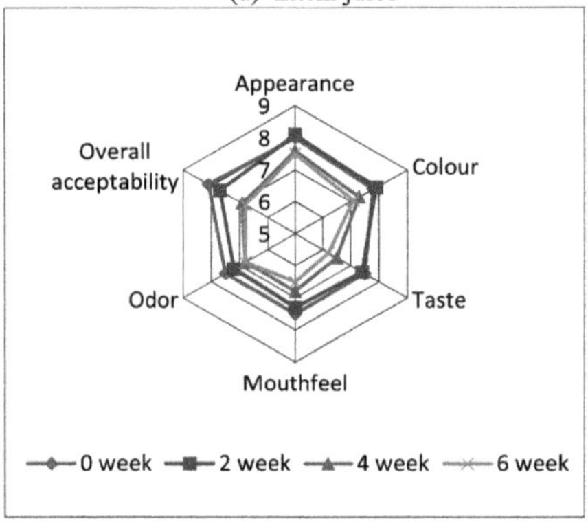

(c) Orange juice

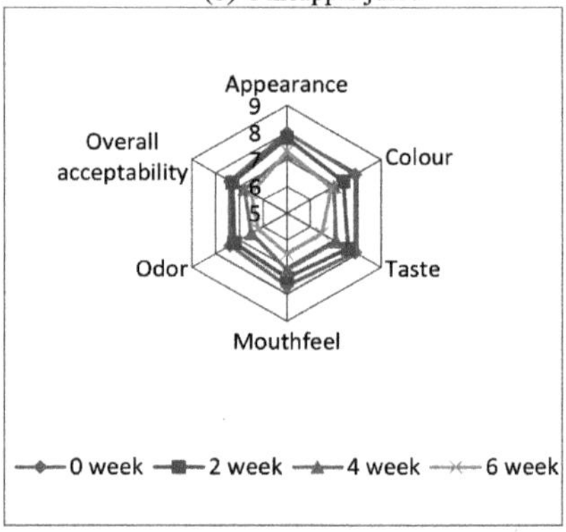

(d) Guava juice

FIGURE 9.5 Sensory score for probiotic fruit juices juice inoculated with *L. acidophilus* during 6 weeks of storage.

9.3.2.5 PHYTOCHEMICAL AND ANTIOXIDANT CHANGES

The total phenolic content (TPC) was very high in litchi juice (2420 mg GAE/100 mL) in comparison to pineapple juice (92 mg GAE/100 mL), as seen from Table 9.6. TPC was observed to vary depending on the storage period and the type of bacterial strain used for probiotication (Table 9.6). A significant ($p \leq 0.05$) decrease in TPC was observed in the control litchi juice with the storage period. TPC in control litchi juice at the end of the 6th week was 79.7% of 0-day value. Similarly, TPC also decreased significantly on probiotication and at the end of the 6th week, the fall was 3 4.2% for Lp treated juice and 42.9% for Lr treated juice against their 0-day values.

There was a significant decrease of TPC in control pineapple juice with a fall of 12.8% from 0 days. Further, there was a significant decrease of 58.8%

and 36.5% of TPC after 6th week in Lp and Lr treated pineapple juices, respectively. The extent of % decrease on probiotication varied between juices and probiotic species used.

No significant change in total flavonoid content (TFC) was observed in control juices. The TFC in fermented litchi and pineapple juice lowered from 11.58 to 3.88 mg QE/100 mL and 4.25 to 1.85 mg QE/100 mL, respectively, after the addition of Lr (Table 9.6). TFC was also found to decrease from 12.56 to 9.58 mg QE/100 mL and 4.12 to 2.12 mg QE/100 mL in litchi and pineapple juice fortified with Lp in 6 weeks refrigerated storage.

The FRAP value of probiotic litchi and pineapple juices showed a declining trend (Figures 9.6 and 9.7) for both strains on storage. However, the highest decrease in litchi was observed in Lp fortified juice while in pineapple, Lr fortified juice showed the highest decrease.

The DPPH radical scavenging activity was higher in litchi juice than pineapple juice and the scavenging activity also varied on storage among the probiotic strain used (Figures 9.6 and 9.7). The DPPH activity was lowest in litchi juice with Lr, it reduced from 88.71% to 13.17% after 6th week. In probiotic pineapple juice with Lr, DPPH activity ranged reduced from 24.8% to 7.8%. Probiotication, therefore, had an adverse effect on radical scavenging activity.

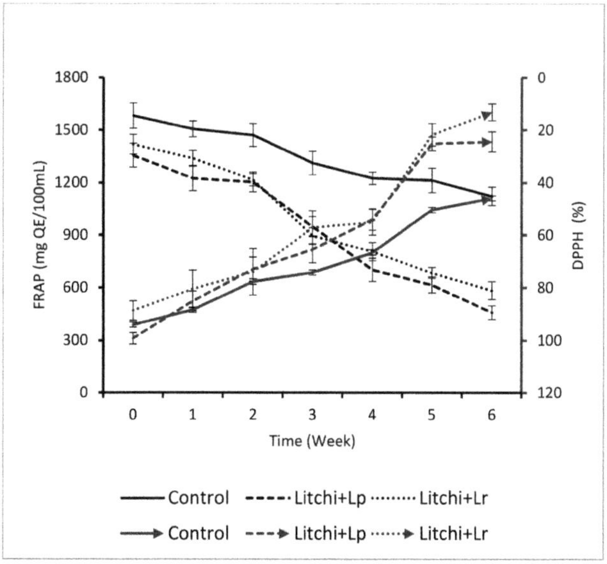

FIGURE 9.6 Changes in ferric reducing antioxidant property (FRAP) and DPPH radical scavenging activity of the probiotic litchi juice during storage at refrigerated condition (4 ± 1 °C).

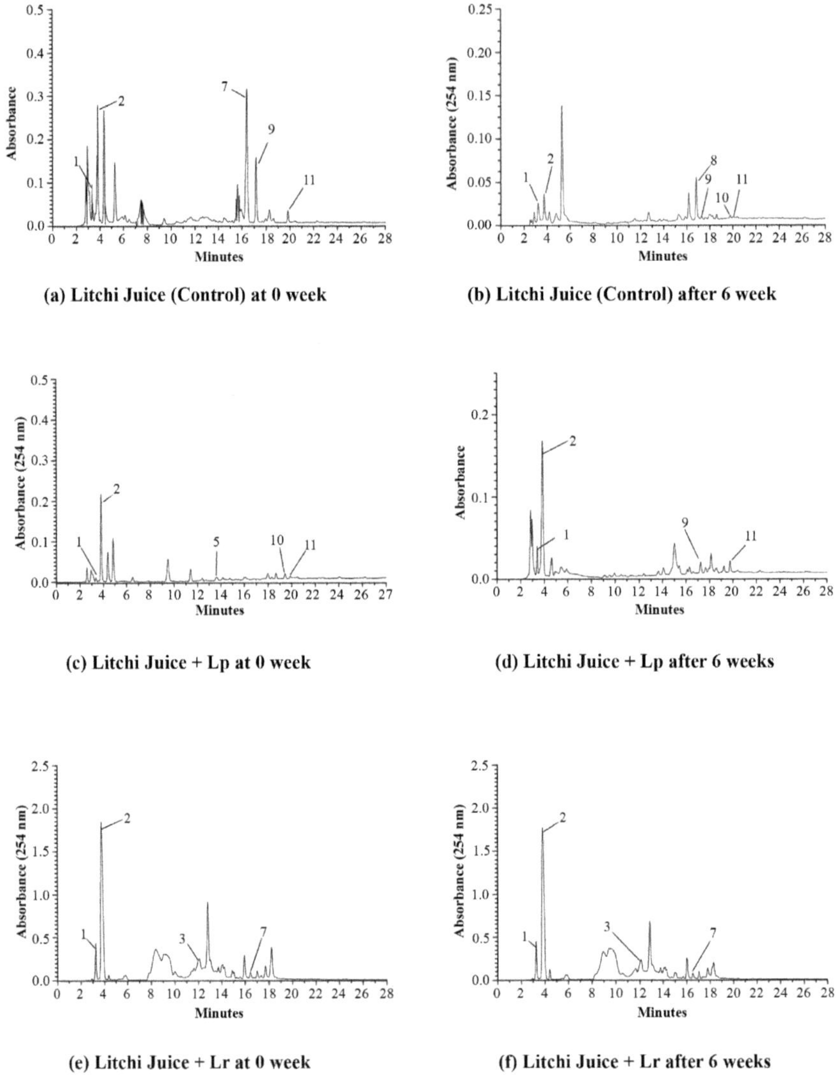

FIGURE 9.7 Changes in ferric reducing antioxidant property (FRAP) and DPPH radical scavenging activity of the probiotic pineapple juice during storage at refrigerated condition (4 ± 1 °C).

The decrease in phytochemicals and antioxidant activity in the probiotic juices could be due to changes in chemical composition and structural change of phenolics. The antioxidant capacity depends on the structural conformation of phenolic compounds (King et al., 2007). The decreasing trend of polyphenols and antioxidant capacity could be understood by the fact that antioxidant capacity of the food depends on the synergistic and redox interactions among the different compounds present in the fruits. The reduction in one group of compounds may lead to the loss in functionality against a certain type of free radicals without changing its functionality toward other radicals (King et al., 2007)..

Phenolic compounds have a tendency to undergo some kind of structural rearrangement that could lead to either increased or decreased antioxidant activities. But mainly, the increase or decrease in phenolic content depends on the overall composition and types of individual phenolic acid present in the maximum in the concerned fruit juice.

9.3.3 HPLC DETERMINATION OF THE PHENOLIC ACIDS CONTENT IN THE PROBIOTIC JUICE SAMPLES

The phenolic acids detected are given in Table 9.6. The following phenolic acids were identified at 254 nm by comparing their known standards in the probiotic fruit juice samples at 0 weeks and after 6 weeks of storages. Gallic acid (RT = 3.23 min), catechin (RT = 11.89 min), chlorogenic acid (RT = 13.54 min), caffeic acid (RT = 14.49 min), syringic acid (RT = 14.73 min), ferulic acid (RT = 16.55), coumaric acid (RT = 16.72 min), rutin (RT = 17.31 min), kaempferol (RT = 19.61 min), and quercetin (RT = 19.89 min) were detected. The phenolic acids in both probiotic juice samples showed a decrease or complete destruction with storage time while, in some cases, an increase or appearance of new phenolic acid originally not detected in the fresh juice was observed. In litchi juice, the addition of probiotic bacteria changed the phenolic acid composition as compared to the control one. The HPLC chromatograms (Figure 9.8) showed the presence of gallic acid, catechin, quercetin in both control and fermented litchi juice. On probiotication of the litchi juice with Lp, the presence of new phenolic compounds like syringic acid and coumaric acid were detected which were not present in the control juice. Moreover, fermentation of litchi juice with Lr did not show any major change in the phenolic acid profile. Similar results were also observed in case of pineapple juice with Lr (Figure 9.9). Storage time also affected the phenolic profile of both the probiotic juices by the destruction of some phenolic acids and the development of new phenolic compounds. In litchi juice, the addition of probiotic bacteria caused a decrease in gallic acid and quercetin content and an increase in catechin and coumaric acid. A similar trend was also observed for gallic acid and ferulic acid in litchi fermented with Lr.

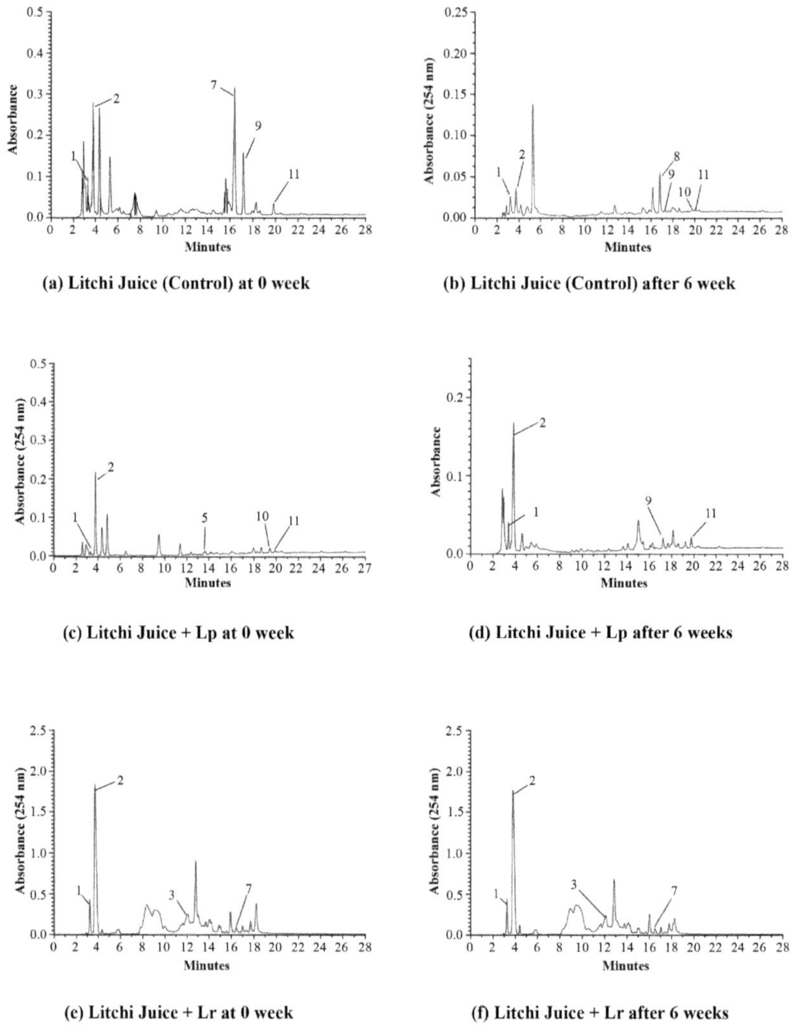

FIGURE 9.8 RP-HPLC chromatogram of the phenolic compounds in probiotic litchi juice with *L. plantarum* (Lp) and *L. rhamnosus* (Lr). [1 = gallic acid (GA); 2 = ascorbic acid (AA); 3 = catechin (CHT); 4 = caffeic acid (CFA); 5 = chlorogenic acid (CGA); 6 = syringic acid (SA); 7 = ferulic acid (FA); 8 = coumaric acid (CMA); 9 = rutin hydrate (RTH); 10 = kaempferol (KF) and 11 = quercetin (QCT)].

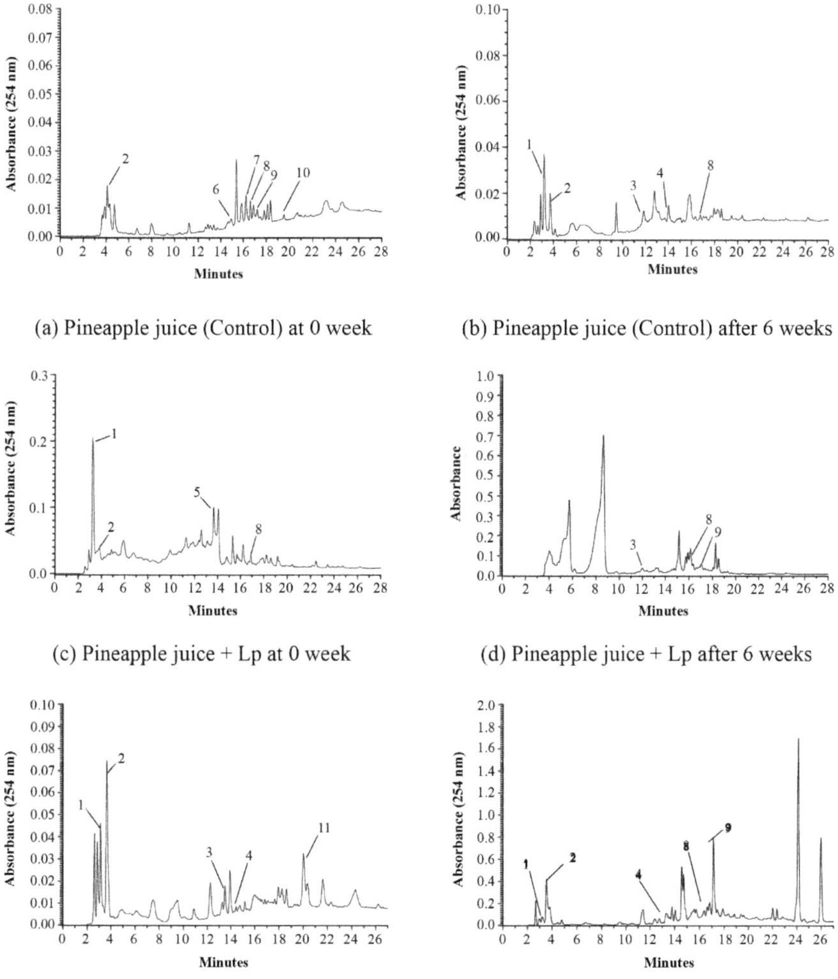

(a) Pineapple juice (Control) at 0 week

(b) Pineapple juice (Control) after 6 weeks

(c) Pineapple juice + Lp at 0 week

(d) Pineapple juice + Lp after 6 weeks

(e) Pineapple juice + Lr at 0 week

(f) Pineapple juice + Lr after 6 weeks storage

FIGURE 9.9 RP-HPLC chromatogram of the phenolic compounds in probiotic pineapple juice with *L. plantarum* (Lp) and *L. rhamnosus* (Lr). 1 = gallic acid (GA); 2 = ascorbic acid (AA); 3 = catechin (CHT); 4 = caffeic acid (CFA); 5 = chlorogenic acid (CGA); 6 = syringic acid (SA); 7 = ferulic acid (FA); 8 = coumaric acid (CMA); 9 = rutin hydrate (RTH); 10 = kaempferol(KF) and 11 = quercetin (QCT).

The destruction of phenolics in most of the cases could be due to oxidation of the phenolic acid due to other factors like light and oxygen (Yoon et al., 2004). Similarly, the increase and detection of new phenolic acids originally absent in the fresh and control samples could be the result of the release of the bound phenolics. The phenolic acids comprise of both free and bound phenolic acids. The bound phenolic acids remain bound to some structural carbohydrate and protein either through ester linkage with carboxylic groups or ether linkages with lignin through their hydroxyl groups in the aromatic ring or acetal bonds. (Nualkaekul and Charalampopoulos, 2011; Saarela et al., 2006; Yoon et al., 2006; Boylston et al., 2004). An increase in the content of some phenolic acid and their antioxidant activity after the addition of probiotics has been reported by Jaiswal et al. (Mättö et al., 2004). Similarly, Kusznierewicz et al. (Vinderola et al., 2002) and Othman et al. (Do Espírito Santo et al., 2011) reported an increase in some phenolic acids such as catechin and caffeic acid in some cases after probiocation. They also reported a decrease in the phenolic acid content with the application of heating, storage time, and storage temperature. This might be the result of the cleavage of the esterified bond between sugar glycoside and phenolic acids.

Another probable reason for the increase in phenolic content could be due to degradation and molecular rearrangements of the existing phenolic acids during the processing like pasteurization of juices (Nagpal et al., 2012). Application of heat may break these bonds and cause their release due to cell disruption and rupture of the food matrix which in turn facilitates their release into the liquid medium (Nagpal et al., 2012).

Overall, probiotic litchi juice showed good content of ferulic acid. Also, probiotic pineapple juices are rich in gallic acid as well as rutin and coumaric acid. Depending on the juice type and phenolic acid compositions, it was found that the addition of probiotic bacteria had both positive as well as negative impacts on the phenolic acid composition of the fruit juice. Moreover, the degradation of large polymeric phenolic by some enzymes of fermented fruit juices could increase the content of total phenolics (Pereira et al., 2011) (Rodríguez et al., 2009). Further investigation of the enzymes related to these biochemical changes would be the focus of future studies.

9.3.4 HPLC DETERMINATION OF THE ORGANIC ACIDS IN THE PROBIOTIC FRUIT JUICES

Organic acids play an important role in taste, flavor and consumer acceptance of fruit beverages. The results are shown in Table 9.7. Lactic acid was recognized as the main metabolite produced by both strains of *Lactobacillus*.

TABLE 9.6 Phenolic Acid Content of the Probiotic Fruit Juices stored at Refrigerated Condition (4 ± 1 °C) Determined by RP-HPLC Expressed in mg/100 mL

Sample	GA	CTH	CFA	CGA	SA	FA	CMA	RTH	KF	QCT
Litchi										
Co_0week	15.25 ± 0.11	ND	ND	ND	ND	2.15 ± 0.12	ND	3.84 ± 0.14	ND	3.18 ± 0.02
Co_6week	12.45 ± 0.24	4.17 ± 0.08	ND	ND	ND	ND	3.17 ± 0.28	1.49 ± 0.09	1.49 ± 0.05	2.63 ± 0.17
Lp_0week	20.97 ± 0.14	ND	ND	ND	1.16 ± 0.10	ND	ND	ND	0.82 ± 0.02	0.78 ± 0.04
Lp_6week	17.24 ± 0.17	6.77 ± 0.12	ND	ND	ND	ND	5.77 ± 0.12	ND	ND	1.08 ± 0.03
Lr_0week	21.01 ± 0.05	ND	ND	ND	ND	20.08 ± 0.08	ND	ND	ND	ND
Lr_6week	16.18 ± 0.04	6.98 ± 0.04	ND	ND	ND	13.51 ± 0.05	ND	ND	ND	ND
Pineapple										
Co_0week	12.50 ± 0.02	ND	ND	ND	1.12 ± 0.15	0.91 ± 0.07	1.09 ± 0.04	2.12 ± 0.16	2.44 ± 0.06	1.18 ± 0.07
Co_6week	22.61 ± 0.11	1.13 ± 0.02	1.05 ± 0.04	ND	ND	ND	1.13 ± 0.02	ND	ND	ND
Lp_0 week	40.72 ± 0.14	ND	ND	ND	3.86 ± 0.11	ND	0.94 ± 0.03	ND	ND	ND
Lp_6week	15.45 ± 0.07	0.85 ± 0.01	ND	ND	1.78 ± 0.07	ND	0.73 ± 0.11	ND	ND	ND
Lr_0week	19.65 ± 0.23	ND	1.23 ± 0.07	ND	ND	ND	ND	ND	ND	0.55 ± 0.04
Lr_6week	28.91 ± 0.03	1.08 ± 0.10	1.46 ± 0.02	ND	ND	ND	ND	ND	ND	0.46 ± 0.02

Results (mg/100 mL) are mean ± S.D of triplicate values; Co: Control; Lp: *L. plantarum* and Lr: *L. rhamnosus*; ND: Contents below the detection limit.

GA, gallic acid; CTH, catechin; CGA, chlorogenic acid; CFA, caffeic acid; SA, syringic acid; FA, ferulic acid; CMA, coumaric acid; RTH, rutin hydrate; KF, kaempferol; QCT, quercetin.

After 6 weeks of storage, lactic acid concentrations in probiotic litchi plus Lp and litchi plus Lr juices reached 16.8 and 15.2 g/L, respectively. Malic acid in probiotic litchi juice for both strains was completely consumed after 6 weeks. The malolactic fermentation of *Lactobacillus* had been reported in the literature (Shah and Jelen, 1990; Pereira et al., 2013). After 6 weeks of storage, slight changes in tartaric, acetic, and citric acid contents also were observed in probiotic litchi and pineapple juices. In this study, after the storage of 6 weeks at 4 ± 1 °C, both lactic acid contents in fermented litchi and pineapple juices increased significantly, and no significant changes ($p < 0.05$) in the content of tartaric, acetic, and citric acids were observed. The malic acid content was below the detectable limit after fermentation. Similar results were also reported by Zheng et al. (Krasaekoopt and Kitsawad, 2010) in fermented litchi juice using a high hydrostatic pressure treatment.

TABLE 9.7 Organic Acid Content of the Probiotic Fruit Juices Stored at Refrigerated Condition (4 ± 1 °C) Determined by RP-HPLC Expressed in g/L

Sample	**Organic Acid (g/L)**				
	Citric				**Lactic Acid**
Litchi					
Co_0 week	$4.15^{a} \pm 0.06$	$4.65^{a} \pm 0.28$	$0.55^{a} \pm 0.12$	$3.65^{b} \pm 0.21$	N.D.^{a}
Co_6 week	$4.65^{a} \pm 0.21$	$4.41^{a} \pm 0.13$	$0.50^{a} \pm 0.10$	$3.54^{b} \pm 0.14$	N.D.^{a}
Lp_0 week	$4.25^{a} \pm 0.13$	$4.55^{a} \pm 0.21$	$0.53^{a} \pm 0.22$	$3.56^{b} \pm 0.24$	$2.12^{b} \pm 0.24$
Lp_6 week	$5.98^{b} \pm 0.25$	$4.18^{a} \pm 0.09$	$0.48^{a} \pm 0.21$	N.D.^{a}	$15.20^{c} \pm 0.29$
Lr_0 week	$4.58^{a} \pm 0.14$	$4.60^{a} \pm 0.15$	$0.54^{a} \pm 0.15$	$3.62^{b} \pm 0.15$	$3.16^{b} \pm 0.15$
Lr_6 week	$6.12^{b} \pm 0.13$	$4.08^{a} \pm 0.01$	$0.47^{a} \pm 0.09$	N.D.^{a}	$16.80^{c} \pm 0.16$
Pineapple					
Co_0 week	$5.18^{a} \pm 0.08$	$4.72^{a} \pm 0.12$	$0.57^{a} \pm 0.04$	$2.76^{c} \pm 0.11$	N.D.^{a}
Co_6 week	$5.23^{a} \pm 0.18$	$5.07^{a} \pm 0.09$	$0.42^{b} \pm 0.30$	$2.41^{c} \pm 0.21$	N.D.^{a}
Lp_0 week	$5.70^{a} \pm 0.15$	$5.43^{b} \pm 0.02$	$0.55^{a} \pm 0.13$	$1.68^{b} \pm 0.19$	$10.68^{b} \pm 0.09$
Lp_6 week	$7.65^{b} \pm 0.10$	$5.66^{b} \pm 0.12$	$0.45^{b} \pm 0.18$	$0.14^{a} \pm 0.02$	$46.74^{c} \pm 0.27$
Lr_0 week	$5.70^{a} \pm 0.07$	$5.45^{b} \pm 0.10$	$0.68^{a} \pm 0.06$	$1.25^{b} \pm 0.08$	$13.25^{b} \pm 0.08$
Lr_6 week	$6.23^{b} \pm 0.22$	$5.62^{b} \pm 0.16$	$0.56^{a} \pm 0.15$	$0.23^{a} \pm 0.06$	$51.23^{b} \pm 0.26$

N.D. Contents below the detection limit. $^{a, b, c, d}$ Different letters represented a significant difference within the same column ($p < 0.05$).

In case of probiotic pineapple juice, increasing trend in lactic acid production was similar to that of litchi juice that reached 46.74 and 51.23 g/L in Lp and Lr fortified pineapple juices, respectively by the end of storage period (Table 9.8). No significant change in citric acid, acetic acid, and tartaric acid was observed on the addition of probiotics but malic acid content decreased significantly when juice was probiocated with LAB. Saradhuldhat et al.and Hong et al. also observed an increase in acid production after fermentation of pineapple juice. There is scant literature on the behavior of organic acids after the addition of probiotics and during storage in fruit juices. Randhawa et al. found that citric acid contents decreased and malic acid contents increased in fermented citrus juices throughout the storage period.

9.4 CONCLUSION

Preliminary probiotic characteristics (acid and bile tolerance and antibiotic sensitivity) were found superior in *L. rhamnosus* followed by *L. plantarum* and *L. acidophilus*. All the three strains of *Lactobacillus* ie, *L. plantarum* (Lp), *L. rhamnosus* (Lr), and *L. acidophilus* (La) have good capacity to survive in the fruit juices studied. *L. plantarum* (Lp) had maintained the required $\log 10^8$ count in litchi and pineapple juices up to 6 weeks in refrigerated condition (4 ± 1 °C). The change in the total soluble sugar, pH, and titratable acidity was minimum in fruit juices fortified with *L. plantarum* (Lp) than other fortified juices. Juice of litchi and pineapple appeared to be better carriers compared to orange and guava for delivery of probiotics due to acceptable changes in their pH, TSS, and titratable acidity on probiotication. The color of the two fermented juices viz. litchi and pineapple were found to be stable during refrigerated storage for up to 6 weeks. Overall acceptability of juices was found to be in the range between 7.4 and 7.6. All other sensory parameters were also found to be in the acceptable range between 5.5 and 7.8. After 6 weeks of storage, the fruit juices were found to have lower acceptability due to changes in appearance, color, taste, and mouthfeel. *L. plantarum* (Lp) was found to be the superior species and litchi juice was found to be the suitable carrier for probiotic bacteria for developing health-promoting functional fruit drink.

KEYWORDS

- **probiotic**
- **fruit juices**
- **litchi**
- **pineapple**
- **total phenolic content**
- **antioxidant activity**

REFERENCES

Boylston, T. D., Vinderola, C. G., Ghoddusi, H. B., and Reinheimer, J. A. Incorporation of Bifidobacteria into Cheeses: Challenges and Rewards. *Int. Dairy J.*, **14**, 375–387, 2004.

Corbo, M. R. Albenzio, M., De Angelis, M., Sevi, A., and Gobbetti, M. Functional Beverages: The Emerging Side of Functional Foods Commercial Trends, Research, and Health Implications. *Compr. Rev. Food Sci. Food Saf.,* **13**, 1192–1206. 2014.

Do Espírito Santo, A. P., Patrizia P., Attilio C., and Maricê N. Oliveira. Influence of Food Matrices on Probiotic Viability—A Review Focusing on the Fruity Bases. *Trends Food Sci. Technol.*, **22**, 377–385, 2011.

Fenech, M. Baghurst, P. , Luderer, W., Turner, J., Record, S., Ceppi, M., and Bonassi S. Low Intake of Calcium, Folate, Nicotinic Acid, Vitamin E, Retinol, b-Carotene and High Intake of Pantothenic Acid, Biotin and Riboflavin are Significantly Associated with Increased Genome Instability-Results from a Dietary Intake and Micronucleus Index Survey in South Australia. *Carcinogenesis*, **26**, 991–999, 2005.

Granato, D. Perotti, F., Masserey, I., Rouvet, M., Golliard, M., and Servin, A. Functional Foods and Nondairy Probiotic Food Development: Trends, Concepts, and Products. *Compr. Rev. Food Sci. Food Saf.,* **9**, 292–302, 2010.

Jankovic, I. Phothirath, P., Ananta, E., and Mercenier, A. Application of Probiotics in Food Products-challenges and New Approaches. *Curr. Opin. Biotechnol.,* 21, 175–181, 2010.

King, V. A. Huang, HY, and Tsen, JH. Fermentation of Tomato Juice by cell Immobilized Lactobacillus acidophilus. *Mid-Taiwan J. Med.*, **12**, 1–7, 2007.

Krasaekoopt, W., and Kitsawad, K., Sensory Characteristics and Consumer Acceptance of Fruit Juice Containing Probiotic Beads in Thailand, *AU.J. T.*, **14(1)**, 33–38, 2010.

Lorca, G. L. and de Valdez, G. F. A Low-pH-Inducible, Stationary-Phase Acid Tolerance Response in Lactobacillus acidophilus CRL639. *Curr. Micro.,* **42**, 21–25. 2001.

Luckow, T. Sheehan, V., Fitzgerald, G., and Delahunty, C. Exposure, Health Information and Flavour-Masking Strategies for Improving the Sensory Quality of Probiotic Juice. *Appetite*, **47**, 315–323, 2006.

Mättö, J., Malinen, E. , Suihko, M. L. , Alander, M. Palva, A., and Saarela, M. Genetic Heterogeneity and Technological Properties of Intestinal Bifidobacteria. *J. Appl. Microb.*, **97**, 459–470, 2004.

Mitsuoka, T. The Human Gastrointestinal Tract. *Lactic Acid Bacteria* **1**, 69–114, 1992.

Nagpal, R., Kumar, A., and Kumar, M. Fortification and Fermentation of Fruit Juices with Probiotic Lactobacilli. *Ann. Microb.,* **62**, 1573–1578, 2012.

Nualkaekul, S. and Charalampopoulos, D. (2011) Survival of Lactobacillus Plantarum in Model Solutions and Fruit Juices. *Int. J. Food Microb.* **146**, 111–117, 2011.

Pereira, A.L.F., Maciel, T.C., and Rodrigues, S. Storage Stability and Acceptance of Probiotic Beverages from Cashew Apple Juice. *Food Bio. Technol.*, **6**, 3115–3165, 2013.

Pereira, A.L.F., Maciel, T.C. and Rodrigues, S. Probiotic Cashew Apple Juice. *Food Res. Int.*, **44**, 1276–1283, 2011.

Peres, C.M. Peres, C., Hernández-Mendoza, A., Xavier Malcata, F. Review on Fermented Plant Materials as Carriers and Sources of Potentially Probiotic Lactic Acid Bacteria-With an Emphasis on Table Olives. *Trends Food Sci. Technol.* **26**, 31–42. 2012.

Perricone, M. Bevilacqua, A., Corbo, M. R., and Sinigaglia, M. Technological Characterization and Probiotic Traits of Yeasts Isolated from Altamura Sourdough to Select Promising Microorganisms as Functional Starter Cultures for Cereal-based Products. *Food Microbiol.*, **38**, 26–35, 2014.

Perricone, M. Corbo, M. R., Sinigaglia, M., Speranza, B., and Bevilacqua, A. Viability of Lactobacillus reuteri in Fruit Juices. *J. Funct. Foods.,* **10**, 421–426, 2014.

Ranadheera, C.S. Prasanna, P.H.P., and Vidanarachchi , J.K. Fruit juice as probiotic carriers. In Fruit Juices: Types, Nutritional Composition and Health Benefits, 1st ed.; Elder, K.E., Ed.; Nova Science Publishers: Hauppauge, NY, USA, 2014; pp. 1–19.

Ranadheera, R.D.C.S. Baines, S.K., and Adams, M. C. Importance of Food in Probiotic Efficacy. *Food Res. Int.*, **43**, 1–7, 2010.

Ross R.P. Desmond, C., Fitzgerald, G.F., and Stanton, C. Overcoming the Technological Hurdles in the Development of Probiotic Foods. *J. Appl. Micro.* **98**, 1410–1417, 2005.

Saarela, M. Virkajarvi, I., Alakomi, H. L., Sigvart-Mattila, P., and Matto, J. Stability and Functionality of Freeze-dried Probiotic Bifidobacterium Cells During Storage in Juice and Milk. *Int. Dairy J.* **16**, 1477–1482, 2006.

Saradhuldhat P., Paull, RE. Pineapple Organic Acid Metabolism and Accumulation During Fruit Development, *Sci. Hort.*, **112**, 297–303, 2007.

Shah, N. P., and Jelen, P. Survival of Lactic Acid Bacteria and their Lactases Under Acidic Conditions. *J. Food Sci.*, **55**, 506–52, 1990.

Sheehan, V.M. Ross, P., and Fitzgerald, G. F. Assessing the Acid Tolerance and the Technological Robustness of Probiotic Cultures for Fortification in Fruit Juices. *Innov. Food Sci. & Emerg. Technol.,* **8**, 279–284. 2007.

Sutton, K.H. Considerations for the Successful Development and Launch of Personalised Nutrigenomic Foods. *Mutat. Res.*, **622**, 117–121, 2007.

Tang H. Jing, Y., XIE, C., Hong, WEI. Antibiotic susceptibility of strains in Chinese medical probiotic products. *J. Medi. Coll. PLA* **22**, 149–152. 2007.

Tannock G.W. Munro, K., Harmsen, H. J. M. , Welling, G. W., Smart, J., and Gopal, P. K. Analysis of the Fecal Microflora of Human Subjects Consuming a Probiotic Product Containing Lactobacillus rhamnosus DR20. *Appl. Environ. Micro.,* **66**, 2578–2588, 2000.

Teixeira P., Castro H., Mohácsi-Farkas C., Kirby R. Identification of Sites of Injury in Lactobacillus Bulgaricus During Heat Stress. *J. Appl. Microb.,* **83**, 219–226. 1997.

Tripathi, M.K. and Giri, S.K. Probiotic Functional Foods: Survival of Probiotics During Processing and Storage. *J. Funct. Foods.* **9**, 225–241. 2014.

Van de Guchte M., Serror P., Chervaux C., Smokvina T., Ehrlich S.D., Maguin E. Stress Responses in lactic Acid Bacteria. Antonie van Leeuwenhoek, *Int. J. Gen. Mol. Micro.*, **82**, 187–216. 2002

Vinderola, C. G., Costa, G.A., Regenhardt, S., and Reinheimer, J. A. Influence of Compounds Associated with Fermented Dairy Products on the Growth of Lactic Acid Starter and Probiotic Bacteria. *Int. Dairy J.*, **12**, 579–589, 2002.

Yoon KY, Woodamns EE, Hang YD Probiotication of Tomato Juice by LAB. *J. Microbiol.* **42**, 315–318, 2004.

Yoon, K. Y. Woodams, E. E., and Hang, Y. D. Production of Probiotic Cabbage Juice by Lactic Acid Bacteria. *Biores. Technol.*, **97**(12), 1427−1430, 2006.

CHAPTER 10

Enthalpy–Entropy Compensation and Adsorption Characteristics of Legumes Using ANN Modeling

PREETISAGAR TALUKDAR[1*], PRANJAL PRATIM DAS[1], and MANUJ KUMAR HAZARIKA[2]

[1]*Department of Chemical Engineering, IIT Guwahati, Assam, India*

[2]*Department of Food Engineering and Technology, Tezpur University, Assam, India*

[*]*Corresponding author. E-mail: preetisagar1891@gmail.com*

ABSTRACT

This work concentrates on moisture sorption isotherm modeling by the use of GAB equation and artificial neural network (ANN) in legumes. The equilibrium moisture content for adsorption was determined by static gravimetric technique. The experiments for legumes were carried out in saturated salt solutions at 30, 40, and 50 °C until the equilibrium moisture content (EMC) was obtained. The moisture adsorption data obtained were fitted in GAB equations and ANN by using MATLAB. For the ANN modeling seven input neurons corresponding to the seven input variables viz., a_w, temperature, ash content, dietary fiber, crude protein, fat content, and carbohydrates were considered whereas the output neuron represents the EMC. The highest R-value and the least mean square error (MSE) value correspond to the hidden neurons of five (for chickpeas) and nine (black-eyed peas, mung beans, and Bengal grams) with the neural network of 7-5-1 and 7-9-1, respectively. At given moisture content, water activity value is estimated by the fitted GAB model in the moisture range from 5 to 20%, at all the temperatures. The generalized isotherm model involving isokinetic temperature and two other parameters, namely, K_1 and K_2 were developed. The overall values were K_1 =

6237.907 and $K_2 = 0.8325$. These two parameters were used for the hydration kinetics study of the same product. For the kinetics of hydration of Bengal gram, black-eyed peas, chickpeas, and mung beans at different temperatures, studies were carried out with measurement of moisture content at known intervals considering the principle of one-dimensional mass transfer to the spherical body. For expressing the dependence of the diffusivity value on temperature and moisture, the model involved an expression containing the parameters K_1 and K_2 and heat of sorption. Simulations were carried out to estimate the value of the pre-exponent of diffusivity expression (D_0).

10.1 INTRODUCTION

In today's world, the legumes' importance has grown (Chenoll et al., 2009). Legumes are replacing meat as a source of dietary protein. Legumes consist of a seed coat and cotyledons, within the cotyledons, the starch granules are present in the protein matrix (Sayar et al., 2001). On soaking legumes in water, a film layer is formed on its surface, which provides resistance to the water molecules. The water molecules overcome this resistance and travel to the cotyledons where they react with the starch granules. During this reaction, a temperature below gelatinization temperature gives swelled starch granules and a temperature above gelatinization temperature gives gelatinized starch (Hoseney, 1994).

Different condition of soaking in grains leads to a different rate of water adsorption and capacities of water adsorptions (Shefaei et al., 2014). Water adsorption affects the final quality of the soaked product (Turhan et al., 2002). Legumes' water adsorption depends on water temperature and duration of soaking (Rajbari et al., 2011); with an increase in water temperature there is a decrease in soaking time, as high water temperature gives high moisture diffusivity (Kashaninejad et al., 2009).

Swelling which takes place during soaking along with moisture diffusion affects the rate of water adsorption. In the soaking process, swelling consideration is difficult as there is a lack of knowledge on swelling velocity in relation to moisture diffusivity. Therefore, a swelling study is carried out to understand the soaking process (Bello et al., 2010). Fick's second law of diffusion is used to study water adsorption of mung beans and to model the characteristic of water diffusivity (Yildirim et al., 2011).

In most literature, mathematical descriptions mentioned are based on data-driven regression model, which describes hydration kinetics or Fickian diffusion. In legumes, a description of diffusions cannot be done by a concentration depended form of Fick's second law of diffusion, since the legumes undergo swelling on soaking (Chapwanya and Misra, 2015).

According to McMinn et al. (2005) the enthalpy–entropy compensation theory, has been analyzed for a different variation of physical and chemical processes. The theory stated that due to changes in the nature of interaction among the solute and solvent, which are responsible for the reaction; the compensation arises. For a specific reaction, the relationship between entropy and enthalpy is linear. In the case of a linear enthalpy and entropy relation, the isokinetic temperature (T_β) was determined from the slope of the line and, if the theory is valid, it should be constant at any point (Heyrovsky, 1970). It represented that the temperature at which all the reactions in the series proceed were at the same rate (Heyrovsky, 1970). The existence of the compensation theory, or the isokinetic relationship, implies that only one reaction mechanism is followed by all members of the reaction series and, therefore, a reliable evaluation of the isokinetic relationship aids elucidation of the reaction mechanisms.

Water sorption of food materials is governed by a complex function between the chemical and physical characteristics of the material. In fruits or legumes, the major water-binding compounds are carbohydrates including both complex biopolymers and simple sugars. Water sorption isotherm is affected by these simple sugars and complex biopolymers. Modeling by incorporating both physical characteristics and chemical composition may give better results than those modeling based only on physical data. Therefore, the usage of artificial neural networks (ANN) helps to incorporate water activity, chemical composition, physical characteristics, and temperature in a single model (Myhara et al., 1998).

10.2 MATERIALS AND METHODS

10.2.1 RAW MATERIALS

Legumes used are mung beans, Bengal grams, black-eyed peas, and chickpeas which were purchased from the local store.

10.2.2 MOISTURE SORPTION ESTIMATION

For adsorption of legumes, equilibrium moisture content (EMC) is determined by using the static gravimetric technique, which based on the isopiestic transfer of water vapor. The relative humidity of each salt solution at the temperature of 30, 40 and 50 °C is in the range of LiCl = 10%, CH_3COOK = 20%, $MgCl_2$ = 30%, K_2CO_3 = 40%, $Mg(NO_3)_2$ = 50%, NaCl = 70%, KCl = 80%, and KNO_3 = 90%. In a low-temperature chamber that has temperature-controlled cabinets, the desiccators are kept at 30, 40, and 50 °C (Thermotech, Sanco, Hopkins, MN, USA) replicated thrice. Each desiccator was filled with a different saturated salt solution to prevail a constant relative humidity environment. Samples are weighted with digital balance (least count: 0.0001 g) after every 2 days (Menkov, 2000).

10.2.3 MOISTURE CONTENT

A hot air oven (SELEG TC344, model no: 1458) at 105 ± 2 °C for 2 h was used to determine moisture content (MC). The samples were taken in triplicates for analyzes and the average moisture content was recorded. Loss in weight was taken as the moisture content of the sample (Prakash et al., 2004).

$$M.C.\ (\%) = \frac{\text{Weight loss}}{\text{Initial weight}} \times 100 \qquad (10.1)$$

10.2.4 ANALYSIS OF DATA

The description of relations between water activity (a_w), temperature, and equilibrium moisture content were verified according to the GAB model.

$$X = \frac{X_m \times C \times K \times a_w}{(1 - K \times a_w) \times (1 - K \times a_w + C \times K \times a_w)} \qquad (10.2)$$

where, X_m = monolayer value (g water/g solids), X = water content (g water/g solids), a_w = water activity, C = monolayer heat sorption constant and K = multilayer heat sorption constant (Chen, 2003).

10.2.5 ARTIFICIAL NEURAL NETWORK DATA PREPARATION

The data of the sample were selected for modeling seven input crude protein, fat, ash content, dietary fiber, carbohydrate, a_w, and temperature and the output variables EMC. The data were available from experimental results and literature review without any missing values. The ANN modeling was carried out in MATLAB using the data available from the moisture sorption values (Myhara et al., 1998).

10.2.6 ESTIMATION OF ENTHALPY–ENTROPY COMPENSATION IN SORPTION

To estimate the enthalpy–entropy compensation data, the method of linear regression is used and applied to the adsorption sorption values obtained from the sorption method of legumes. MATLAB was used to estimate the constant parameters *A*, *B*, and *C* in the following equation:

$$MC = \frac{a_w}{A*(a_w)^2 + B*a_w + C} \tag{10.3}$$

where *A*, *B*, and *C* are the constant parameters and are estimated in MATLAB by the process of curve fitting. The moisture content of the legumes is taken to be constant at 5%, 7.5%, 10%, 12.5%, 15%, 17.5%, and 20%. The corresponding water activity is estimated using the MATLAB. The obtained water activity values are used in excel sheets to obtained the T_β, isokinetic temperature. The formula $\psi T*(\ln a_w) = K_1 K_2^m$ is used to obtain a graph between EMC and $\ln(\ln a_w * \psi T)$; from the graph the K_1 and K_2 values are obtained (Heyrovsky, 1970).

10.2.7 HYDRATION BEHAVIOR

The weighted legumes were taken in a beaker containing 200 ml of distilled water. The beaker containing the sample is placed in a water bath (Lab companion, Model number BW-20G) which was heated to the required temperature. The experiment was conducted at various temperatures of 30, 40, and 50 °C. The hydration behavior is continued until the samples attain a stable weight. The final weight of the legumes was measured and taken into account (Chenoll et al., 2009).

10.2.8 HYDRATION MODELING TO ESTIMATE D_0

The hydration modeling was done using the data obtained from the soaking treatment and the K_1 and K_2 values obtained in the estimation of enthalpy–entropy compensation in sorption isotherm. The diffusivity of the sample was estimated using Equation (10.4)

$$D(\rho_A) = D_0 \exp\left[-\frac{\left(R_g K_1 K_2^{\frac{u}{u_i}} + \lambda(\mathrm{T})\right)}{2R_g T}\right] \quad (10.4)$$

where D_0 is a constant (equivalent to the diffusion coefficient at infinitely high temperature), T is the absolute temperature and R_g is the gas constant, K_1 and K_2 are constants to be calculated from sorption equilibrium data and u_i is the monolayer moisture content, local moisture content in dry basis, u, equation ρ_A represents the local volumetric concentration of water and $D(\rho A)$ is the diffusion coefficient

10.3 RESULT AND DISCUSSION

10.3.1 MOISTURE SORPTION MODELING USING ARTIFICIAL NEURAL NETWORK

For adsorption of mung beans, Bengal gram, chickpeas, and black-eyed beans equilibrium moisture content is determined by using a static gravimetric technique, which is based on an isopiestic transfer of water vapor. Preparation of saturated salt solutions of NaCl, LiCl, $MgCl_2$, Mg $(NO_3)_2$, KNO_3, K_2CO_3, KCl, and CH_3COOK were done to prevail a constant relative humidity (RH) environment at 30 °C, 40 °C, and 50 °C. From Labuza (1984), water activities of the above-saturated salts solution at different temperatures were taken. The experiment was carried out in saturated salt solutions at the temperature of 30 °C, 40 °C, and 50 °C till the equilibrium moisture content was obtained. The moisture content values were obtained by following the procedures in Sections 10.2.2 and 10.2.3. The moisture adsorption graphs at the three temperatures were given in Figures 10.1–10.4.

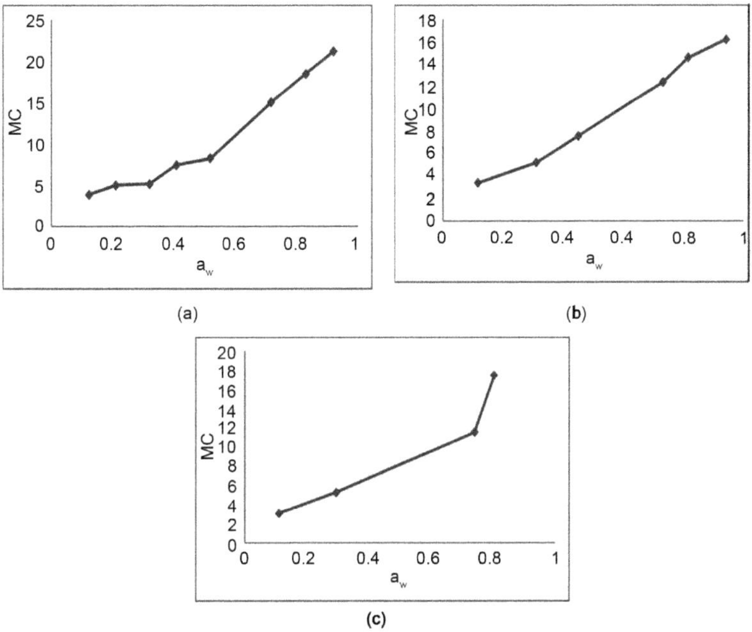

FIGURE 10.1 The adsorption graphs of Bengal gram at (a) 30 °C, (b) 40 °C, and (c) 50 °C.

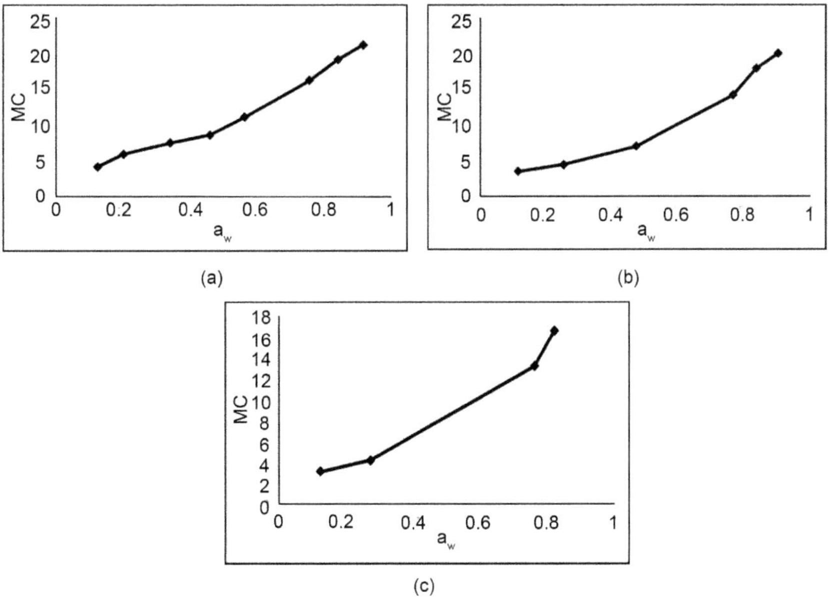

FIGURE 10.2 The adsorption graphs of black-eyed peas at (a) 30 °C, (b) 40 °C, and (c) 50 °C

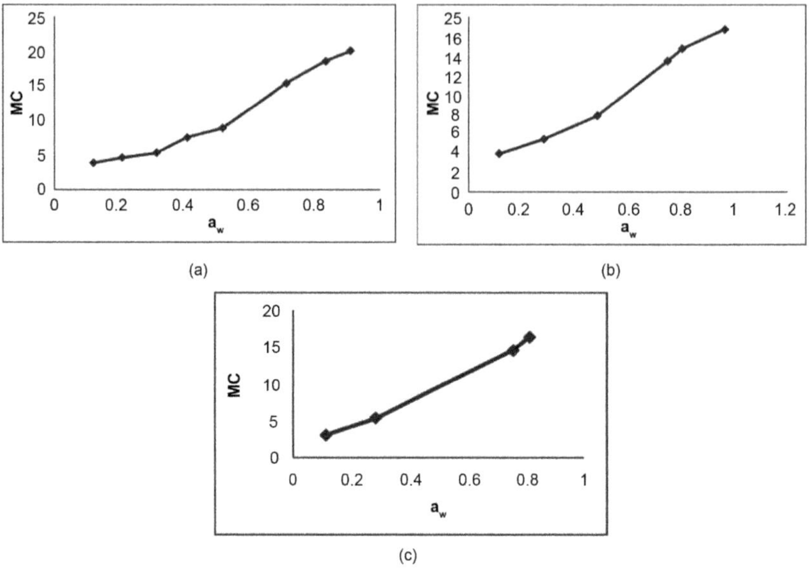

FIGURE 10.3 The adsorption graphs of chickpeas at (a) 30 °C, (b) 40 °C, and (c) 50 °C.

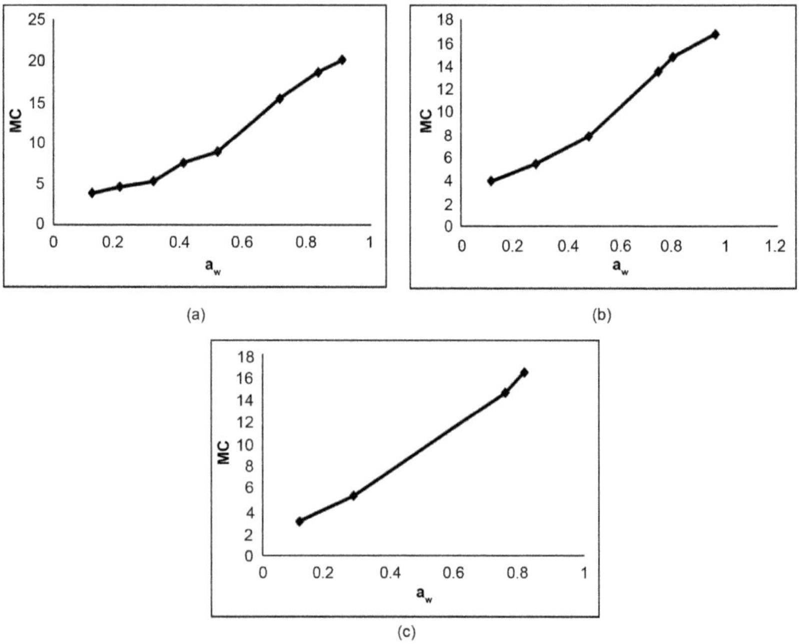

FIGURE 10.4 The adsorption graphs of mung beans at (a) 30 °C, (b) 40 °C, and (c) 50 °C.

The above moisture adsorption data obtained were fitted in the GAB equation (Equation (10.2)) by using MATLAB. The constant parameters were estimated from the curve fitting tool and the RMSE and R^2 values are represented in Tables 10.1–10.4.

TABLE 10.1 The Values of GAB Equation of Bengal Grams

Parameters	50 °C	40 °C	30 °C
X	1.34	10.71	8.21
C	1.30	4.42	4.46
K	1.15	0.51	0.71
RMSE	2.33	0.80	0.95
R^2	0.95	0.98	0.97

TABLE 10.2 The Values of GAB Equation of Black-eyed Peas

Parameters	50 °C	40 °C	30 °C
X	3.60	5.81	7.68
C	22.65	7.34	9.38
K	0.95	0.81	0.72
RMSE	0.17	0.90	0.64
R^2	0.99	0.99	0.99

TABLE 10.3 The Values of GAB Equation of Chickpeas

Parameters	50 °C	40 °C	30 °C
X	4.17	9.96	11.40
C	1.22	5.83	3.12
K	0.92	0.52	0.61
RMSE	1.85	1.11	1.09
R^2	0.97	0.97	0.98

TABLE 10.4 The Values of GAB Equation of Mung Beans

Parameters	50 °C	40 °C	30 °C
X	1.23	11.87	8.13
C	1.20	5.47	5.08
K	1.11	0.59	0.86
RMSE	2.56	0.72	0.85
R^2	0.90	0.95	0.96

Multilayer neural networks including output, input, and hidden layers in variation of 1–10 were studied. Seven neurons formed the input layer corresponding to seven input variables of components: temperature, a_w, ash content, dietary fiber, crude protein, fat content, and carbohydrates. The composition of raw Bengal gram, black eyed pea, chick pea, and mung beans are given in Table 10.5. The output layer had one neuron, which represented the equilibrium moisture content. The optimal configuration was based on minimizing the difference between values predicted by the neural networks and the desired outputs. The optimal value of the ANN process is the number of hidden layers and neural networks which corresponds to the highest *R*-value and the least MSE value. The seven input neurons correspond to seven input variables: a_w, temperature, ash content, dietary fiber, crude protein, fat content, and carbohydrates, and the output neuron represents the EMC. The data set of 26 cases obtained from experiments and was divided into three sets of training, validation, and testing. The training set consisted of 18 (70%) cases, the validation set consisted of 4 (15%) cases, and the testing consisted of 4 (15%) cases. The cases for the three sets, training, validation, and testing were chosen randomly from the set of 26 cases. The number of hidden layers and neural networks that correspond to the highest *R*-value and the least MSE value is the optimal value of the ANN process.

TABLE 10.5 Compositions of the Raw Samples

Samples	Crude Protein	Fat	Fiber	Ash	Carbohydrates
Bengal gram	19.00	2.10	6.00	3.14	56.00
Black-eyed pea	22.30	2.10	4.10	3.77	55.07
Chickpea	23.60	3.10	4.61	4.42	50.00
Mung beans	21.80	2.40	4.20	3.20	51.30

TABLE 10.6 The ANN Values of Bengal Gram

Number of Hidden Layers	Neural Network	MSE	*R*
1	7-1-1	14.3231	0.7517
2	7-2-1	1.3029	0.9763
3	7-3-1	0.6047	0.9925
4	7-4-1	0.7132	0.9906
5	7-5-1	0.7943	0.9869
6	7-6-1	59.5205	0.6588
7	7-7-1	0.1230	0.9978
8	7-8-1	0.6901	0.9939
9	7-9-1	0.0593	0.9991
10	7-10-1	0.1731	0.9701

From Table 10.6 it can be seen that both the highest *R*-value and the least MSE value corresponds to the hidden neurons of nine with the neural network of 7-9-1. Hence the best neural network is 7-9-1 with hidden layers of nine. The best neural network figure is given in Figure 10.5.

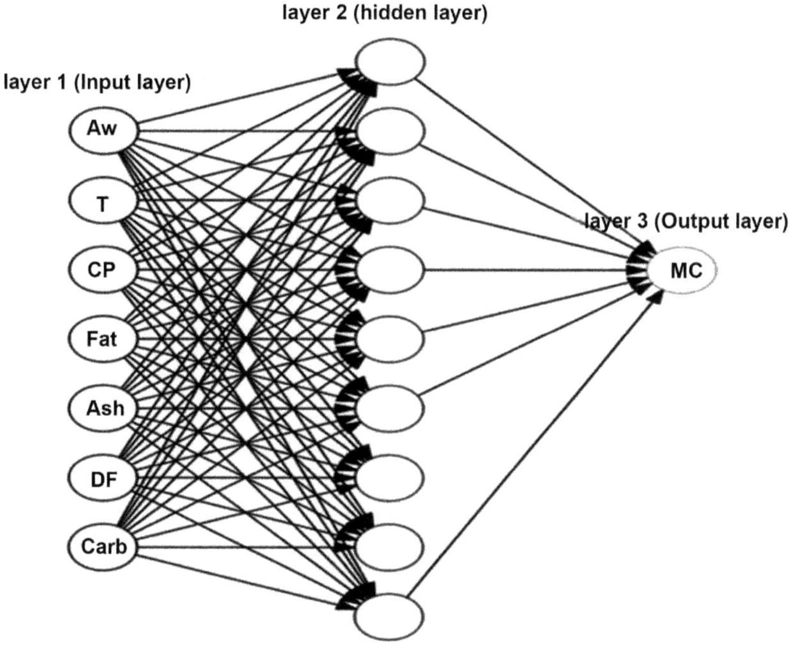

FIGURE 10.5 The neural network of nine hidden layers.

TABLE 10.7 The ANN Values of Black-Eyed Peas

Number of Hidden Layers	Neural Network	MSE	*R*
1	7-1-1	0.4999	0.9940
2	7-2-1	0.4234	0.9942
3	7-3-1	0.2485	0.9967
4	7-4-1	0.5018	0.9947
5	7-5-1	0.0835	0.9990
6	7-6-1	0.0690	0.9991
7	7-7-1	0.0543	0.9989
8	7-8-1	0.0575	0.9993
9	7-9-1	0.0187	0.9998
10	7-10-1	3.1689	0.9631

From Table 10.7, it is seen that both the highest R-value and the least MSE value corresponds to the hidden neurons of nine with the neural network of 7-9-1. Hence the best neural network is 7-9-1 with hidden layers of nine. The best neural network figure is given in Figure 10.6.

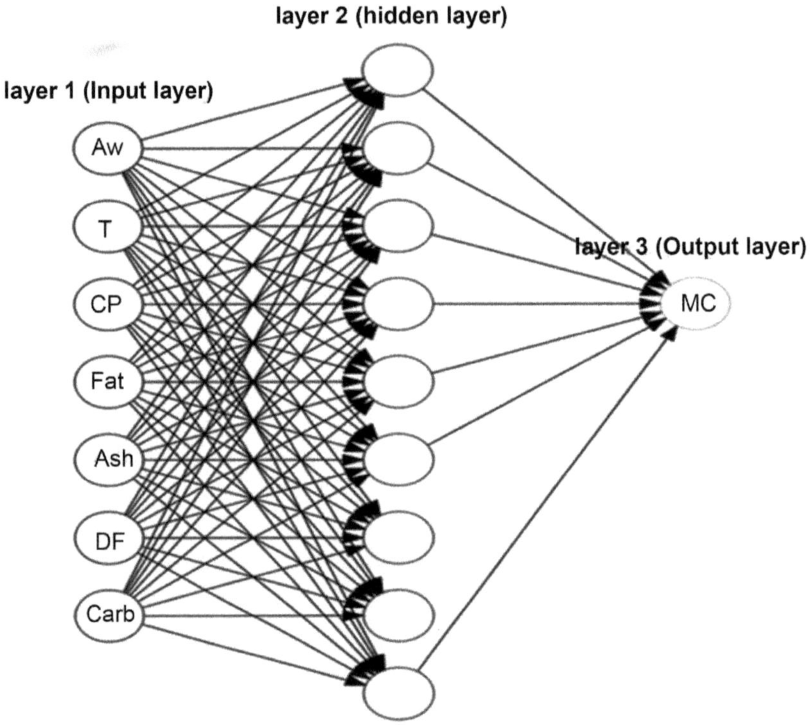

FIGURE 10.6 The neural network of nine hidden layers.

TABLE 10.8 The ANN Values of Chickpeas

Number of Hidden Layers	Neural Network	MSE	*R*
1	7-1-1	0.9656	0.9823
2	7-2-1	0.1415	0.9833
3	7-3-1	0.6375	0.9381
4	7-4-1	0.1550	0.9976
5	7-5-1	0.1371	0.9979
6	7-6-1	0.1707	0.9766
7	7-7-1	0.1745	0.9974
8	7-8-1	0.5095	0.9956
9	7-9-1	3.5761	0.9396
10	7-10-1	3.2569	0.9510

From Table 10.8, it is seen that both the highest R-value and the least MSE value corresponds to the hidden neurons of five with the neural network of 7-5-1. Hence the best neural network is 7-5-1 with hidden layers of five. The best neural network figure is given in Figure 10.7.

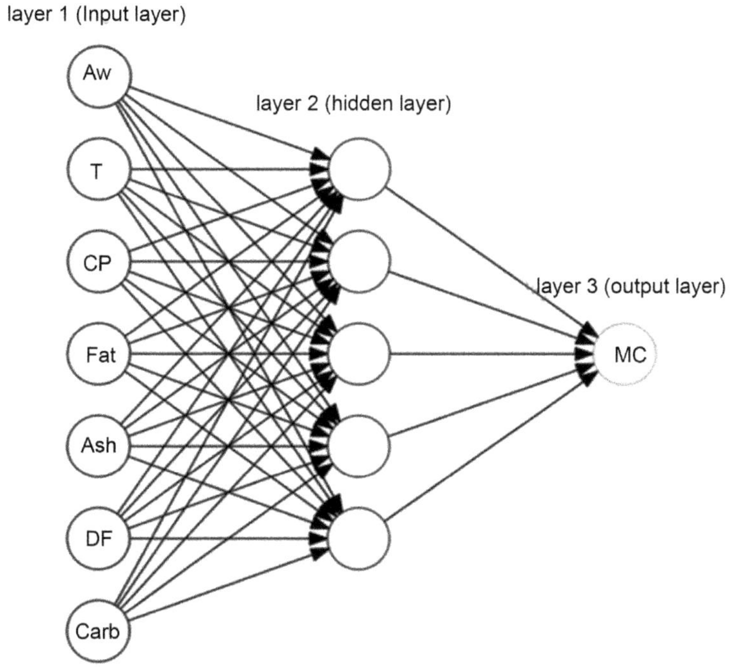

FIGURE 10.7 The neural network of five hidden layers.

TABLE 10.9 The ANN Values of Mung Beans

Number of Hidden Layers	Neural Network	MSE	*R*
1	7-1-1	13.3767	0.8112
2	7-2-1	1.2187	0.9087
3	7-3-1	0.6986	0.9555
4	7-4-1	0.7111	0.9199
5	7-5-1	0.7456	0.9876
6	7-6-1	0.9098	0.7109
7	7-7-1	0.2234	0.9810
8	7-8-1	0.6871	0.9204
9	7-9-1	0.1129	0.9897
10	7-10-1	0.1322	0.9581

From Table 10.9, it is seen that both the highest R-value and the least MSE value corresponds to the hidden neurons of nine with the neural network of 7-9-1. Hence the best neural network is 7-9-1 with hidden layers of nine. The best neural network figure is given in Figure 10.8, the neural network of nine hidden layers.

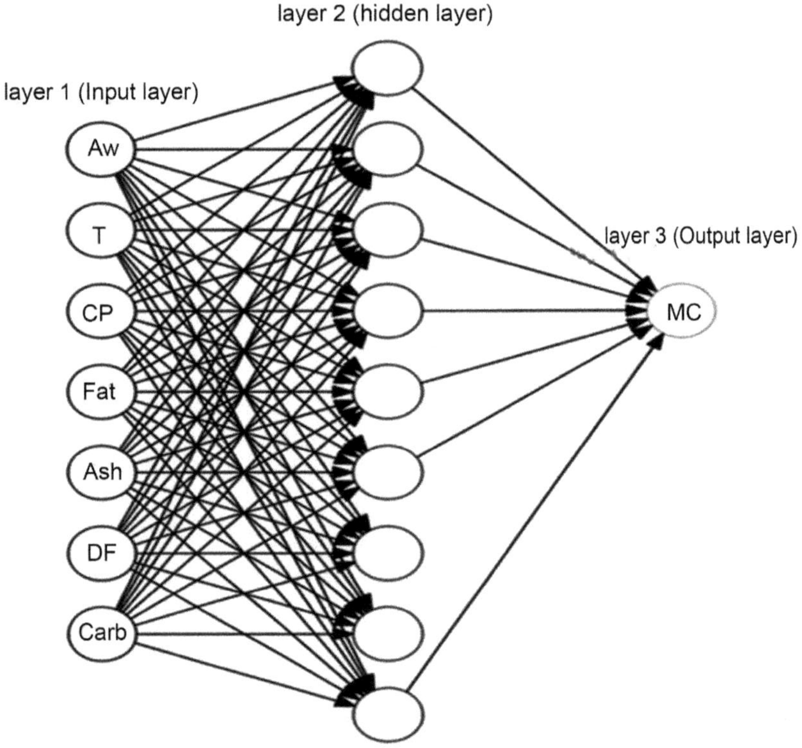

FIGURE 10.8 The neural network of nine hidden layers.

10.3.2 ENTHALPY-ENTROPY COMPENSATION IN SORPTION PHENOMENA

The understanding of water properties with its relationship to a biological system is classified into three categories: thermodynamic, dynamic, and structural. The understanding of water equilibrium at a particular temperature and relative humidity is given by the thermodynamic approach. The thermodynamic functions can be calculated from sorption isotherms (McMinn et al., 2005).

During the sorption process, the energy variation that occurs on mixing water molecules with sorbent is measured as enthalpy change (ΔH). The binding or repulsive forces in the system is given as entropy (ΔS), it is associated with the spatial arrangement of the water–sorbent interface. For a linear enthalpy and entropy relation, the isokinetic temperature (T_β) can be determined from the slope of the line and, if the theory is valid, should be constant at any point. It represents the temperature at which all the reactions in the series proceed at the same rate (Heyrovsky, 1970).

The existence of the compensation theory, or the isokinetic relationship, implies that only one reaction mechanism is followed by all members of the reaction series and, therefore, a reliable evaluation of the isokinetic relationship aids elucidation of the reaction mechanisms.

The initial step is to calculate the constant parameters of Equation (10.3) using the MATLAB (Table 10.10). After this step, the constant parameters were taken into account and the corresponding water activity for constant moisture content of 5%, 7.5%, 10%, 12.5%, 15%, 17.5%, and 20% were calculated in MATLAB (Table 10.11).

10.3.2.1 RESULTS AND GRAPHS OF BENGAL GRAM

The calculation of the constant parameters and predicted water activity for constant moisture content for black eyed peas are given in Table 10.12 and Table 10.13, respectively.

TABLE 10.10 The Calculated Constant Parameters of Equation (10.3) for Bengal Gram

Temperature (°C)	*A*	*B*	*C*
50	−0.3702	0.3589	−0.0024
40	−0.0369	0.0511	0.0412
30	−0.0679	0.0672	0.0379

TABLE 10.11 The Predicted Water Activity of Bengal Gram

Moisture Content (%)	50 °C	40 °C	30 °C
5	0.3053	0.2604	0.2526
7.5	0.5982	0.4221	0.4051
10	0.6897	0.5854	0.5440
12.5	0.7440	0.3610	0.6591
15	0.7809	0.7673	0.7514
17.5	0.9069	0.8686	0.8051
20	0.9769	0.9547	0.9247

The predicted water activity is taken into consideration to calculate the $1/T$ and ln(1/ERH). The plots between $1/T$ and ln(1/ERH) are given in Figure 10.9.

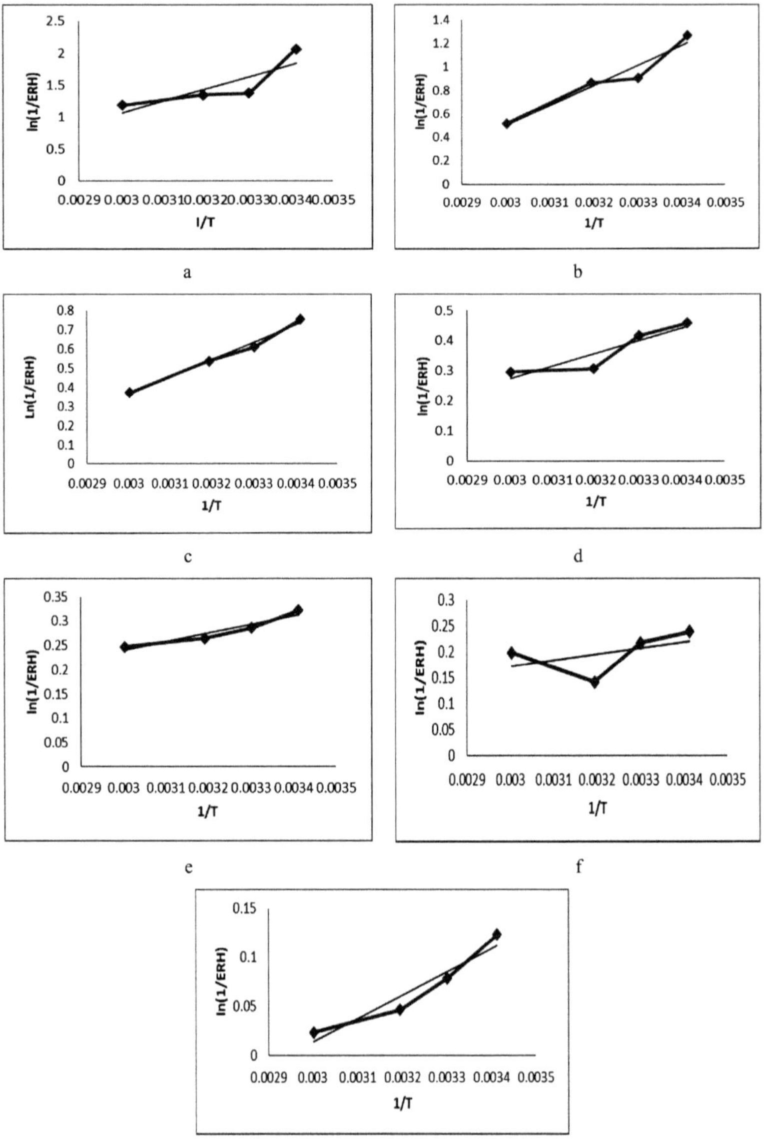

g

FIGURE 10.9 The graphs of $1/T$ vs ln(1/ERH) at (a) 5%, (b) 7.5%, (c) 10%, (d) 12.5%, (e) 15%, (f) 17.5%, and (g) 20% MC.

The intercept and the slope of the above graphs were used to calculate the del H of enthalpy and del S of entropy. The following formulae were used for calculation: del H = slope*R and del S = intercept*R. The calculated del S and del H were plotted in a graph, from the slope of the graph the value of T_β is obtained. T_β = the slope of the graph is given in Figure 10.10.

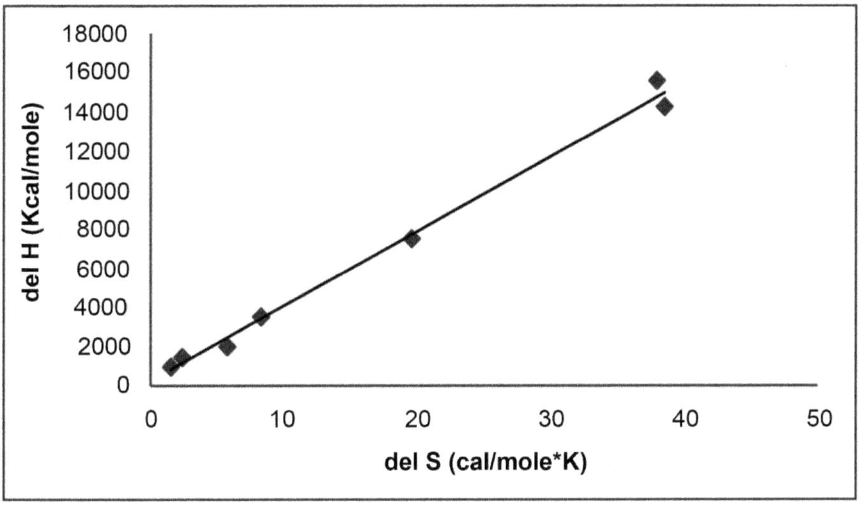

FIGURE 10.10 del S vs del H.

From the graph in Figure 10.10, the value of T_β is obtained. T_β = 381.52 K. The obtained value is nearer to the value obtained by Aguerre et al. (1986). Therefore, it can be deduced that the method used to obtain the value of T_β is valid. The next step is to calculate the value of $\psi(T) = 1/(1/T_\beta - 1/T)$ for the temperature of 30, 40, and 50 °C. The obtained $\psi(T)$ is used to calculate the values of $\ln(\ln a_w * \psi(T))$. A graph is plotted between $\ln(\ln a_w * \psi(T))$ and EMC to obtained the values of K_1 and K_2 of Equation (10.4). The K_1 and K_2 values were obtained from the exponential of the intercept and slope of the graph plotted between $\ln(\ln a_w * \psi(T))$ and EMC, that is, K_1 = exp(intercept) and K_2 = exp(slope). From the graph given in Figure 10.11, the values of K_1 and K_2 were obtained in the mentioned methods.

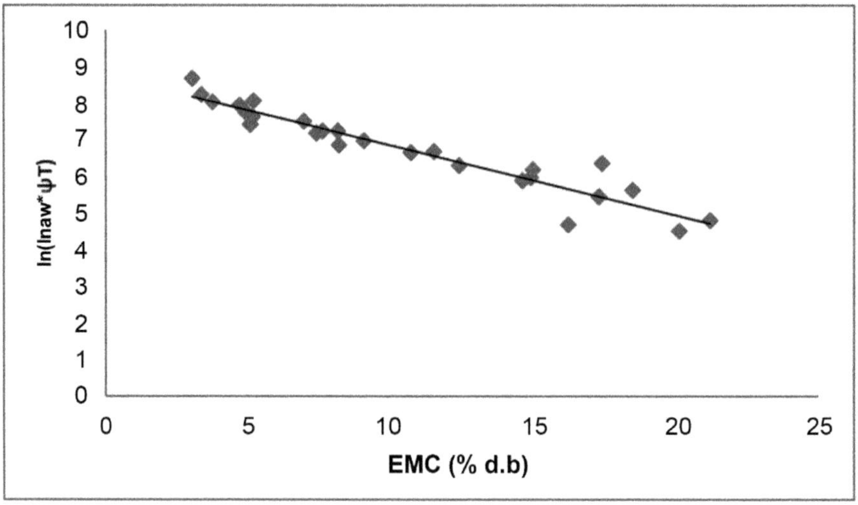

FIGURE 10.11 ln (ln a_w*ψ(T)) vs EMC.

The values of K_1 and K_2 were obtained as K_1 = 6436.3107 and K_2 = 0.8258.

10.3.2.2 RESULTS AND GRAPHS OF BLACK-EYED PEAS

TABLE 10.12 The Calculated Constant Parameters of Equation (10.4) for Black-Eyed Peas

Temperature (°C)	*A*	*B*	*C*
50	−0.2535	0.2528	0.0128
40	−0.1207	0.1253	0.02888
30	−0.08435	0.1024	0.01909

TABLE 10.13 The Predicted Water Activity of Black-Eyed Peas

Moisture Content (%)	50 °C	40 °C	30 °C
5	0.3518	0.2694	0.1705
7.5	0.5612	0.4570	0.3265
10	0.6773	0.6051	0.4942
12.5	0.7491	0.7116	0.6267
15	0.7976	0.7890	0.7326
17.5	0.8325	0.8471	0.8144
20	0.8568	0.8921	0.8788

The predicted water activity is taken into consideration to calculate the $1/T$ and ln(1/ERH). The plots between $1/T$ and ln(1/ERH) are given in Figure 10.12, Tables 10.12, and 10.13.

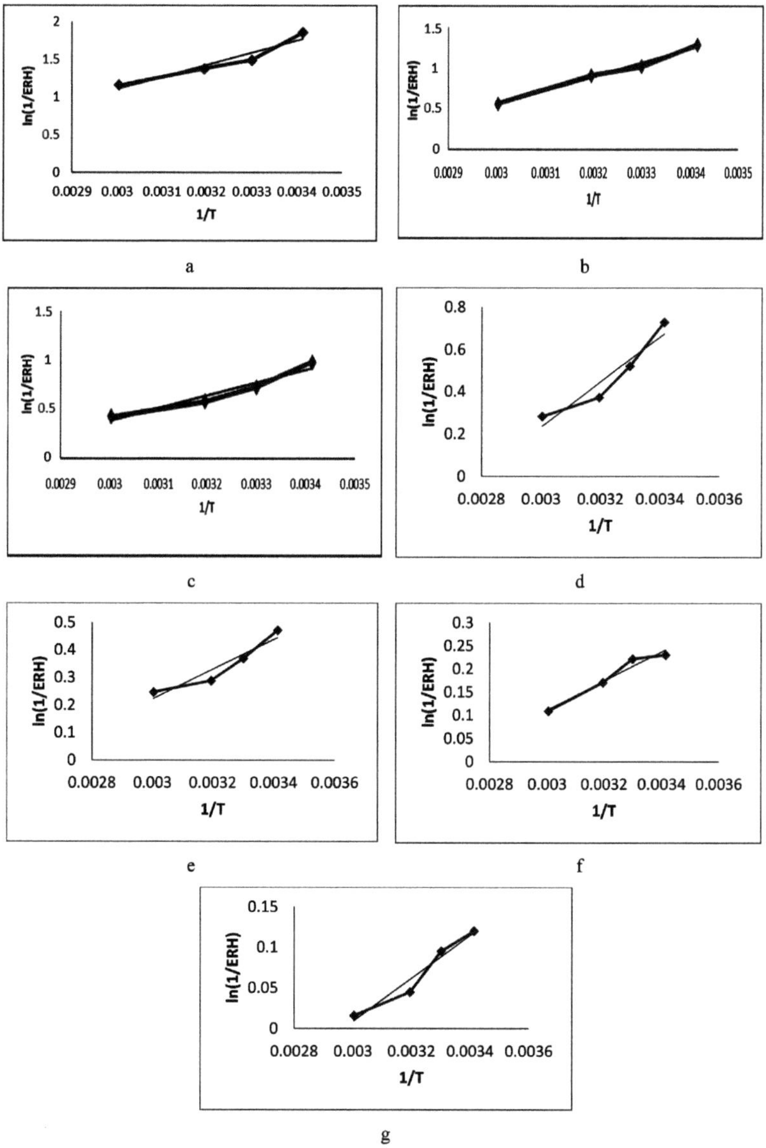

FIGURE 10.12 The graphs of $1/T$ vs ln(1/ERH) at (a) 5%, (b) 7.5%, (c) 10%, (d) 12.5%, (e) 15%, (f) 17.5%, and (g) 20% MC.

The calculations of del S and del H were done as mentioned earlier and were plotted in a graph, from the slope of the graph (Figure 10.13) the value of T_β is obtained.

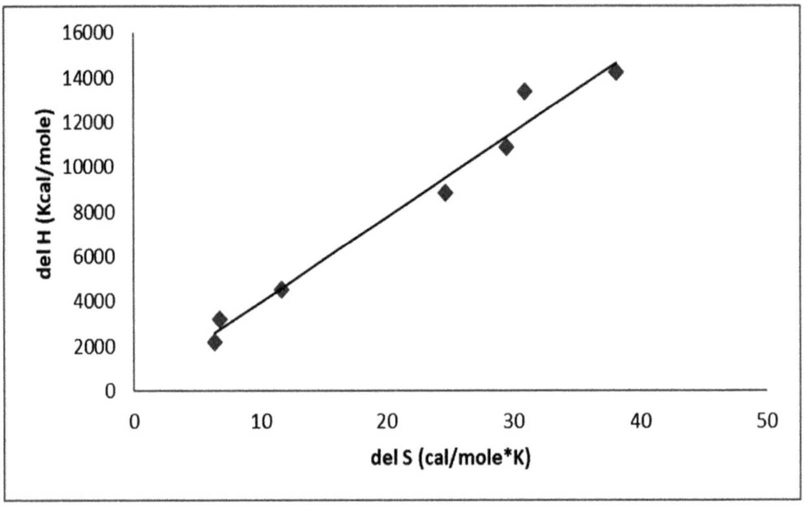

FIGURE 10.13 del S vs del H.

The value of $T_\beta = 382.91$ K. For the values of K_1 and K_2, the same procedure of calculations that is done for Bengal gram is followed for black-eyed peas.

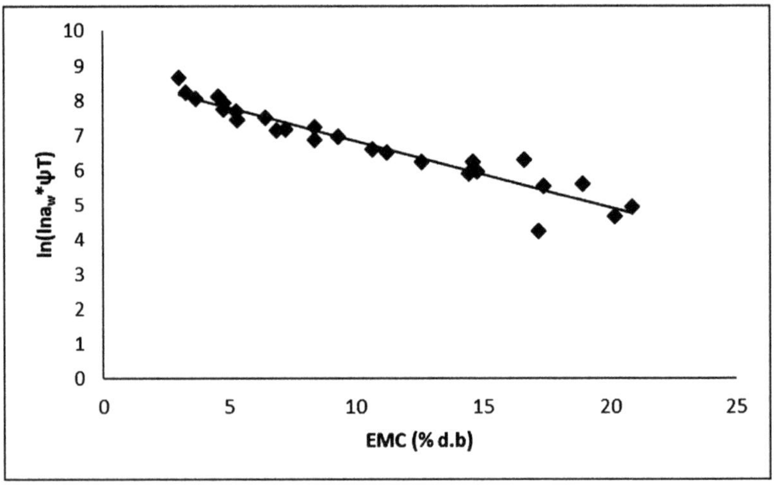

FIGURE 10.14 ln (ln a_w*$\psi(T)$) vs EMC.

The values of K_1 and K_2 were obtained as $K_1 = 6086.935$ and $K_2 = 0.8402$ (Figure 10.14).

10.3.2.3 RESULTS AND GRAPHS OF CHICKPEAS

TABLE 10.14 The Calculated Constant Parameters of Equation (10.3) for Chickpeas

Temperature (°C)	*A*	*B*	*C*
50	−0.1067	0.1134	0.0274
40	−0.0438	0.0659	0.0326
30	−0.0367	0.0316	0.0455

TABLE 10.15 The Predicted Water Activity of Chickpeas

Moisture Content (%)	50 °C	40 °C	30 °C
5	0.2437	0.2014	0.1614
7.5	0.4222	0.3873	0.3524
10	0.5728	0.5583	0.5106
12.5	0.7073	0.6772	0.6352
15	0.7914	0.7543	0.7343
17.5	0.8852	0.8014	0.7816
20	0.9524	0.9202	0.8516

The predicted water activity is taken into consideration to calculate the $1/T$ and ln(1/ERH). The plots between $1/T$ and ln(1/ERH) are given in Figure 10.15, Tables 10.14 and 10.15.

The calculations of del S and del H were done as mentioned earlier and were plotted in a graph, from the slope of the graph (Figure 10.16) the value of T_β is obtained.

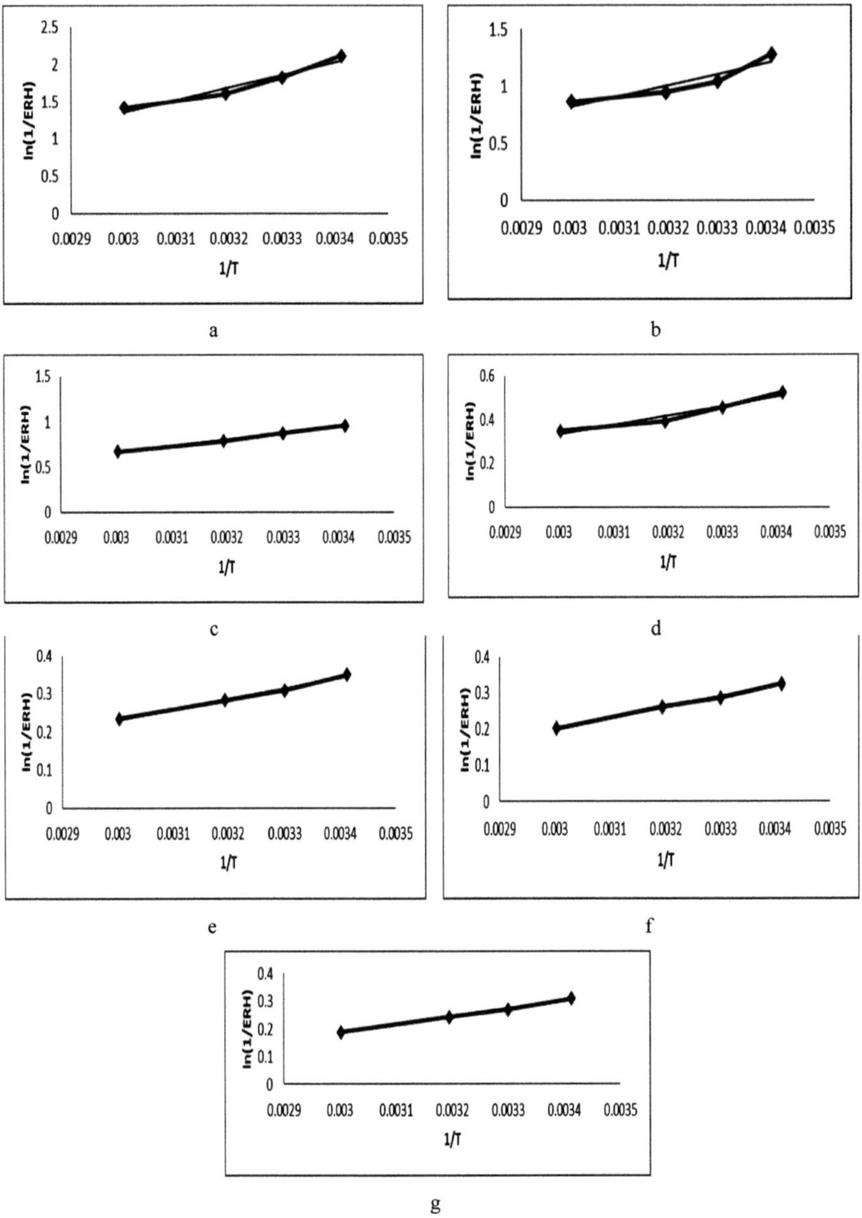

FIGURE 10.15 The graphs of $1/T$ vs ln(1/ERH) at (a) 5%, (b) 7.5%, (c) 10%, (d) 12.5%, (e) 15%, (f) 17.5%, and (g) 20% MC.

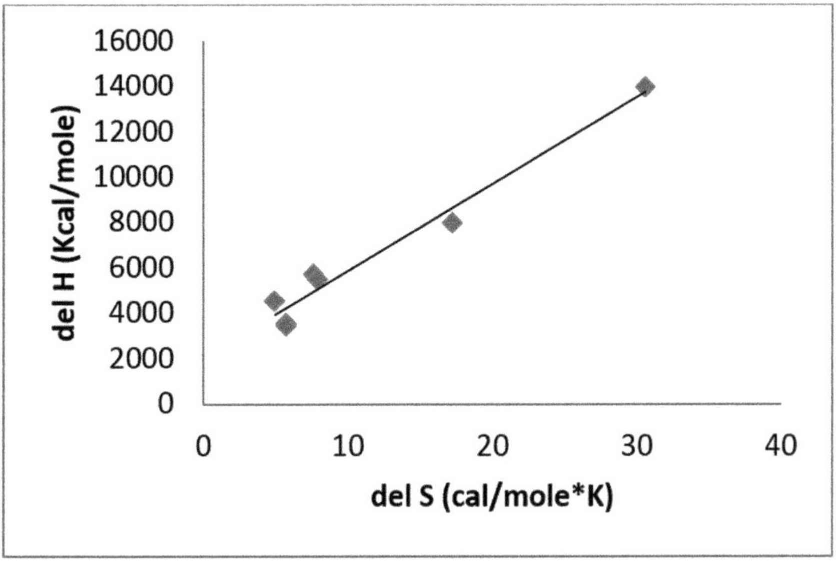

FIGURE 10.16 del *S* vs del *H*.

The value of $T_\beta = 381.61$ K. For the values of K_1 and K_2, the same procedure of calculations that is done for Bengal gram is followed for chickpeas.

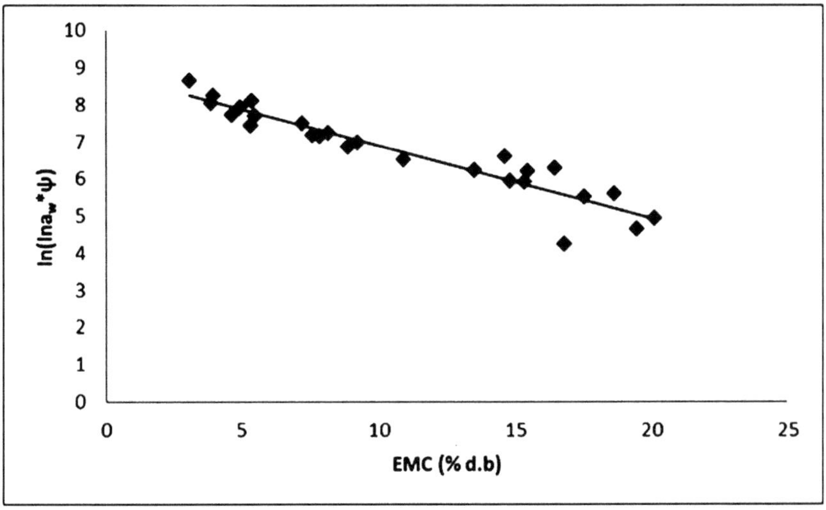

FIGURE 10.17 $\ln(\ln a_w{*}\psi(T))$ vs EMC.

The values of K_1 and K_2 were obtained as $K_1 = 6887.751$ and $K_2 = 0.8226$ (Figure 10.17).

10.3.2.4 RESULTS AND GRAPHS OF MUNG BEANS

The mung beans' constant parameters and the predicted water activity for constant moisture content are given in Table 10.16 and Table 10.17, respectively.

TABLE 10.16 The Calculated Constant Parameters of Equation (10.3) for Mung Beans

Temperature (°C)	*A*	*B*	*C*
50	−0.2438	0.2476	0.0116
40	−0.0729	0.0645	0.0457
30	−0.0281	0.0366	0.0332

TABLE 10.17 The Predicted Water Activity of Mung Beans

Moisture Content (%)	50 °C	40 °C	30 °C
5	0.3369	0.2918	0.1969
7.5	0.5548	0.4502	0.3151
10	0.6760	0.5855	0.4393
12.5	0.7510	0.6932	0.5622
15	0.8010	0.7778	0.6777
17.5	0.8682	0.8346	0.7826
20	0.9457	0.8982	0.8758

The predicted water activity is taken into consideration to calculate the $1/T$ and ln(1/ERH). The plots between $1/T$ and ln(1/ERH) are given in Figure 10.18, Tables 10.16 and 10.17.

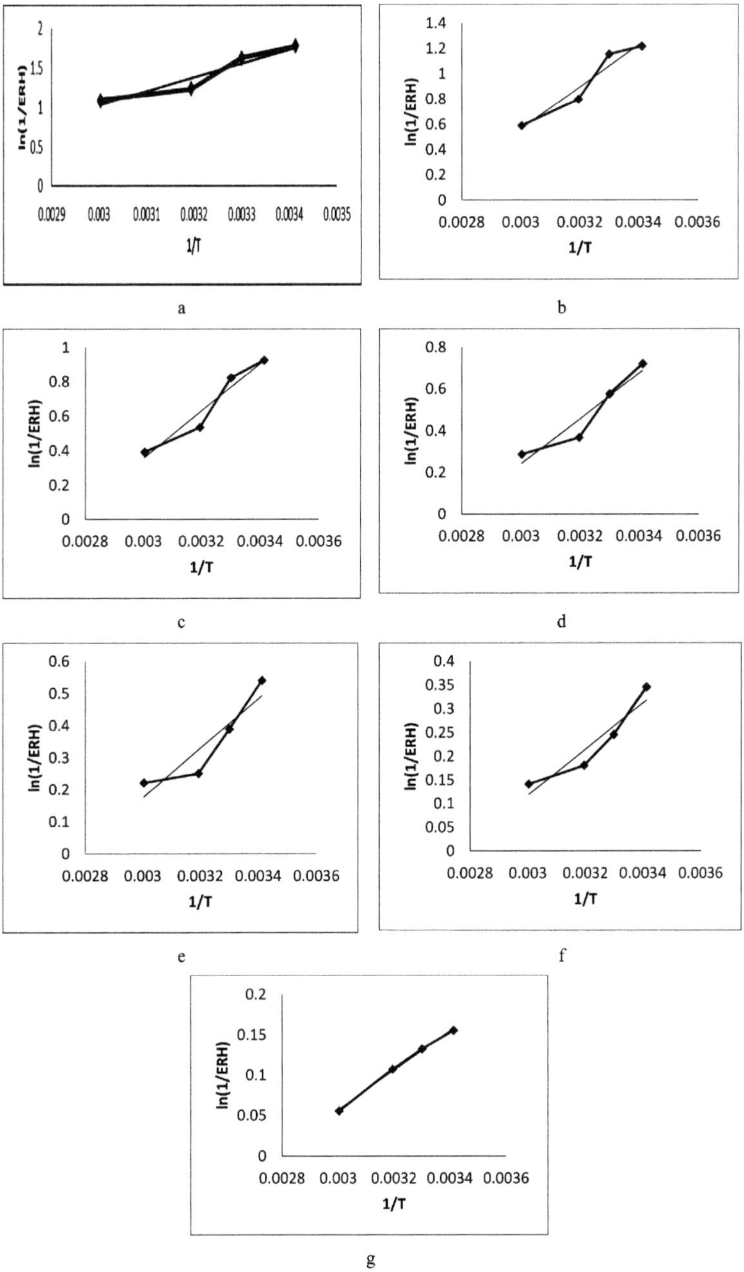

FIGURE 10.18 The graphs of $1/T$ vs ln(1/ERH) at (a) 5%, (b) 7.5%, (c) 10%, (d) 12.5%, (e) 15%, (f) 17.5%, and (g) 20% MC.

The calculations of del S and del H were done as mentioned earlier and were plotted in a graph, from the slope of the graph (Figure 10.19) the value of T_β is obtained.

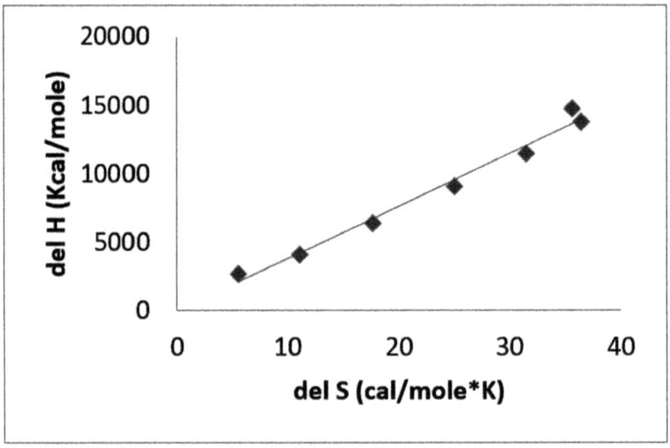

FIGURE 10.19 del S vs del H.

The value of $T_\beta = 383.89$ K. For the values of K_1 and K_2, the same procedure of calculations that is done for Bengal gram is followed for mung beans.

The values of K_1 and K_2 were obtained as $K_1 = 6160.418$ and $K_2 = 0.8395$ (Figure 10.20).

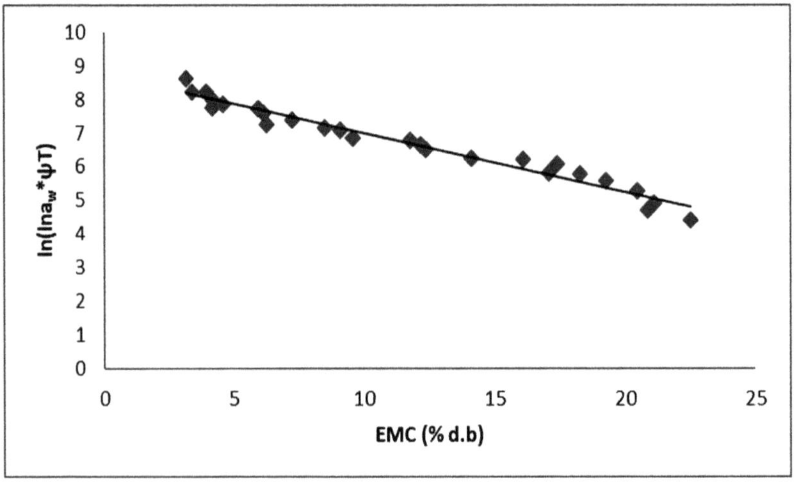

FIGURE 10.20 ln (ln a_w*$\psi(T)$) vs EMC.

These values of K_1 and K_2 were substituted in Equation (10.4) for finding the diffusivity of the legumes.

The overall graph of all the samples to determine the K_1 and K_2 is shown in Figure 10.21.

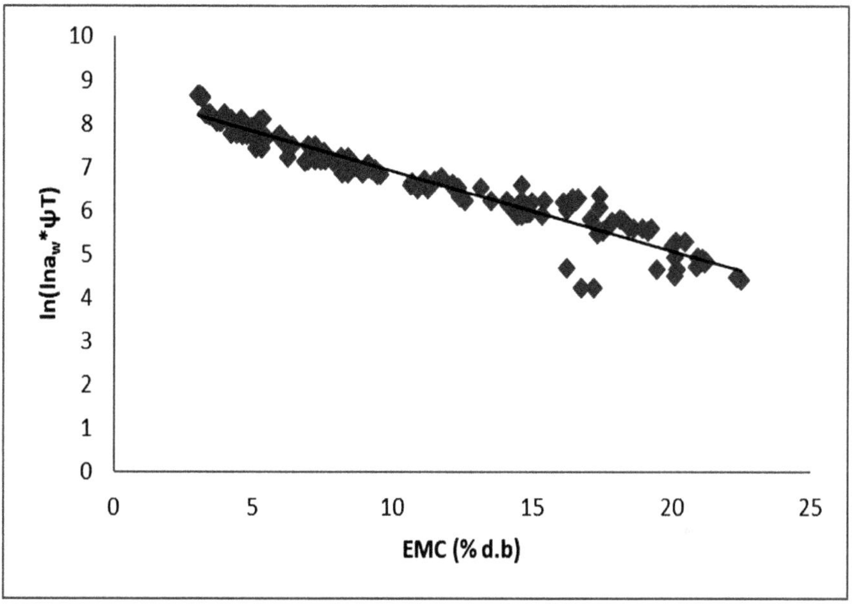

FIGURE 10.21 The K_1 and K_2 graph of all samples.

The values of K_1 and K_2 were obtained as $K_1 = 6237.907$ and $K_2 = 0.8325$.

10.3.3 HYDRATION MODELING BASED ON ENTHALPY-ENTROPY COMPENSATION DATA

The diffusion coefficient in starchy foods depends on moisture concentration; hence, it is necessary to calculate the moisture content of a sample with respect to time to determine the diffusivity of the sample. For obtaining the moisture content of a sample with time the process of hydration at a specific temperature must be done. The hydration treatments of samples were done by following the procedure as mentioned in Section 10.2.7 at 30 °C, 40 °C, and 50 °C. From this method, the moisture content of the samples with respect to time is obtained in a uniform time interval. These values were further evaluated in MATLAB to calculate the diffusivity of the samples. The values of hydration with time are shown in a graph (Figure 10.22).

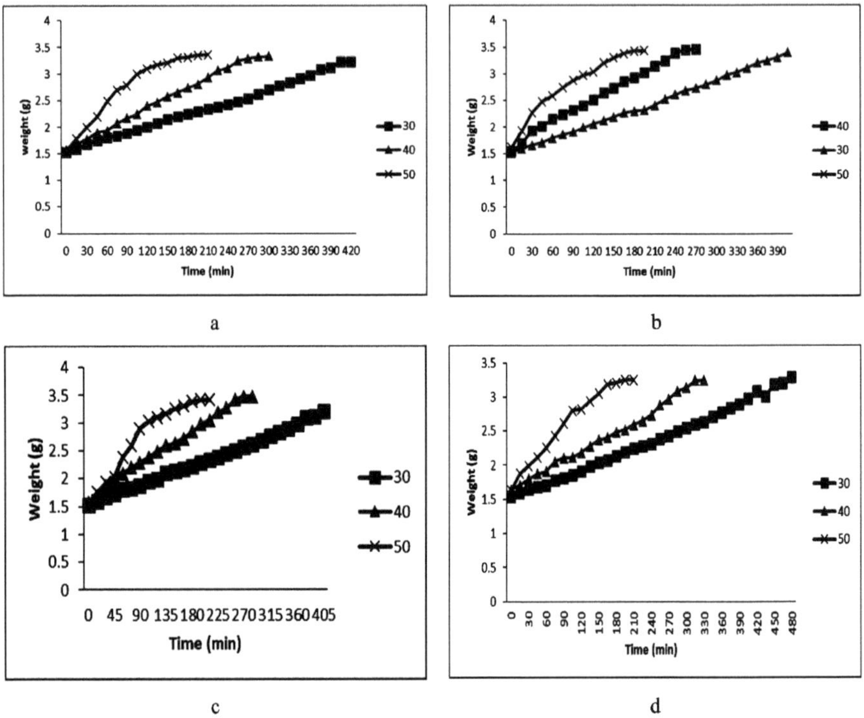

FIGURE 10.22 Graphs of hydration behavior for (a) Bengal grams, (b) black-eyed peas, (c) chickpeas, and (d) mung beans at 30 °C, 40 °C, and 50 °C.

From the graphs in Figure 10.22, it can be concluded that hydration at 50 °C is fastest as compared to hydration at 30 °C which is slowest, it is also seen that hydration time decreases with an increase in time.

The initial moisture content (IMC) was estimated by the method described in Section 2.3 using the formula in Equation (10.1), the equivalent radius (ER), the dry density (ρ_d), and θ (saturation moisture fraction) are estimated by the following formula:

$$ER = \left(\frac{G_m + S_m + A_m}{6} \right)^{1/3} \tag{5}$$

where, $G_m = (2abc)^{1/3}$, $A_m = \left(\frac{2a+2b+2c}{3}\right)$ and $S_m = \left(\frac{4ab+4bc+4ca}{3}\right)^{\frac{1}{3}}$ also: a = length/2,

b = breath/2, and c = thickness/2.

$$\rho_d = \rho(1+MC) - (1000 \times MC) \quad (6)$$

$$\theta = \frac{\frac{\text{density of solute}}{\text{density of water}} + \text{saturation moisture content}}{1 + \left(\frac{\text{density of solute}}{\text{density of water}} + \text{saturation moisture content}\right)} \quad (7)$$

TABLE 10.18 Physical Properties of the Four Samples

Parameters	Bengal Gram	Black-eyed Peas	Chickpeas	Mung Beans
ER (mm)	3.12	2.76	4.02	4.60
IMC	0.13	0.12	0.12	0.15
Θ	0.87	0.87	0.88	0.87
ρd (kg/m^3)	1264.88	1170.28	1404.80	1456.70

The moisture content of the samples is determined at 30, 40, and 50 °C. The obtained moisture content is used along with the K_1 and K_2 and the values given in Table 10.18 in Equation (10.4) to estimate the D_0 value (Table 10.19). For the estimation of D_0 MATLAB is used for the analysis of the obtained and calculated values. The depiction of the relation between predicted v/s estimated moisture content graphs of (a) Bengal grams (b) black eyed peas (c) chick peas, and (d) mung beans is shown in Figure 10.23.

TABLE 10.19 D_0 of the Legumes

Samples	D_0 (m^2/s)
Bengal gram	2.670×10^{-7}
Black-eyed peas	2.644×10^{-7}
Chickpeas	6.370×10^{-7}
Mung beans	4.744×10^{-7}

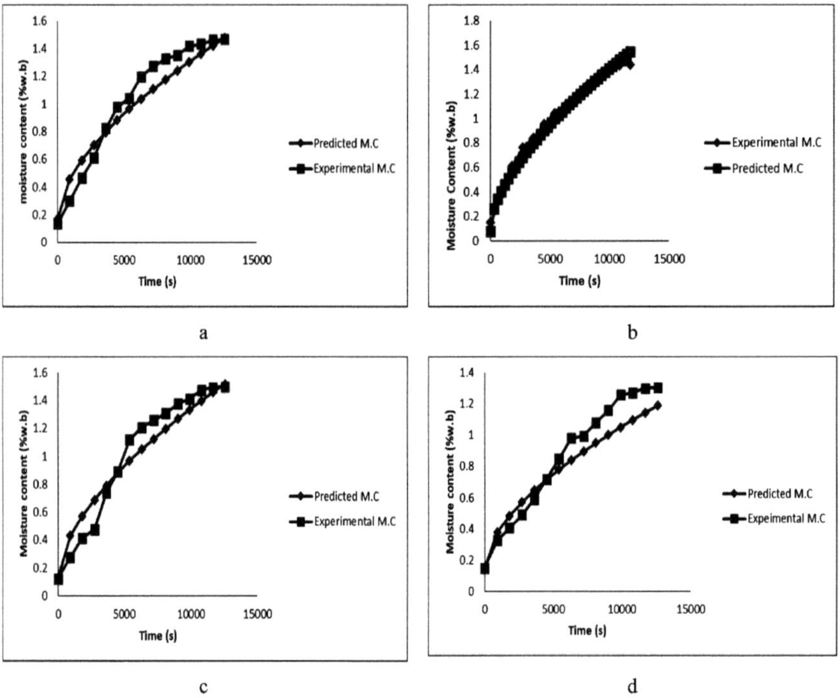

FIGURE 10.23 Predicted vs estimated moisture content graphs of (a) Bengal grams, (b) black-eyed peas, (c) chickpeas, and (d) mung beans.

10.4 CONCLUSION

Experiments were carried out in the Department of Food Engineering and Technology of Tezpur University to estimate the equilibrium moisture content of Bengal gram, black-eyed peas, chickpeas, and mung beans at known relative humidity by the chemical method, at temperatures from 30 °C to 50 °C. The equilibrium moisture data is related to storage relative humidity by the GAB equation as well as by ANN model. The seven input neurons corresponded to the seven input variables, a_w, temperature, ash content, dietary fiber, crude protein, fat content, and carbohydrates while the output neuron represented the EMC. The highest R-value and the least MSE value corresponding to the hidden neurons with the neural network are obtained.

At given moisture content, water activity value is estimated by the fitted GAB model in the moisture range from 5% to 20%, at all the temperatures.

Based on these predicted data of water activity, enthalpy–entropy compensation analysis of the four samples were carried out to estimate isokinetic temperature and develop the generalized isotherm model involving isokinetic temperature and two other parameters viz., K_1 and K_2. These two parameters are later on used for hydration kinetics study of the same product. Also, an overall K_1 and K_2 containing all the data of the four samples together was calculated and obtained.

Finally, for studying the kinetics of hydration of Bengal gram, black-eyed peas, chickpeas, and mung beans soaking experiments at different temperatures were carried out with measurement of moisture content at known intervals. For explaining the moisture migration during the soaking process, the principle of one-dimensional mass transfer to the spherical body was considered with the use of equivalent diameter. For expressing the dependence of the diffusivity value on temperature and moisture, the model involved an expression containing the parameters K_1 and K_2 and heat of sorption. Simulations were carried out to estimate the value of the pre-exponent of diffusivity expression (D_0).

This work validated the application of the approach of enthalpy–entropy compensation analysis of sorption behavior to analyze hydration behavior of the product under consideration. The developed model is applicable to predict the endpoint of soaking to obtain the desired final moisture content.

KEYWORDS

- **hydration kinetics**
- **enthalpy-entropy compensation data**
- **isokinetic temperature**
- **GAB equation**
- **ANN modeling**
- **equilibrium moisture content**

REFERENCES

Aguerre, R. J., Gabitto, J. F., & Chirife, J. (1985). Utilization of Fick's Second Law for the Evaluation of Diffusion Coefficients in Food Processes Controlled by Internal Diffusion. *International Journal of Food Science & Technology*, *20*(5), 623–629.

Bello, M., Tolaba, M. P., Aguerre, R. J., & Suarez, C. (2010). Modelling Water Uptake in a Cereal Grain During Soaking. *Journal of Food Engineering*, *97*(1), 95–100.

Chapwanya, M., & Misra, N. N. (2015). A Soft Condensed Matter Approach Towards Mathematical Modelling of Mass Transport and Swelling in Food Grains. *Journal of Food Engineering*, *145*, 37–44.

Chen, C. (2003). Moisture Sorption Isotherms of Pea Seeds. *Journal of Food Engineering*, *58*(1), 45–51.

Chenoll, C., Betoret, N., & Fito, P. (2009). Analysis of Chickpea (var.“*Blanco lechoso*”) Rehydration. Part I. Physicochemical and Texture Analysis. *Journal of Food Engineering*, *95*(2), 352–358.

Heyrovsky, J. (1970). Determination of Isokinetic Temperature. *Nature*, *227*, 66–67.

Hoseney, R. C. (1994). *Principles of Cereal Science and Technology* (No. Ed. 2). American Association of Cereal Chemists (AACC).

Kashaninejad, M., Dehghani, A. A., & Kashiri, M. (2009). Modelling of Wheat Soaking Using Two Artificial Neural Networks (MLP and RBF). *Journal of Food Engineering*, *91*(4), 602–607.

Labuza, T. P. (1984). Application of Chemical Kinetics to Deterioration of Foods. *Journal of Chemical Education*, *61*(4), 348–358.

McMinn, W. A. M., Al-Muhtaseb, A. H., & Magee, T. R. A. (2005). Enthalpy–Entropy Compensation in Sorption Phenomena of Starch Materials. *Food Research International*, *38*(5), 505–510.

Menkov, N. D. (2000). Moisture Sorption Isotherms of Chickpea Seeds at Several Temperatures. *Journal of Food Engineering*, *45*(4), 189–194.

Myhara, R. M., Sablani, S. S., Al-Alawi, S. M., & Taylor, M. S. (1998). Water Sorption Isotherms of Dates: Modelling Using GAB Equation and Artificial Neural Network Approaches. *LWT-Food Science and Technology*, *31*(7–8), 699–706.

Prakash, S., Jha, S. K., & Datta, N. (2004). Performance Evaluation of Blanched Carrots Dried by Three Different Driers. *Journal of Food Engineering*, *62*(3), 305–313.

Sayar, S., Turhan, M., & Gunasekaran, S. (2001). Analysis of Chickpea Soaking by Simultaneous Water Transfer and Water–Starch Reaction. *Journal of food Engineering*, *50*(2), 91–98.

Turhan, M., Sayar, S., & Gunasekaran, S. (2002). Application of Peleg Model to Study Water Absorption in Chickpea During Soaking. *Journal of Food Engineering*, *53*(2), 153–159.

Yildirim, A., Oner, M. D., & Bayram, M. (2011). Fitting Fick’s Model to Analyze Water Diffusion into Chickpeas During Soaking with Ultrasound Treatment. *Journal of Food Engineering*, *104*(1), 134–142.

CHAPTER 11

Enzymatic Production of Chito-Oligosaccharides and D-glucosamine by Fungal Chitosanases from *Aspergillus* spp.: A Review

CARLOS NEFTALÍ CANO-GONZÁLEZ[1], ENA DEYLA BOLAINA-LORENZO[1], ALICIA COUTO[2], FAUSTINO LARA[3], RAÚL RODRÍGUEZ-HERRERA[1], and JUAN CARLOS CONTRERAS-ESQUIVEL[1*]

[1]*Food Research Department, School of Chemistry, Universidad Autonoma de Coahuila, Saltillo City 25250, Coahuila State, Mexico*

[2]*Universidad de Buenos Aires, FCEN, Departamento de Química Orgánica—CONICET, CIHIDECAR, Intendente Güiraldes 2160, C1428GA, Ciudad Universitaria, Buenos Aires, Argentina*

[3]*CISEF, Saltillo, Coahuila, Mexico*

**Corresponding author. E-mail: carlos.contreras@uadec.edu.mx*

ABSTRACT

Research related to chitosan is primarily focused as a bioactive compound. However, its high molecular weight and viscosity reduce its solubility in water, which makes it usefulness in food and difficult in biomedical applications. Therefore, research has focused on the generation of hydrolyzed products, such as the chito-oligosaccharides (COS) and D-glucosamine. The enzymatic method is of great importance since the percentage of contaminant residues in the products generated is minimal or null, making the use of chitosanase very attractive. Fungi as an enzymatic source are a great alternative, among them one of the most used genus is *Aspergillus*. Interestingly, the fungal chitosan has demonstrated its great potential for application in industry and diagnosis in medicine. The enzyme showed a great ability to hydrolyze chitosan to obtain the DP 2-7 COS. A recent report has shown that *Aspergillus* chitosanase was successfully applied to a pilot study of chitosan hydrolysis at the 200 kg scale.

11.1 INTRODUCTION

Chitosan is a linear polysaccharide composed mainly of β-(1,4)-2-deoxy-2-amino-D-glucan and constitutes a prominent polysaccharide due to its biocompatibility. Chitosan is composed of less than 20% β-(1,4)-2-acetamido-D-glucopyranose and more than 80% β-(1,4)-2-amino-D-glucopyranose (Zou et al., 2016). Consequently, chitosan can be considered a partially deacetylated derivative of chitin. Chitosan structure contains heteromonomer chains of *N*-acetylglucosamine (GlcNAc)–glucosamine (GlcN) and GlcN–GlcNAc, as well as homomonomers of GlcNAc–GlcNAc and GlcN–GlcN (Wang et al., 2012). It is often referred to as a cationic polymer, which can easily bind to negatively charged molecules.

The products generated by the enzymatic hydrolysis of chitosan are the chito-oligosaccharides (COS) and D-glucosamine (GlcN). COS have a low molecular weight compared to chitosan and are soluble in aqueous solutions, they play a key role at the cellular and molecular level (Du et al., 2010). COS develop antioxidant, antimicrobial, immunostimulant, and antitumor activities. In addition, COS and D-glucosamine are biocompatible, nontoxic, and nonallergic to living tissues. D-glucosamine is the main monosaccharide of chitosan and it is obtained from the hydrolysis generated by chitosanase type exo. This molecule has benefits in health as GlcN-based products show positive effects in the liver, bones, and membrane cell (Sun et al., 2013). The biological activities of COS and D-glucosamine have been extensively studied and potential applications have been found in food, pharmaceutical, agricultural, and environmental industries. The decomposition of chitosan by enzymatic route using chitosanase constitutes a green alternative to obtain its oligosaccharides. This enzyme is synthesized by different microorganisms by fungi, *Aspergillus* spp. *Trichoderma* spp., and bacteria, including *Bacillus* spp., *Paenibacillus* spp, and *Streptomyces* spp. (Jeon and Kim, 2000). Fungi are a great alternative due to their capability of enzyme production, for example, *Aspergillus* is one of the most commonly used genera for enzymatic production.

Aspergillus is a genus of filamentous fungi whose species are found throughout the world in different habitats, comprising more than 250 species. Some species generate deterioration in food while others produce mycotoxins that are pathogens for humans and animals (Kaya et al., 2015). Species of this genus are used in biotechnology to produce metabolites such as antibiotics, organic acids, drugs or enzymes, and agents for food fermentations (Samson et al., 2014). *Aspergillus* is one of the most studied genera of great economic importance due to its good performance producing enzymes for applications in the areas of biomedicine, food, pharmaceutical, and so forth. In fact, *Aspergillus* spp. has been used to produce chitosanases in several research works (Thadathil and Velappan, 2014).

The aim of this chapter is to inform about the production process of COS and D-glucosamine through the enzymatic pathway involving the *Aspergillus* genus.

11.2 CHITOSAN

Chitosan is a polysaccharide with biological activity useful in medical applications. However, its high molecular weight reduces its solubility in water. This disadvantage has promoted the use of chitosan-oligosaccharides since they have an improved water solubility which makes the product easy to manage by the food and pharmaceutical industry. More importantly, chitosan-oligosaccharides with antimicrobial, antitumor, antioxidant, and immunomodulatory activities are even better than chitosan (Younes et al., 2014). There are different methods for the degradation of the linkages of chitosan resulting in different numbers and sequences of GlcN and GlcNAc, as well as different degrees of polymerization, among which are acid hydrolysis (Tsao et al., 2011), enzymatic hydrolysis (Wu et al., 2011), oxidative degradation (Xia et al., 2013), and sonic degradation (Liu et al., 2006). Acid hydrolysis and enzymatic hydrolysis are the most commonly used methods to degrade chitosan. Although the chemical method is the most used in the industry, the products cannot be used as bioactive materials because there is a big risk of contamination with acid traces. On the other hand, the yield of chitosan-oligosaccharide production is low, without mentioning the problems of environmental contamination generated by this method. According to these results, enzymatic hydrolysis has been preferred as a method to obtain chitosan-oligosaccharides. It is a green method free from toxic solvents and chemical compounds (Jeon and Kim, 2000).

11.3 CHITO-OLIGOSACCHARIDES AND D-GLUCOSAMINE

COS and D-glucosamine contain three types of reactive functional groups: an amino/acetamido group at C-2 and two hydroxyl groups at C-6 and C-3. The amino content, the degree of deacetylation, and molecular weight are the main reasons for the differences between the physicochemical properties. These characteristics also correlate with biological functions as antimicrobial (Jeon and Kim, 2000), antitumor, antioxidant, and immunomodulators (Younes et al., 2014; Mengíbar et al., 2013). Reaching an adequate molecular weight, COS will be absorbed through the intestine and enter the bloodstream undergoing different biological processes in the human body (Mengíbar et al., 2013).

By elimination of acetyl groups, COS becomes positively charged and then can bind to microbial cell walls that lead to growth inhibition. This physicochemical property, as well as molecular weight and immune stimulation, influence antitumor activity (Guo et al., 2006). COS can stimulate the production of cytokines such as interleukin-1 (IL-1) and the tumor necrosis factor-alpha (TNF-a).

D-glucosamine that occurs at a percentage higher than 80% in the chitosan can be purified by hydrolytic methods (Zou et al., 2016). Pharmaceutical products containing GlcN have reported positive effects such as the protection of liver cells and the stabilizing activity of the cell membrane; GlcN has been used in the treatment of osteoporosis and for the healing of wounds because it promotes the synthesis of hyaluronic acid. Recently, GlcN is receiving more attention due to its easy absorption by the human body and excellent biological and stimulating properties (Sun et al., 2013).

11.4 CHITOSANASES

Chitosanases (EC 3.2.1.132) are enzymes that catalyze the hydrolysis of the glycosidic bonds of chitosan. Therefore, the search for efficient and stable chitosanases to produce specific COS and D-glucosamine products is very important scientifically. Based on the amino acid sequence, the chitosanases belong to the family of glycoside hydrolases (GH) families GH5, GH7, GH8, GH46, GH75, and GH80. Currently, the GH46, GH75, and GH80 families contain exclusively chitosanases. The GH 75 family stands out for covering only fungal chitosanases (Cheng et al., 2006). Fungal chitosanases have shown the capacity to hydrolyze chitosan to obtain the DP 2-7 COS (Zhang et al., 2015). This chitosan degrading enzymes are useful in pharmaceutical and other industries.

Based on the mechanism of action on chitosan, there are two types of chitosanases, exo-chitosanase (EC 3.2.1.165) and endo-chitosanase (EC 3.2.1.132); the former cuts the polysaccharide at the end of the chain producing monomers of D-glucosamine and the later cuts at any part of the polysaccharide chain producing chito-oligosaccharides (Figure 11.1) (Chen et al., 2014).

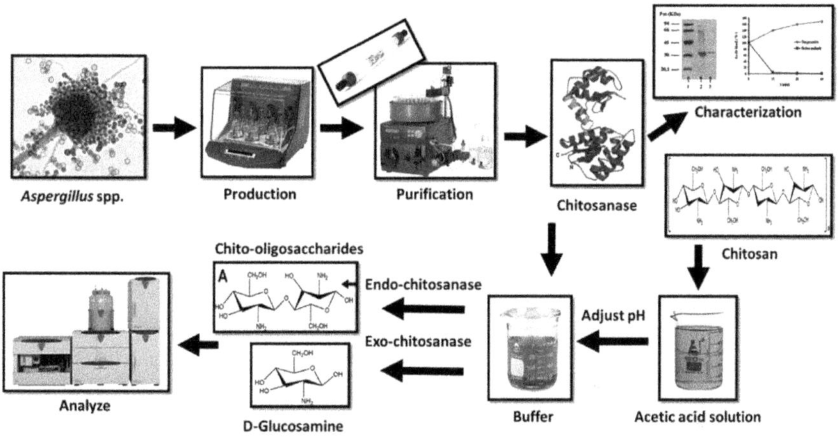

FIGURE 11.1 Chitosan hydrolysis by endo- and exochitosanases to produce (A) chitosan oligosaccharides and (B) D-glucosamine molecules.

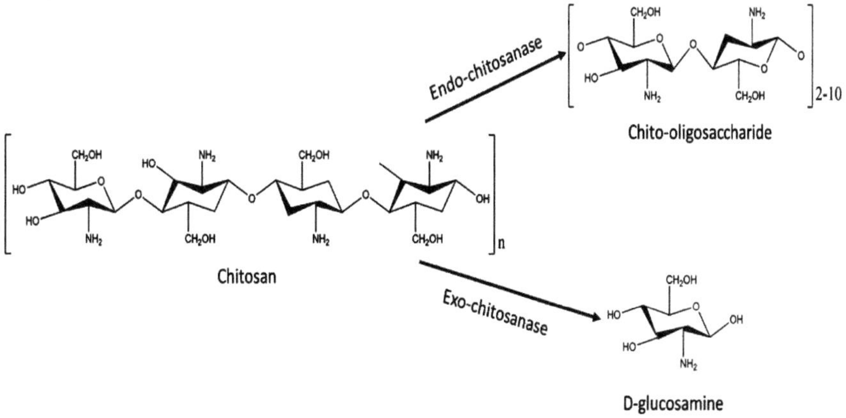

FIGURE 11.2 Process of production of COS and D-glucosamine by the enzymatic route.

11.5 PRODUCTION

COS and GlcN obtained by acidic hydrolysis on an industrial scale are not considered as bioactive materials due to the possibility of contamination by chemical compounds. In recent years, the enzymatic hydrolysis has been proposed as an alternative method for the production of bioactive COS (Figure 11.2). Chitosanases from different nature have particular catalytic action mainly depending on the degree of deacetylation of chitosan. In fact, it has been generally reported that chitosanases produced by microorganisms generate a high yield of COS compared to chitosanases from other sources (Table 11.1) (Kim and Rajapakse, 2005).

TABLE 11.1 COS Production by Enzymatic Route

Aspergillus spp.	Time (h)	pH	Temp (°C)	Concentration (%)	Major End Products (DP)	References
sp. Y2K	68	5.3	45 – 50	5.5	3–5	Cheng and Li (2000)
fumigatus KH-94	NR	5.5	NR	NR	2–4	Kim et al. (1998)
sp. W-2	20	6.0	55	NR	2–4	Zhang et al. (2015)
fumigatus Y2K	24	NR	60	10	2, 3, 6	Chen et al. (2012)
sp. EGY1	44	NR	25	NR	2–11	Embaby et al. (2018)
sp. CJ22-326	NR	6.0	65	NR	3–6	Chen et al. (2004)

Note: NR = Not reported.

Cheng et al. dissolved chitosan in diluted acetic acid at 5.5% w/v and then 5 units of chitosanase per 5 L of chitosan were added. The reaction was left for 68 h at 45–50 °C with a pH of 5.3. The generated products were chito-oligosaccharides with different degrees of polymerization labeled as DP3, DP4, DP5, and DP6, with approximately 28%, 30%, 22%, and 7% of yield, respectively. The remaining percentages are oligomers of high molecular weight and dimers. This chitosanase is considered to be very useful for industrial applications of chito-oligosaccharide preparation (Cheng and Li, 2000).

Notably, the most outstanding work on COS production at a large scale was the biotransformation of 200 kg of chitosan to COS, with only 1.6% of remaining insoluble chitosan. Chitosan (10% w/v) was prepared in acetic acid solution at 2.6% with 3 g of chitosanase (25,000,000 U). The hydrolysis reaction was carried out at 60 °C for 24 h with stirring. According to the thin layer chromatography (TLC) analysis, the generated products were mainly chitotriose to chitohexaose with a small amount of chitobiose (Chen et al., 2012).

In general, the production of GlcN is based on the action of exo-chitosanase. Chitosan is hydrolyzed at the nonreducing end of the chain generating only monomers.

Purification of exo-chitosanase in the presence of chito-oligomers (DP 2-7) as substrate resulted in GlcN and the shortened oligomers of a single residue in the early stage of the reaction. This enzyme generates only GlcN from soluble chitosan, but not chitobiose, chitotriose, or chitotetraose, even after prolonged reactions. The reaction product was confirmed by HPLC demonstrating that the purified enzyme produced only GlcN. The methods such as TLC present fractionation in the run and by viscosimetry the change is not significant. The combination of the endo and exo chitosanases generates a heterologous group of oligomers and monomers, but with time prolongation, it ends in only the presence of GlcN (Jung et al., 2006; Zhang et al., 2000).

11.6 APPLICATION

There is a great demand for COS and D-glucosamine due to their bioactive potential specifically in the food industry and biomedicine fields. Several methods for the production of COS have been reported using different reactors, for example, batch reactors, reactors of column with immobilized enzymes, membrane reactors; continuous COS production by the dual reactor system (Kim and Rajapakse, 2005; Ming et al., 2006). In this context, the immobilization of enzymes is of great importance in order to allow its repetitive use. Immobilization prevents deactivation of the enzyme by agents that can affect or inhibit enzymatic activity. Economically, the immobilization of chitosanase is very feasible since its production process is very expensive (Ming et al., 2006).

Polyacrylonitrile nanofibrous (PAN) membranes were prepared by electrospinning from 10% by weight of PAN solution and their surface was modified by amidination reaction. Then, chitosanase was covalently immobilized in the PAN membranes. The 80% immobilization efficiency was achieved. Factors such as the optimal temperature and pH for the immobilized enzyme were 50 °C and 5.8 °C, respectively. The immobilized chitosanase maintained approximately 70% enzymatic activity after ten discontinuous reactions. Storage was up to 60 days at 4 °C with a nonsignificant loss of activity (Sinha et al., 2012).

The amino content, degree of deacetylation, and molecular weight are the main reasons for the differences between the structures and the physicochemical properties of COS. These features also correlate with their biological functions as antimicrobial agent. COS can alter the membrane permeability of the microbial cell, preventing the entry of materials or the release of cellular components, and ultimately leading to microbial death. Another mechanism involves the absorption of COS by the cells until they penetrate the bacterial DNA, blocking RNA transcription. The positively charged nature of COS is responsible for their binding to the bacterial cell wall. The amino group positively charged at the C-2 position of the GlcN monomer, interacts with the negatively charged carboxylic acid group of macromolecules of the bacterial cell wall (Jeon and Kim, 2000). In addition, COS exhibits antitumoral activity. The antitumor mechanisms of COS are probably related to the induction of lymphocyte factors, increasing proliferation of T cells to produce tumor inhibitory effect. The antitumor mechanisms of COS were acquired to improve immunity by accelerating T-cell differentiation to increase cytotoxicity and maintain T-cell activity (Suzuki et al., 1986). COS also presents immunomodulatory activity which is thought to be responsible for antitumor activity (Xu and Du, 2003). Also, COS can stimulate the production of cytokines such as (IL-1β and TNF-α (Feng et al., 2004). Another feature of COS is related to their capacity as free radical scavengers; it means that it can break down the oxidative sequence at any level (Mengíbar et al., 2013). The degree of polymerization determines the capacity of free radical elimination. Other studies have shown that COS can reduce the damage caused by oxidative stress induced by H_2O_2 in cells (Zou et al., 2016). It also has an immunomodulatory activity which is thought to be responsible for antitumor activity (Xu and Du, 2003).

11.7 CONCLUSION

COS are biopolymers that present anticancer, antioxidant, antiglycemic bioactivity, inflammation-oriented diseases. The bioactivity depends directly on the degree of acetylation and polymerization and the ionization pattern of GlcN that makes COS totally soluble with low viscosity, absorbable by the human body, and permeable by micro-organism cells.

COS production routes present disadvantages for the application on the biomedical, pharmaceutical, and food industries. Enzymatic hydrolysis presents better characteristics for the application of COS, without the presence of contaminants and friendly to the environment. In that sense, microbial chitosanase presents a higher COS yield than chitosanase from other sources.

KEYWORDS

- **chito-oligosaccharides**
- **D-glucosamine**
- **bioactive compound**
- **enzymatic**

REFERENCES

Chen, X.; Xia, W.; Yu, X. Purification and Characterization of Two Types of Chitosanase from *Aspergillus* sp. CJ22-326. *Food Res. Int.* 2004, 38, 315–322.

Chen, X.; Zhai, C.; Kang, L.; Li, C.; Yan, H.; Zhou, Y.; Yu, X.; Ma, L. High-level Expression and Characterization of a Highly Thermostable Chitosanase from *Aspergillus* Fumigatus in *Pichia pastoris*. *Biotechnol. Lett.* 2012, 34, 689–694.

Cheng, C.; Chang, C.; Wu, Y.; Li, Y. Exploration of Glycosyl Hydrolase Family 75, a Chitosanase from *Aspergillus fumigatus*. *J. Biol. Chem.* 2006, 281, 3137–3144.

Cheng, C.; Li, Y. An *Aspergillus* Chitosanase with Potential for Large Scale Preparation of Chitosan Oligosaccharides. *Biotechnol. Appl. Biochem.* 2000, 32, 197–203.

Du, Y.; Ying, X.; Wang, L.; Zhai, Y.; Yuan, H.; Yu, R.; Hu, F. Sustained Release of ATP Encapsulated in Chitosan Oligosaccharide Nanoparticles. *Int. J. Pharm.* 2010, 392, 164–169.

Embaby, R.; Melika, A.; Hussein, A.; El-Kamel, H.; Marey. Biosynthesis of Chitosan-oligosaccharides (COS) by Non-aflatoxigenic *Aspergillus* sp. Strain EGY1 DSM 101520: A Robust Biotechnological Approach. *Process Biochem.* 2018, 64, 16–30.

Feng, J.; Zhao, L.; & Yu, Q. Receptor-mediated Stimulatory Effect of Oligochitosan in Macrophages. *Biochem. Biophys. Res. Commun.* 2004, 317, 414–420.

Guo, Z.; Chen, R.; Xing, R.; Liu, S.; Yu, H.; Wang, P.; Li, C.; Li, P. Novel Derivatives of Chitosan and Their Antifungal Activities In Vitro. *Carbohyd. Res.* 2006, 3, 352–354.

Jeon, Y.; Kim, S. Production of Chitooligosaccharides Using an Ultrafiltration Membrane Reactor and Their Antibacterial Activity. *Carbohyd. Polym.* 2000, 41, 133–141.

Jung, W.; Kuk, J.; Kim, K.; Jung, K.; Park, R. Purification and Characterization of exo-β-glucosaminidase from *Aspergillus fumigatus* S-26. *Protein Expr. Purif.* 2006, 45, 125–131.

Kaya, H.; Kumar, P.; Kaaya, A. MiD-infrared Spectroscopy for Discrimination and Classification of *Aspergillus* spp. Contamination in Peanuts. *Food Control.* 2015, 52, 103–111.

Kim, S.; Rajapakse, N. Enzymatic Production and Biological Activitis of Chitosan Oligosaccharides (COS): A Review. *Carbohyd. Polym.* 2005, 62, 357–368.

Kim, S.; Shon, D.; Lee, K. Purification and Characteristics of Two Types of Chitosanases from *Aspergillus fumigatas* KH-94. *J. Microbiol. Biotechnol.* 1998, 8, 568–574.

Liu, H.; Bao, J.; Du, Y.; Zhou, X.; Kennedy, J. Effect of Ultrasonic Treatment on the Biochemphysical Properties of Chitosan. *Carbohyd. Polym.* 2006, 64, 553–559.

Mengíbar, M.; Mateos, I.; Miralles, B.; Heras, A. Influence of the Physico-chemical Characteristics of Chito-oligosaccharides (COS) on Antioxidant Activity. *Carbohyd. Polym.* 2013, 97, 776–782.

Ming, M.; Kuroiwa, T.; Ichikawa, S.; Sato, S.; Mukataka, S. Production of Chitosan Oligosaccharides by Chitosanase Directly Immobilized on an Agar Gel Coated Multidisk Impeller. *Biochem. Eng. J.* 2006, 28, 289–294.

Samson, R.; Visagie, C.; Houbraken, J.; Hong, S.; Hubka, V.; Klaassen, C.; Perrone, G.; Seifert, K.; Susca, A.; Tanney, J.; Varga, J.; Kocsubé, S.; Szigeti, G.; Yaguchi, T.; Frisvad, J. Phylogeny, Identification and Nomenclature of the Genus *Aspergillus*. *Stud. Mycol.* 2014, 78, 141–173.

Sinha, S.; Dhakate, S.; Kumar, P.; Mathur, R.; Tripathi, P.; Chand, S. Electrospun Polyacrylonitrile Nanofibrous Membranes for Chitosanase Immobilization and its Application in Selective Production of Chitooligosaccharides. *Bioresour. Technol.* 2012, 115, 152–157.

Sun, Y.; Zhang, J.; Wu, S.; Wang, S. Preparation of D-glucosamine by Hydrolysis of Chitosan with Chitosanase and β-D-glucosaminidase. *Int. J. Biol. Macromol.* 2013, 61; 160–163.

Suzuki, K.; Mikami, T.; Okawa, Y.; Tokoro, A.; Suzuki, S.; Suzuki, M. Antitumor Effect of Hexa-N-acetylchitohexaose and Chitohexaose. *Carbohyd Res.* 1986, 151, 403–408.

Thadathil, N.; Velappan, S. Recent Developments in Chitosanase Research and its Biotechnological Applications: A Review. *Food Chem.* 2014, 150, 392–399.

Tsao, C.; Chang, C.; Lin, Y.; Wu, M.; Han, J.; Hsieh, K. Kinetic Study of Acid Depolymerization of Chitosan and Effects of Low Molecular Weight Chitosan on Erythrocyte Rouleaux Formation. *Carbohyd. Res.* 2011, 346, 94–102.

Wang, S.; Liu, C.; Liang, T. Fermented and Enzymatic Production of Chitin/chitosan Oligosaccharides by Extracelular Chitinases from *Bacillus cereus* TKU027. *Carbohyd. Polym.* 2012, 90, 1305–1313.

Wu, J.; Tang, C.; Wu, M.; Li, J.; Chen, W.; Shi, H. Method for Producing Chitosanase by Using *Aspergillus niger*, Cn. patent. CN102154241-A. 2011.

Xia, Z.; Wua, S.; Chen J. Preparation of Water Soluble Chitosan by Hydrolysis Using Hydrogen Peroxide. *Int. J. Biol. Macromol.* 2013, 59, 242–245.

Xu, Y.; Du, Y. Effect of Molecular Structure of Chitosan on Protein Delivery Properties of Chitosan Nanoparticles. *Int. J. Pharm.* 2003, 250, 215–226.

Younes, I.; Hajji, S.; Frachet, V.; Rinaudoc, M.; Jelloulia, K.; Nasri, M. Chitin Extraction from Shrimp Shell Using Enzymatic Treatment. Antitumor, Antioxidant and Antimicrobial Activities of chitosan. *Int. J. Biol. Macromol.* 2014, 69, 489–498.

Zhang, J.; Cao, H.; Lia, S.; Zhao, Y.; Wang, W.; Xu, Q.; Du, Y.; Yin, H. Characterization of a New Family 75 Chitosanase from *Aspergillus* sp. W-2. *Int. J. Biol. Macromol.* 2015, 81, 362–369.

Zhang, X.; Dai, A.; Zhang, X.; Kuroiwa, K.; Kodaira, R.; Shimosaka, M.; Okazaki, M. Purification and Characterization of Chitosanase and exo-β-D-glucosaminidase from Koji Mold, *Aspergillus oryzae* IAM2660. *Biosci. Biotech. Bioch.* 2000, 64, 1896–1902.

Zou, P.; Yang, X.; Wanga, J.; Li, Y.; Yu, H.; Zhang, Y.; Liu, G. Advances in Characterization and Biological Activities of Chitosan and Chitosan Oligosaccharides. *Food Chem.* 2016, 190, 1174–1181.

CHAPTER 12

Isomaltulose: The Next Sweetener, A Quick Review

JUAN PABLO BRACHO-OLIVEROS[1], ANDREA CAROLINA RAMIREZ-GUTIERREZ[1], GASTON EZEQUIEL-ORTIZ[1], JUAN CARLOS CONTRERAS-ESQUIVEL[2], and SEBASTIAN FERNANDO CAVALITTO[1*]

[1]*Research and Development Center for Industrial Fermentations, CINDEFI (CONICET, La Plata, UNLP), Calle 47 y 115 B1900ASH, La Plata, Argentina*

[2]*Laboratory of Applied Glycobiotechnology, Food Research Department, School of Chemistry, Universidad Autonoma de Coahuila, Saltillo 25280, Coahuila, Mexico*

[*]*Corresponding author. E-mail: cavalitto@quimica.unlp.edu.ar*

ABSTRACT

The social behavior of new generations consume highly processed foods and high levels of fat and sugar, such as glucose or high-fructose syrup (corn syrup), commonly used in the manufacture of hard candies, have generated a growing increase in cases of diabetes and nutrition problems. In this context, we present the isomaltulose as a substitute potential for sweeteners, due to its property of maintaining stable glycemic levels, dosing energy to the organism in a long-term way, promoting the fat oxidation and it is a noncariogenic food. There is just one company that distributes isomaltulose in an industry level, under the name of Palatinose™ (Beneo-Palatinti, Sudzücker Group, Mannheim, Germany) and already incorporated isomaltulose products in the food processing as beverage, bakery, energy drinks, dry-powder drinks, chewing gum, hard candies, and others. But it is necessary to produce quantities of isomaltulose that will supply the demand of the food industry. The latest advances in biotechnology provide tools to efficiently produce any product that may result from the metabolism of micro-organisms. But first, it is important to know the global aspects of the enzyme, the reaction that catalyzes and its products to determine their chemical and biological parameters. This work aims to make a quick review of all these aspects that allow us to understand and start the biotechnological path for the design and development of future foods based on isomaltulose.

12.1 INTRODUCTION

Currently, the social behavior of new generations who consume highly processed foods, increased sedentary lifestyle, and increased consumption of fat and sugars, such as glucose, or high fructose (corn syrup) used in the manufacture of foods and drinks, have generated a growing increase in cases of diabetes, 422 million in 2014 compared to 108 million in the 1980s, according to the global report on diabetes made by the world health organization (WHO, 2016). These numbers have motivated the institutions to promote the awakening of the scientists' interest in the development of foods that are healthy and have nutritional value. In that way, isomaltulose has the potential to become a substitute for sucrose in processed foods due to its property of not increasing glycemic levels in its consumers and the ability to dose energy to the organism in a long term, promoting the oxidation of fats. At the moment, it is an interesting option used in the baking products industry, cereal bars, dairy products, powder drinks, and energy drinks for athletes. According to the Food and Agricultural Organization and Codex Alimentarius, it can be used as an additive of several products in its hydrogenated form (isomaltose, SIN-953). Due to its molecular structure, its crystalline form is very similar to sucrose but it is commercialized in three presentations according to the particle size distribution (PSD), a large one (similar to table sugar), <0.71 mm PSD) and a thinner (<0.05 mm DTP) used for compressed products or as a cover (Beneo, Sudzücker, Mannheim, Germany). In the food industry, it is also used as an anticaking agent, due to its low hygroscopicity, thickener and/or stabilizer; its taste and palatability are very familiar to sucrose, even in numerous trials it is evident that the subjects hardly differentiate the foods that were sweetened with sucrose from those that were sweetened with isomaltulose (Oosthuyse et al., 2013; Stevenson et al., 2017) nor does it produce the bitter taste generated by the use of synthetic sweeteners such as Aspartame or Acesulfame (E951 and E950) (García-Almeida, 2013; Wee et al., 2018) commonly used in the industry, besides, the consumption of isomaltulose it is not related to any disease, such as some synthetic sweeteners does (Corzo et al., 2015; Dahlqvist et al., 1963; García-Almeida, 2013; Rubio-Arraez et al., 2016) Another novel feature of isomaltulose is its prebiotic activity and nutritional functionality(Corzo et al., 2015). With all these characteristics, the food industry is gradually incorporating more foods with isomaltulose additives.

12.2 ISOMALTULOSE HEADLINES

Isomaltulose is a carbohydrate resulting from the isomerase activity of sucrose isomerase, present in fungal, bacteria, and plants. Its chemical name is 6-O-α-D-glucopyranosil-D-fructofuranose, composed of glucose and fructose, linked through the α-1, 6 bond (different from sucrose, α-1,2) (Mu et al., 2014). This isomer has physical, viscous, and organoleptic characteristics similar to those of sucrose. Compared with sucrose, its melting point is lower, 120–128 °C versus 160–185 °C.(Zhang et al., 2003). In terms of stability, α-1,6 binding makes isomaltulose stable at acidic pH values without hydrolysis, for example, when incubating isomaltulose in 10% solution and pH 1 and 95 °C, it remains stable for more than 30 min, whereas sucrose under the same conditions was hydrolyzed almost completely in the first 15 min of incubation (Campbell et al. 2018). Isomaltulose has a caramelization point lower than sucrose, it is completely digested in the small intestine and provides the same amount of energy as sucrose, about 4 kcal/g; however, the incorporation of glucose and fructose monosaccharides, resulting from digestion, occurs more slowly than in the case of sucrose (Dahlqvist et al., 1963). This feature allows to this sugar provide energy, in the form of glucose and fructose, in extended periods of time, classifying it as a slow-digesting sugar (Lina et al., 2002; Mu et al., 2014). This phenomenon also promotes the oxidation of fats (fatty acids) found in the body (Campbell et al. 2018; Dahlqvist et al., 1963). Sucrose, on the other hand, is a fast-digesting sugar; hydrolysis occurs completely in the first region of the intestine, so the energy provided by it is used by the body in short periods of time; the excess is stored in the form of fats and adipose tissues.(Holub et al., 2010; Stevenson et al., 2017). In industrials, the solubility of isomaltulose is approximately 29% at 20 °C and, as a very important aspect, it is less hygroscopic than sucrose in the solid state, this is an excellent attribute in the design of foods and powders since it would act as a sweetener and anticaking agent; its sweetening power is 50%–75% relative to sucrose.(Moraes et al., 2005; Mu et al., 2014).

12.3 DIGESTION

Due to its physicochemical characteristics, hydrolysis in fructose and glucose occurs at a very low rate, in this way the available glucose is also slowly incorporated into the blood, generating as a response a low glycemic index and low insulinemic response in the individual. This reveals a great substitute to be incorporated in diets for diabetic patients and athletes, although, it is also being incorporated in dietary regimens for healthy patients or with obesity problems, due to the ability to promote the oxidation of fats. This "antiobesity" effect occurs in the process of digestion of isomaltulose in the small intestine when the low availability of glucose in the blood triggers downstream regulation of the receptor that is responsible for activating the expression of the α and γ genes of the peroxisome proliferator-activated receptor in the liver (PPAR). (Campbell et al. 2018; Dahlqvist et al., 1963). Another feature that shows the use of this sweetener is that isomaltulose is difficultly fermented by the micro-organisms of the oral/dental cavity, efficiently inhibiting the formation of insoluble glycans that acidify the oral cavity and produce dental caries; therefore, its consumption does not produce cavities. The patients who consumed isomaltulose maintained an oral pH higher than 5.6, whereas those who consumed sucrose, oral pH was less than 5.6, loosing away the tooth enamel and then the occurrence of caries.(Campbell et al. 2018; Ooshima et al., 1983; Rubio-Arraez et al., 2016).

12.4 METABOLISM

Metabolism of isomaltulose occurs in the small intestine slower than sucrose.(van Can et al., 2009). Intestinal cell homogenates were prepared, as a source of enzymes and mixed with solutions of sucrose and isomaltulose, this explains that the metabolism of sucrose released at least six times more glucose molecules than the metabolism of isomaltulose, in the same period of time (Dahlqvist et al., 1963; Lina et al., 2002; Mu et al., 2014). The same effect it is shown when tested in vivo conditions on diabetic and nondiabetic patients that have ingestion of isomaltulose and sucrose, finding that there is no occurrence of critical levels of blood glucose in those individuals who ingested isomaltulose. These results suggest that the consumption of isomaltulose would be safe for patients with diabetic conditions or predisposition to diabetes (Dahlqvist et al., 1963). In mice, the same response is observed, all isomaltulose is absorbed in the small intestine and there are very few residues of the disaccharide in feces and urine. In other trials the purification of the sucrase/isomaltase complex from the small intestine of mice revealed that isomaltulose hydrolysis occurs at rates similar to those seen in the homogenate of human intestines, pointing to this complex as responsible for the absorption and digestion in the small intestine (Campbell et al., 2018). On the other hand, the glycoamylase/maltase complex is not able to hydrolyze isomaltulose, as previously thought. In adult pigs, isomaltulose can be digested by both enzymatic complexes; however, in experiments in vivo and in humans, the inactivation of the enzyme isomaltase is related to the fall in the activity of isomaltulase and the activity of sucrase is undetected. In this way, the activity of the isomaltase and isomaltulase enzymes, of the sucrase/isomaltase complex, are responsible for digesting isomaltulose in the small intestine of humans (Dahlqvist et al., 1963). Also in pigs, fed with high doses of isomaltulose, a very small fraction can be traced in the large intestine that promotes fermentative processes, an effect that is rare and generated by high doses of the diet (20% isomaltulose). When feces are compared with both diets (sucrose and isomaltulose), the fermentation residues that occur in the large intestine were not found to be significant or dangerous and coincide with the results in mice with radiolabeled isomaltulose (0.5 g/kg of body weight) (Campbell et al. 2018; Dahlqvist et al., 1963; Holub et al., 2010).

After hydrolyzing, the resulting glucose and fructose pathways follow the same path as those that are released by sucrose; however, because the process of digestion of isomaltulose is slower, the levels of glucose and fructose in the blood are different (van Can et al., 2009). When mice and dogs that have been fed with consistent doses of sucrose and isomaltulose (5.5 mg/kg of body weight) were studied for 90 minutes and every 15 min, it was found that 83% of isomaltulose was excreted intact in the urine, increasing as the dosage was finished, whereas, conversely, the glucose/fructose levels in the blood and urine remained constant, against 76% of the sucrose effect. Another interesting result of this trial is that at the end of the dosage and after 120 min total, there was no evidence of hypertriglyceridemia, an excessive increase in triglycerides due to the presence of glucose in isomaltulose treatments.(Mu et al., 2014).

12.5 TOXICOLOGY

Toxicologically, isomaltulose does not represent any danger in terms of its consumption. Taking into international guidelines to determine if a food product is toxic for human consumption, the effects of isomaltulose consumption on mice of Wistar and Sprague–Dawley lines were tested without finding any adverse effects or differences with controls fed with sucrose, even those fed high doses of isomaltulose did not present any chronic pathology or abnormalities derived from its consumption, now, differing from the control with high doses of sucrose (Lina et al., 2002; Mu et al., 2014). Isomaltulose did not generate any mutagenic, teratogenic, or embryotoxic effect in mice, validating its safe use as a food additive for humans. In humans, the results are similar to those seen in mice and do no more than reaffirm the difference in the metabolism rates of isomaltulose hydrolysis versus sucrose hydrolysis, showing values in blood tests of insulin, glucose, and fructose lower than those found with the sucrose consumption; already in the first 30 min of the trial, the glycemic levels are higher in sucrose diet than with isomaltulose (Lina et al., 2002). Considering fats, the lipid metabolism was unaltered by the consumption of isomaltulose, and no gastrointestinal damage was observed derived from its consumption, even in the presence of high doses (50 g/day). In contrast, recent studies have shown the appearance of gastrointestinal discomfort when prolonged isomaltulose ingestion and under an intense exercise regime, compared with foods and beverages composed of fast-digesting carbohydrates (maltodextrin and fructose), and high levels of glucose, in healthy athletes (nondiabetics) who consumed isomaltulose on a regular basis during a whole day of high-intensity exercise, which would put on risk diabetic athlete consumers (Oosthuyse et al., 2013). However, at present, the dosage analysis should be deeper to clarify the relationship that exists in prolonged consumption and exercise intensity, since other authors propose combining drinks with fast-digesting carbohydrates with those with slow digestion, such as Isomaltulose, and depending on the energy requirement of the athlete. Even so, the *codex alimentarius* allowed using it as a safe additive and a low toxicity ingredient, using their hydrogenated form known as isomaltose (COD SIN-953).

12.6 TECHNOLOGICAL APPLICATION

In 2015, trials were conducted to include isomaltulose as a substitute for sucrose in the production of marshmallows. The physicochemical processes and qualitative traits were analyzed on the final product, being the best formulation the mixture of isomaltulose: fructose (1:1), according to its mechanical properties and organoleptic and sensorial evaluations, compared with those that were made with a mixture of sucrose and high fructose syrups (corn syrup). Six mixtures of sweeteners and one control were designed with 40 g of sucrose and 60 g of corn syrup, which corresponds to 100% of the sugars present in the marshmallows. The mixtures tested were alternated between corn syrup, isomaltulose, and fructose. The addition of fructose to the test aims to improve the sweetening power of the marshmallow because isomaltulose has only one-third of the sweetening power of sucrose and high fructose syrup. The proportions of isomaltulose and fructose range from 30:70 to 70:30, respectively. The physical, mechanical, and sensory parameters were measured such as texture, color, hardness, and physicochemical parameters (water activity, pH, °Brix). Sensory evaluation values were appearance, color, taste, chewiness, elasticity, and buy intention.(Periche et al., 2015).

Another use of isomaltulose in the food industry occurs in the optimization of osmotic extraction (extraction of plum juice). Osmotic extraction is an industrial process of preparation whose objective is the dehydration of fruits to be processed or conserved. So far it is common for the industry to use sucrose to increase the osmotic pressure in the solution where the fruits are deposited. In Korea, the use of this method on an industrial scale for the conservation of fresh plums and plum juice making is very common, maybe, also means significant energy savings by not requiring thermal processes to dehydrate plums (Kim et al., 2018). The presence of sucrose in this process generates concern among worried consumers because its excessive consumption could trigger diabetes or another disease-related. This is why researchers have experimented with replacing the use of sucrose with isomaltulose sugar, obtaining interesting results. Mixtures of sugars (fructose and isomaltulose and glucose and isomaltulose) were used and each one of them independently, having a total of five samples: sucrose, fructose, isomaltulose, glucose-isomaltulose, and fructose-isomaltulose. The control was a 20 mM solution of sucrose, which is used for osmotic dehydration. The best extraction and color, which one it's very important in the plum juice industry, (measured in ° Brix and Browning scale) were found for the mixtures tested with isomaltulose; by itself, extraction with only isomaltulose did not generate the best extraction values. As for the amount of blood glucose of the individuals who consumed the extract, isomaltulose was the one that obtained the lowest value. Although isomaltulose extraction was not the best, it was the most stable in time, maintaining the initial quality of the extract for longer than the others, a factor that is very attractive for industries considering the time of packaging and distribution. For this last attribute, the extraction industry has shown great interest and has begun the incorporation of isomaltulose into osmotic extraction processes.(Kim et al., 2018).

As we have already explained, the most relevant characteristics of isomaltulose are related to the slow rate of hydrolysis by the small intestine, and it is this what has most interested the developers of food and energy drinks. This ability to deliver energy in the form of glucose for a long time has been the focus of the design of energy drinks for athletes. The recent research has considered the need for energy when practicing high-intensity sports, such as soccer or cycling. However, it is common to evaluate the phenomenon of glucose absorption and subsequent energy recovery but not the performance of the athlete as the supplement is ingested (Oosthuyse et al., 2013). The effect produced by the ingestion of isomaltulose on cyclists at steady state and on a 16-km route time trial was tested. For this, energy drinks based on fast-hydrolyzing carbohydrates (a mixture of fructose and maltodextrin) were compared, which are typically formulated. Placebo was given as a control, containing noncaloric flavored water and two options with 7% carbohydrates, one of slow digestion (isomaltulose), and the other of rapid digestion (fructose and maltodextrin). The availability of energy in both trials was similar, compared to placebo; however, the behavior of the body at continuous doses and under intense exercise produced a drop in the performance of the athletes, even generating the impossibility of concluding the day; even so, isomaltulose maintained stable glycemic levels, as common energy drinks do during exercise but promoting the oxidation of fatty acids, a unique and novel effect for carbohydrate-based food supplements (Oosthuyse et al., 2013). Another effect shown in soccer players is the decrease in the levels of epinephrine, the hormone responsible for the acceleration of the heart rate and the contraction of the blood vessels, during the execution of high-intensity exercises in prolonged time, so it is recommended intake of drinks based on isomaltulose in the phase prior to the execution of the physical activity, that is, the warming up and physical preparation, which would substantially reduce the use of energy drinks based on carbohydrates with the high glycemic index, for example, those composed of fructose and maltodextrin (Stevenson et al., 2017).

12.7 BIOTECHNOLOGY

12.7.1 WHERE TO FIND?

In the isomaltulose nature, it is found in vegetables such as sugar beet and honey, but it is a small amount and it is difficult to synthesize it artificially. (Mu et al., 2014; Xu et al., 2013). The use of biotechnology to produce the enzyme sucrose isomerase, responsible for the isomerization of sucrose to isomaltulose, has been widely studied in recent years. There are several microorganisms capable of isomerizing sucrose, forming mainly isomaltulose and trehalulose, in smaller quantities; for example *Protaminobacter rubrum* (Lee et al., 2008), *Erwinia rhapontici* (Cheetham, 1984; Kawaguti et al., 2006; Moraes et al., 2005; Wu and Birch, 2011) , *Klebsiella sp* (Li et al., 2004; Zhang et al., 2002; Zhang et al., 2003) , *Serratia plymuthica* (Véronèse and Perlot, 1999) *Pseudomonas mesoacidophilus* (Mu et al., 2014) and *Agrobacterium radiobacter* (Mu et al., 2014). This enzyme, also referred to as isomaltulose synthase, has been purified from these organisms for its biochemical characterization (Cheetham, 1984; Ravaud et al., 2009). It produces mainly isomaltulose (from 75% to 85%) with traces of trehalulose and in some cases glucose and fructose (Cheetham, 1984); *Serratia plymuthica* in addition produces isomaltose and isomalezitose (Véronèse and Perlot, 1999). In 1995, the first expression of the recombinant enzyme of *P. rubrum* in *Escherichia coli* was reported. Currently, at least 9 organisms express the recombinant enzyme in heterologous systems through *E. coli, Lactococcus lactis* and *Sacharomyces cerevisiae* (De Costa et al., 2003; Mu et al., 2014; Soliman, 2018; Wu and Birch, 2005).

12.7.2 SUCROSE ISOMERASE

Sucrose isomerase, synonymous with isomaltulose synthase, sucrose α-glucosyltransferase and trehalulosa synthase (when the production of trehalulose exceeds the production of isomaltulose), catalyzes the isomerization of the α-1,2 bond between glucose and fructose n sucrose at a type of α-bond -1.6 (isomaltulose) or α-1,4 (trehalulose); in the process also small amounts of glucose and fructose are released product of the hydrolysis of sucrose (Cheetham, 1984). The catalytic mechanism consists of an acid–base equilibrium between the active site of the enzyme and sucrose (Ravaud et al., 2009). The domain of the enzyme is one of the treophosphate isomerase types whose active site belongs to the glycoside hydrolase family; this family consists of three domains A (N-terminal), B (loop of type β-α barrel), and C (C-terminal) (Ravaud et al., 2009; Ravaud et al., 2006). Unlike the glucoside-hydrolases enzymes, sucrose isomerase has a specialized motif that is responsible for isomerization, whose sequence is RLDRD and is highly conserved among all enzymes. This characterization was done with the *Klebsiella* sp enzyme (Cheetham, 1984; Contesini et al., 2013; Zhang et al., 2003)

The pH of catalytic activity of this enzyme is in the range of pH 5–6, with some specific exceptions (Silva et al., 2005); even so, some authors report activity in pH values close to 7 but due to the mechanism of catalysis it is necessary that the pH is slightly acid so that it protons the oxygen of Asp (Rubio-Arraez et al., 2016; Campbell et al., 2018) and thus can act as a nucleophile and attack de sucrose molecule.(Queneau et al., 2008; Ravaud et al., 2009; Ravaud et al., 2006). These considerations depend directly on the resulting product from the catalysis, for example, at high pH increases the production of trehalulose and low pH increases the amount of monosaccharides (glucose and fructose) in *K. planticola*.(Zhang et al., 2002; Zhang et al., 2003) This suggests that at a low-pH (acid) hydrolysis occurs mediated by the sucrose isomerase and as the pH becomes alkaline, the isomers show up becoming the main product of the reaction; is what could be observed in *Pantoea dispersa* (Wu and Birch, 2005; Wu and Birch, 2011). An effect similar to pH occurs at the optimum reaction temperature. The optimal catalytic activity is between 20 °C and 40 °C of the sucrose isomerase of several organisms, which results in a great industrial potential because other isomerases require high temperatures(Queneau et al., 2008; Silva et al., 2005); when temperatures above 50 °C are applied, no isomaltulose appears as a result of the catalysis, suggesting that the enzyme is thermolabile; however, have been also reported to have activity in short periods of time at high temperatures (Silva et al., 2005). Again, when temperature values are analyzed below and above the optimum in *K. planticola*, it produces a greater amount of isomaltulose at 30 °C but decreases when the temperature is below the optimum, increasing the trehalulose levels produced (Wu and Birch, 2011). In *P. dispersa*, in the range of 20–40 °C produces isomaltulose: trehalulose levels of 21:1 to 8:1; however, above 40 °C the amounts of isomaltulose are lower (Wu and Birch, 2005; Wu and Birch, 2011). Sucrose isomerase of *S. plymuthica* at 30 °C produces isomaltulose, at 25 °C trehalulose, at 45 °C monosaccharides, and at 30 °C isomelezitosa, its production optimum being 50 °C (Véronèse and Perlot, 1999).

12.7.3 PRODUCING SUCROSE ISOMERASE

Optimizing the industrial production of sucrose isomerase using microorganisms implies to optimize culture media to upgrade the growth conditions for it, obtaining attractive levels of catalytic activity. Several researchers have proposed culture media that optimize the growth and production of the enzyme sucrose isomerase at the level of bioreactors. In 2005, a group of Brazilian researchers proposed to optimize the culture medium for *Erwinia* sp in fermenters, adding 12% molasses, different from the classic culture medium containing sucrose as a source of energy and carbon, obtaining a maximum of 15.6 U/ml of sucrose isomerase activity, against the 10 U/ml obtained with *K. planticola* in a medium with sucrose.(Moraes et al., 2005; Mu et al., 2014; Wang et al., 2019). They evaluated the growth conditions and the effects on enzyme production on variable molasses concentrations from 0% to 10%, as control the same concentrations (0%–10%) were used but sucrose (Moraes et al., 2005; Wang et al., 2019). The effect of concentration of peptone was also considered, keeping constant the concentration of carbohydrate (molasses, 12%) and testing different concentrations values (Moraes et al., 2005). Due to the thermal sensitivity of the sucrose isomerase, different temperatures at different pH were also considered as variables. After microbial growth, the cells were immobilized using sodium alginate, a flag methodology for the biotechnological utilization of recombinant enzymes in terms of production on an industrial scale (Contesini et al., 2013; Kawaguti et al., 2011; Moraes et al., 2005). The addition of molasses as a source of energy and carbon to culture media is a great advantage, due to its economic accessibility and because its presence stimulates the glycosyltransferase activity of sucrose isomerase (Wang et al., 2019).

More recently, Kawaguti (2016) reported activity levels of 14.6 U/ml in shaking flasks at 30 °C, using a concentration of 160 g/l (16%) of molasses instead, keeping the concentration of peptone and yeast extract (20 and 15 g/l), for 10 h. At 8 h of culture, there was an activity of 7.47 U/ml, suggesting a maximum of production in the exponential growth phase.(Kawaguti et al., 2011; Kawaguti et al., 2006). These results of optimization of growth conditions were validated through the response surface methodology that allows analyzing the interactions of one variable influenced by others, using *Erwinia* sp, as a model organism in the production of sucrose isomerase (Kawaguti et al., 2006; Wang et al., 2019)

Other living systems have also been used for the production of sucrose isomerase, using genetic engineering. In 2018, Soliman et al. achieved the expression of the sucrose isomerase gene in transgenic potato plants, transformed by *Agrobacterium tumefasciens*. The gene coding for sucrose isomerase is *pal* I, which has a size of ≈1694 bp in *E. rhapontici,* expressed in *E. coli* was previously reported (De Costa et al., 2003; Fernie et al., 2001; Soliman, 2018). This work was aimed at achieving the expression of the sucrose isomerase gene in the apoplastic region, the aqueous space between the cell membrane and the cell wall of plants, of potato tuber, *A. tumefaciens* mediated, using the pBinAR-pal I plasmid (De Costa et al., 2003; Fernie et al., 2001; Soliman, 2018). Until now, only the sucrose isomerase gene had been expressed in the apoplastic region of *Nicotiana tabaco*, the *tobacco* plant but the transformants showed physiological and growth deficiencies, such as the impossibility of differentiating cells or the difficulty to intake water, so seedlings finished showing symptoms of chlorosis and necrosis.(Fernie et al., 2001; Soliman, 2018). Determining the tubers of transgenic plants using HPLC techniques, the production of isomaltulose in the apoplastic space reached 19.5 mol/g as opposed to zero from the nontransformed plants. This would be the first time that transgenic plants for sucrose isomerase would be completely viable, maintaining the apoplastic production of the enzyme (Soliman, 2018). This extends the frontiers regarding the biotechnological possibilities for the design of new isomaltulose production systems as a substitute for sucrose. Moreover, if the production of isomaltulose of potato tuber (without considering starch) compares with the levels of sucrose or glucose, they are significantly higher (1.4 mol/g of sucrose and 0.48 mol/g of glucose; result of the HPLC analysis), that is, a potato composed mostly of slow-digesting sugars and a low glycemic index (Soliman, 2018; van Can et al., 2009).

12.8 THE NEAR FUTURE

In accordance with the goals established by the World Health Organization, in its objective 3 for 2030, to guarantee a healthy life and well-being. On the other hand, the Food and Agriculture Organization establishes the increase of economic investment in the scientific development of biotechnology to improve the quality of the food industry, making them safer and with a sustainable approach. Isomaltulose is then shown as an interesting option to be considered as a substitute for sucrose, partially or totally, revealing its nutritional and safe characteristics. Even so, we must follow the increasing of research to upgrade the food quality, offering the industry and consumers suitable tools for handling and applying the newest technologies, also attending the economic, social, and environmental needs of the present century. There is now a company that distributes the isomaltulose under the name Palatinose (Beneo Group Südzucker, Mannheim, Germany) and already incorporated in the design and manufacture of food, beverage, bakery, energy and powder drinks, chewing gum, candies, and other foods whose companies are motivated to improve the welfare of their consumers, who, in turn, demand healthier food.

KEYWORDS

- **isomaltulose**
- **sucrose isomerase**
- **biotechnology**

REFERENCES

Campbell, G.; Belobrajdic, D.; Bell-Anderson, K. Determining the Glycaemic Index of Standard and High-Sugar Rodent Diets in C57BL/6 Mice. *Nutrients* **2018**, *10* (7), 856.

Cheetham, P. S. J. The Extraction and Mechanism of a Novel Isomaltulose-Synthesizing Enzyme from *Erwinia Rhapontici*. *Biochem. J.* **1984**, *220* (1), 213–220.

Contesini, F.; de Alencar Figueira, J.; Kawaguti, H.; de Barros Fernandes, P.; de Oliveira Carvalho, P.; da Graça Nascimento, M.; Sato, H. Potential Applications of Carbohydrases Immobilization in the Food Industry. *IJMS* **2013**, *14* (1), 1335–1369.

Corzo, N.; Alonso, J. L.; Azpiroz, F.; Calvo, M. A.; Cirici, M.; Leis, R.; Lombó, F.; Mateos-Aparicio, I.; Plou, F. J.; Ruas-Madiedo, P.; et al. Prebióticos; concepto, propiedades y efectos beneficiosos. *Nutr. Hosp.* **2015**, *31* (1), 21.

Dahlqvist, A.; Auricchio, S.; Prader, A. Human Intestinal Disaccharidases and Hereditary Disaccharide Intolerance. The Hydrolysis of Sucrose, Isomaltose, Palatinose (isomaltulose), and a 1,6-a-oligosaccharide (iso-malto-oligosaccharide) Preparation. *J. Clin. Investig.* **1963**, *42* (4), 7.

De Costa, D. M.; Suzuki, K.; Yoshida, K. Structural and Functional Analysis of a Putative Gene Cluster for Palatinose Transport on the Linear Chromosome of Agrobacterium Tumefaciens MAFF301001. *J. Bacteriol.* **2003**, *185* (7), 2369–2373.

Fernie, A. R.; Roessner, U.; Geigenberger, P. The Sucrose Analog Palatinose Leads to a Stimulation of Sucrose Degradation and Starch Synthesis When Supplied to Discs of Growing Potato Tubers. *Plant Physiol.* **2001**, *125* (4), 1967–1977.

García-Almeida, J. M. Una visión global y actual de los edulcorantes. Aspectos de regulación. *Nutr. Hosp.* **2013**, *28* (4), 16.

Holub, I.; Gostner, A.; Theis, S.; Nosek, L.; Kudlich, T.; Melcher, R.; Scheppach, W. Novel Findings on the Metabolic Effects of the Low Glycaemic Carbohydrate Isomaltulose (Palatinose™). *Br. J. Nutr.* **2010**, *103* (12), 1730–1737.

Kawaguti, H. Y.; Carvalho, P. H.; Figueira, J. A.; Sato, H. H. Immobilization of Erwinia Sp. D12 Cells in Alginate-Gelatin Matrix and Conversion of Sucrose into Isomaltulose Using Response Surface Methodology. *Enzym. Res.* **2011**, *2011*, 1–8.

Kawaguti, H. Y.; Manrich, E.; Sato, H. H. Production of Isomaltulose Using Erwinia Sp. D12 Cells: Culture Medium Optimization and Cell Immobilization in Alginate. *Biochem. Eng. J.* **2006**, *29* (3), 270–277.

Kim, H.-W.; Han, S. H.; Lee, S.-W.; Suh, H. J. Effect of Isomaltulose Used for Osmotic Extraction of Prunus Mume Fruit Juice Substituting Sucrose. *Food Sci. Biotechnol.* **2018**, *27* (6), 1599–1605.

Lee, H. C.; Kim, J. H.; Kim, S. Y.; Lee, J. K. Isomaltose Production by Modification of the Fructose-Binding Site on the Basis of the Predicted Structure of Sucrose Isomerase from "Protaminobacter Rubrum." *Appl. Environ. Microbiol.* **2008**, *74* (16), 5183–5194.

Li, X.; Zhang, D.; Chen, F.; Ma, J.; Dong, Y.; Zhang, L. Klebsiella Singaporensis Sp. Nov., a Novel Isomaltulose-Producing Bacterium. *Int. J. Syst. Evol. Microbiol.* **2004**, *54* (6), 2131–2136.

Lina, B. A. R.; Jonker, D.; Kozianowski, G. Isomaltulose (Palatinose®): A Review of Biological and Toxicological Studies. *Food Chem. Toxicol.* **2002**, *40* (10), 1375–1381.

Moraes, A. L. L.; Steckelberg, C.; Sato, H. H.; Pinheiro, A. Production of Isomaltulose From Enzymatic Transformation of Sucrose, Using Erwinia sp D12 Immobilized with Calcium Alginate. *Ciênc. Tecnol. Aliment.* **2005**, *25* (1), 9.

Mu, W.; Li, W.; Wang, X.; Zhang, T.; Jiang, B. Current Studies on Sucrose Isomerase and Biological Isomaltulose Production Using Sucrose Isomerase. *Appl. Microbiol. Biotechnol.* **2014**, *98* (15), 6569–6582.

Ooshima, T.; Izumitani, A.; Sobue, S.; Okahashi, N.; Hamada, S. Non-Cariogenicity of the Disaccharide Palatinose in Experimental Dental Caries of Rats. *Infect. Immun.* **1983**, *39* (1), 7.

Oosthuyse, T.; Carstens, M.; M. Millen, A. Ingesting Isomaltulose Versus Fructose-Maltodextrin During Prolonged Moderate-Heavy Exercise Increases Fat Oxidation but Impairs Gastrointestinal Comfort and Cycling Performance. *Wilderness Environ. Med.* **2013**, *24* (1), 78–79.

Periche, A.; Heredia, A.; Escriche, I.; Andrés, A.; Castelló, M. L. Potential Use of Isomaltulose to Produce Healthier Marshmallows. *LWT - Food Sci. Technol.* **2015**, *62* (1), 605–612.

Queneau, Y.; Chambert, S.; Besset, C.; Cheaib, R. Recent Progress in the Synthesis of Carbohydrate-Based Amphiphilic Materials: The Examples of Sucrose and Isomaltulose. *Carbohydr. Res.* **2008**, *343* (12), 1999–2009.

Ravaud, S.; Robert, X.; Watzlawick, H.; Haser, R.; Mattes, R.; Aghajari, N. Structural Determinants of Product Specificity of Sucrose Isomerases. *FEBS Lett.* **2009**, *583* (12), 1964–1968.

Ravaud, S.; Watzlawick, H.; Haser, R.; Mattes, R.; Aghajari, N. Overexpression, Purification, Crystallization and Preliminary Diffraction Studies of the *Protaminobacter Rubrum* Sucrose Isomerase SmuA. *Acta Crystallogr. F. Struct. Biol. Cryst. Commun.* **2006**, *62* (1), 74–76.

Rubio-Arraez, S.; Capella, J. V.; Castelló, M. L.; Ortolá, M. D. Physicochemical Characteristics of Citrus Jelly with Non Cariogenic and Functional Sweeteners. *J. Food Sci. Technol.* **2016**, *53* (10), 3642–3650.

Silva, Z.; Sampaio, M.-M.; Henne, A.; Bohm, A.; Gutzat, R.; Boos, W.; da Costa, M. S.; Santos, H. The High-Affinity Maltose/Trehalose ABC Transporter in the Extremely Thermophilic Bacterium Thermus Thermophilus HB27 Also Recognizes Sucrose and Palatinose. *J. Bacteriol.* **2005**, *187* (4), 1210–1218.

Soliman, H. I. A. Molecular Cloning of Sucrose Isomerase Gene and Agrobacterium-Mediated Genetic Transformation of Potato (Solanum Tuberosum L.) Plants. *Int. J. Environ. Agric. Biotechnol.* **2018**, *3* (3), 864–874.

Stevenson, E. J.; Watson, A.; Theis, S.; Holz, A.; Harper, L. D.; Russell, M. A Comparison of Isomaltulose versus Maltodextrin Ingestion during Soccer-Specific Exercise. *Eur. J. Appl. Physiol.* **2017**, *117* (11), 2321–2333.

van Can, J. G. P.; IJzerman, T. H.; van Loon, L. J. C.; Brouns, F.; Blaak, E. E. Reduced Glycaemic and Insulinaemic Responses Following Isomaltulose Ingestion: Implications for Postprandial Substrate Use. *Br. J. Nutr.* **2009**, *102* (10), 1408–1413.

Véronèse, T.; Perlot, P. Mechanism of Sucrose Conversion by the Sucrose Isomerase of Serratia Plymuthica ATCC 15928. *Enzym. Microb. Technol.* **1999**, *24* (5–6), 263–269.

Wang, Z.-P.; Wang, Q.-Q.; Liu, S.; Liu, X.-F.; Yu, X.-J.; Jiang, Y.-L. Efficient Conversion of Cane Molasses Towards High-Purity Isomaltulose and Cellular Lipid Using an Engineered Yarrowia Lipolytica Strain in Fed-Batch Fermentation. *Molecules* **2019**, *24* (7), 1228.

Wee, M.; Tan, V.; Forde, C. A Comparison of Psychophysical Dose-Response Behaviour across 16 Sweeteners. *Nutrients* **2018**, *10* (11), 1632.

Wu, L.; Birch, R. G. Characterization of the Highly Efficient Sucrose Isomerase from Pantoea Dispersa UQ68J and Cloning of the Sucrose Isomerase Gene. *Appl. Environ. Microbiol.* **2005**, *71* (3), 1581–1590.

Wu, L.; Birch, R. G. Isomaltulose Is Actively Metabolized in Plant Cells. *Plant Physiol.* **2011**, *157* (4), 2094–2101.

Xu, Z.; Li, S.; Li, J.; Li, Y.; Feng, X.; Wang, R.; Xu, H.; Zhou, J. The Structural Basis of Erwinia Rhapontici Isomaltulose Synthase. *PLoS One* **2013**, *8* (9), e74788.

Zhang, D.; Li, X.; Zhang, L.-H. Isomaltulose Synthase from Klebsiella Sp. Strain LX3: Gene Cloning and Characterization and Engineering of Thermostability. *Appl. Environ. Microbiol.* **2002**, *68* (6), 2676–2682.

Zhang, D.; Li, N.; Lok, S.-M.; Zhang, L.-H.; Swaminathan, K. Isomaltulose Synthase (PalI) of Klebsiella Sp. LX3: Crystal Structure and Implication of Mechanism. *J. Biol. Chem.* **2003**, *278* (37), 35428–35434.

Zhang, D.; Li, N.; Swaminathan, K.; Zhang, L.-H. A Motif Rich in Charged Residues Determines Product Specificity in Isomaltulose Synthase. *FEBS Lett.* **2003**, *534* (1–3), 151–155.

CHAPTER 13

Going Through Pulsed Electric Field Technology for Food Processing: Assessment of Progress and Achievements

ANA MAYELA RAMOS-DE-LA-PEÑA[1*], MERAB MAGALY RIOS-LICEA[2], EMILIO MENDEZ-MERINO[2], and JUAN CARLOS CONTRERAS-ESQUIVEL[3*]

[1]*Tecnológico de Monterrey, School of Engineering and Sciences, Av. Eugenio Garza Sada 2501 Sur, Monterrey 64849, Nuevo Leon, Mexico*

[2]*Sigma-Alimentos, Torre Sigma, San Pedro Garza Garcia 66269, Nuevo Leon, Mexico*

[3]*Laboratory of Applied Glicobiotechnology, Food Research Department, School of Chemistry, Universidad Autonoma de Coahuila, Blvd. Venustiano Carranza and Jose Cárdenas s/n República Oriente, Saltillo 25280, Coahuila, Mexico*

[*]*Corresponding author.*
E-mail: ramos.amay@tec.mx; carlos.contreras@uadec.edu.mx

ABSTRACT

Pulsed electric field processing (PEF) is a nonthermal technology that enables inactivation of microorganisms and has an impact on enzymes, besides new purposes in the food processing have been found. Until now, this technology has not been yet introduced to all countries and there is a lack of information about the benefits it provides. The aim of this chapter is to provide insights into the main applications of PEF in food science, such as microbial and enzyme inactivation. Moreover, advances in the processing of different products, such as fruit juices, purees and smoothies, milk beverages, oil, plant tissues, and meat, have been included. In addition, results of sensory evaluation of PEF treated products from people from different countries are compared, as well as the products' acceptance. Finally, topics about environmental issues and legislation frameworks are discussed.

13.1 INTRODUCTION

1. Food preservation is required since fresh food always desired, containing the nutrients needed for good health. However, it is not always easy to obtain (Barbosa-Cánovas, 1999). Common practices of food industries, such as thermal processing operations (pasteurization, sterilization, drying, and evaporation), are used to guarantee the microbial safety of the products (Pereira and Vicente, 2010). However, there is a demand for alternative methods for food preservation due to thermal processing has shown to produce color, flavor, and nutrients degradation (Barbosa-Cánovas, 1999). It has been demonstrated that nonthermal processing has a favorable impact on foods such as fresh quality and higher nutritional value due to color and flavor retention, compared with foods subjected to heat treatments (Barbosa-Cánovas and Altunakar, 2006). A nonthermal technology based on the use of electric fields of high voltage is the technology of pulsed electric fields (PEF) (Rajkovik et al., 2010). Inactivation of vegetative cells of bacteria in foods such as liquid eggs, fruit smoothies, milk, and juices can be achieved with the PEF processing due to the changes produced by PEF in microorganisms. These changes include cell electroporation and disruption, which lead to cell swelling, shrinking, and lysis (Barbosa-Cánovas and Altunakar, 2006). Moreover, PEF is also considered a promising nonthermal technology and a potential alternative to traditional solvent extraction of bioactive compounds from fruits and vegetables (Medina-Meza and Barbosa-Cánovas, 2015). Despite all the benefits offered by this technology, information about PEF is not enough and not equally distributed among consumers from different countries. The aim of this chapter is to provide a general perspective for the use of PEF on pasteurization, sterilization, and enzyme inactivation. In addition, other purposes of PEF in food are described, such as extraction, drying, freezing, etc. Available equipment, consumer's point of view, environmental impact, and legislation frameworks are considered to increase the knowledge and interest about the

use of this novel technology and the impact of the final product on consumers, industry, and environment.

13.1.1 PASTEURIZATION, ENZYME INACTIVATION, AND STERILIZATION BY MEANS OF THERMAL TREATMENT

1. Thermal pasteurization is the physical process of food decontamination and enzyme inactivation, which is still used today for being efficient, environmentally friendly, preservative-free, and inexpensive compared to other preservation technologies (Silva and Gibbs, 2009). The use of thermal processing requires the determination of heating conditions to produce microbiologically safe products while maintaining quality and increasing shelf-life. Thermal processing needs a time-temperature profile, indicating heating medium temperature and holding time. Thermal pasteurization temperatures are usually 90–100 °C. A typical industrial pasteurizer for glass jars of an acidic product would have a preheating section (70 °C; 5 min), a pasteurization section (90 °C; 30 min), a first cooling section (50 °C; 7 min), and a second cooling section (20 °C; 7 min) (Álvarez et al., 2006). High-temperature short-time (HTST) pasteurization is the most widely used technique for the heat treatment of fruit juices. For instance, orange juice is pasteurized at 90–95 °C for 15–30 s and apple juice is processed at 77–88 °C for 25–30 s (Ortega-Rivas, 2011).
2. In the case of milk, a range of temperatures between 63°C and 100 °C are used. First, the milk needs to be heated at 63 °C for 30 min, then HTST methods take place, being 15 s at 72 °C, 1 s at 89 °C, 0.5 s at 90 °C, 0.1 s at 94 °C, and 0.01 s at 100 °C (Holsinger et al., 1997).
3. On its part, thermal sterilization of food achieves microbial destruction by means of saturated steam under pressure, causing irreversible denaturation of enzymes and structural proteins. Heat sterilization is usually carried out at 121–124 °C and 200 kPa for 15 min (World Health Organization, 2018).
4. In the thermal sterilization process, a target microorganism needs to be specified, capable of reproducing in food under nonrefrigerated conditions. Temperature profile to which the food has been submitted, obtained at the point of slowest heating in the product during the process has to be registered, being F_0 the unit that expresses the

lethality of the thermal process (Holdsworth and Simpson, 2007; Álvarez et al., 2006). F_0 value is the number of minutes of heating at 121 °C required to achieve the same thermal destruction ratio of a target micro-organism having a *z* value of 10 °C. The *z* value is the temperature increment needed to accelerate the destruction of microorganisms by a factor of 10 (Berk, 2013).

13.1.2 MICROBIAL AND ENZYME INACTIVATION BY ELECTRIC PULSES

1. PEF is one of the alternative technologies capable of inactivating microorganisms at temperatures below those used during thermal processing. Electric fields in the range of 5–50 kV/cm generated by the application of short high voltage pulses (µs) between two electrodes cause microbial inactivation, leading to the permeabilization of microbial membranes (Álvarez et al., 2006). The application of pulsed electric fields of high intensity and short duration (sub µs to 5 µs) may cause temporary or permanent permeabilization of cell membranes. This phenomenon is referred to as electroporation or electropermeabilization (Mañas and Pagán, 2006). Electroporation is caused by externally applied short and intense electric pulses (Donsi et al., 2010). The phenomenon of the membrane damage is caused by the increase of the potential difference (u_m) produced across a cell membrane, which is produced by the increase in the intensity of the applied electric field. When u_m is higher than a value range (0.5–1.5 V), a reversible or irreversible loss of membrane semipermeability takes place (Vorobiev and Lebovka, 2010). The time between pulses is much longer than the pulse width and the generation of pulses involves slow charging and fast discharging of the capacitor. Electric field pulses that are most commonly applied are in the form of exponentially decaying and square waves (Barbosa-Cánovas, 1999). Microbiological inactivation and food preservation effects depend on multiple high strength electric field pulses of relatively short duration (Dunn, 2001). Excellent illustrations of electroporation principles can be found in works such as Donsi et al. (2010), Mahnič-Kalamiza et al. (2014), and Wang et al. (2018).
2. PEF inactivation kinetics is often assessed, microbial evolution is described as a function of time, then parameters appearing in

models are characterized as a function of process parameters and environmental conditions. Finally, both models are combined to predict microbial evolution as a function of process parameters and environmental conditions (Álvarez et al., 2006).

3. Several studies have been conducted to investigate inactivation kinetics of different microorganisms and spores in food. It has been demonstrated a reduction in colony count in a range of 4–6 log cycles, which is comparable to thermal pasteurization (Knorr et al., 2011). In the case of spores, it has been found that a 100% inactivation is not achieved by means of PEF, due to spores have cortex, coat, and exosporium, which provide their membranes with increased resistance (Hamilton and Sale, 1967; Raso et al., 1998). However, complete spore inactivation can be achieved by combining PEF with other technologies, for instance, thermal heating. Table 13.1 enlists some microorganisms which have been investigated for inactivation by PEF.
4. Mechanisms of microbial inactivation and factors affecting microbial inactivation are usually described. Tests consist of the application of an electric field, causing the inactivation of a maximum number of microorganisms without an electrical breakdown of food.
5. In a microbial resistance analysis, Pagán and Mañas (2006) reported that spores are the most resistant to PEF treatment, followed by Gram-positive vegetative cells and Gram-negative cells. On the other hand, yeasts and molds were classified as more susceptible to inactivation. In the case of viruses associated with food diseases (Peng et al., 2017), PEF effects have shown to be negligible (Khadre and Yousef, 2002).
6. Thermal pasteurization and sterilization are fully accepted as effective methods for pathogens destruction with minimal loss of food quality (Peng et al., 2017). On the other hand, nonthermal technologies such as pulsed electric fields have emerged the last 20 years claiming that the application of heat during the processing operation causes not only microbial destruction but also nutrient degradation such as loss of vitamins and volatile flavor compounds, texture changes, enzyme inactivation, and nonenzymatic browning, besides the cooked flavor is perceived (Terefe et al., 2015). Moreover, is emphasized that quality attributes of nonthermal processed food are comparable to fresh food, besides avoiding the use of food additives to extend shelf-life. For instance, after PEF processing no change of

pH, acidity, soluble solids vitamin C content, and color in fruit beverages was observed. In addition, no change in polyphenol contents and antioxidant capacity was detected at low electric fields (1–5 kV/cm). However, when electric fields were increased, changes could be detected (Bi et al., 2013).

7. Effect of PEF on enzymes varies in function of the enzyme, for instance, lipase, glucose oxidase, and α-amylase reduced their activity to 70–85% after PEF treatment (Yang et al., 2004). Other enzymes such as plasmin, lactoperoxidase, polyphenoloxidases, papain, pectinmethylesterase, lactate dehydrogenase, among others, have been studied in terms of inactivation by means of PEF, and different inactivation percentages have been obtained (Yeom et al., 2000; Van Loey et al., 2001; Yeom et al., 2002; Zhong, 2008; Meza-Jiménez et al., 2019). It was found that parameters for enzyme inactivation by PEF are different depending on the enzyme tested.
8. Milk processing by means of PEF is limited to outlet temperatures below 63 °C, which is the minimum low-temperature long-time (30 min) pasteurization conditions at long (10^8 μs) or shorter treatment times (10^5 μs). These conditions are below 72 °C and 15 s (HTST pasteurization conditions) due to excessive heat treatment causes a detrimental effect in milk (Rodriguez-Gonzalez, 2010), for instance, Maillard reactions (Van Boekel, 1998). Moreover, findings about a protective effect of milk fats on bacterial inactivation have been found (Yu et al., 2009), which need to be understood. Key proteins in milk have been denatured successfully by PEF, such as alkaline phosphatase, plasmin, and a protease, from *Pseudomonas fluorescens* M3/6 (Vega-Mercado et al., 2007).
9. The challenge remains in the capacity of PEF to replace thermal technologies to achieve fully microbial decontamination, this goal can be accomplished by the combination of different technologies (Lee et al., 2018) or by the addition of antimicrobial compounds in food, such as *Mentha piperita* L. essential oil (de Carvalho et al., 2018), instead of applying solely PEF. In addition, longer times and high voltages need to be applied if full decontamination is aimed; however, these processing conditions may lead to the undesired food detrimental.

13.2 PEF EQUIPMENT

1. Food can be treated in a batch-wise or continuous mode. The treatment chamber is the component of the PEF unit where the direct application of the electric field occurs. It consists of two electrodes held in position by an insulating material to form an enclosure containing the food. The most commonly used chambers are with parallel plates, with concentric cylinders and coaxial and colinear chambers. Chambers with parallel plates are characterized by uniform electric field strength distribution, whereas a smooth and uniform product flow is commonly reported in concentric cylinders. The coaxial and colinear chambers have less uniform electric field distribution; however, they are easy to clean and have a high load resistance (Barbosa-Cánovas et al., 1999). Continuous and static chambers for PEF processing have been developed; moreover, U-shaped, parallel plate, disk-shaped, wire–cylinder, rod–rod, and sealed are among the static treatment chambers (Huang and Wang, 2009).
2. Ayhan et al. (2001a) developed a glove box as part of an integrated pilot plant scale PEF processing and packaging system to facilitate studies of product shelf-life. PDF treated samples of orange juice (35 kV/cm for 59 μs) were incubated at 22 and 37 °C for 2 weeks and analyzed for microbial growth. Turbidity or microbial growth in the orange juice samples at 22 and 27 °C was not found. The treated orange juice had a shelf life of more than 16 weeks at 4 °C.
3. Jaeger et al. (2009) modified a colinear treatment chamber in a continuous flow-through system to improve microbial inactivation of *Escherichia coli* and to reduce PEF impact on alkaline phosphatase activity in raw milk. The modification consisted of the insertion of stainless steel and polypropylene grids to alter the field strength distribution, to increase turbulence kinetic energy, and to improve temperature homogeneity. An improved electric field strength distribution with increased average electric field strength was achieved. Moreover, the velocity profile was improved, being increased the turbulence kinetic energy due to the insertion of the grids. The temperature of the liquid decreased and alkaline phosphatase inactivation was reduced from 78% to 92% of residual activity and an increase of microbial inactivation of 0.6 log-cycles was observed.
4. The same year, Li et al. (2009) used a coaxial and tube-plate treatment chamber for inactivation of bacteria suspended in solution

by means of PEF and they concluded that the tube-plate treatment chamber was more effective than the coaxial treatment chamber.

5. Due to the presence of bubbles in chambers of PEF may lead to nonuniform treatment, as well as operational and safety problems (Barbosa-Cánovas, 1999), Wang et al. (2009) developed an electrode system to reduce bubbles in chambers of treatment. It allowed the use of the equipment for a long-time avoiding discharging and holding low temperatures. The authors treated orange juice in the PEF chamber, with 12.5 kV/cm for 340 μs. The results showed that this equipment with the modified electrode is able to avoid electrical discharges and that it fulfills the requirements for industrial production.
6. In another study, equipment was developed for the batch processing of energy crop: the KEA-MOBIL (Mobile Karlsruher Elektroporationsanlage, mobile Karlsruhe electroporation device). It consisted of a hydraulic press made from insulating material to apply a pressing force ≥ 11 tons to the piston. The piston and the bottom of the test vessel served as electrodes and the capacity was 50 kg/h with an automated feeding system. An electroporation device for large throughput also was developed. The device KEA was equipped with generators with a charging voltage of 350 kV and 1.2 kJ per pulse operated at 20 Hz (Sack et al., 2010).
7. In 2011, a domestic device for PEF was launched by Nutri-Pulse®, the e-cooker® equipment which requires 220 or 380 V. The vendors claim that it is a standalone system, which has a central programmable logic controller for easy operation and has advanced diagnostic and data logging facilities. This domestic equipment offers new opportunities for cooks in restaurants and food service, for innovative product developers in the food industry and consumers. It is available as NP60 (capacity 30–100 cc) or NP600 (capacity 100–800 cc).
8. Microchips also have found application in PEF sterilization, for instance, Zhu et al. (2017) started from interdigital, planar, and symmetrical comb teeth to design and processing of a microelectrode, which was encapsulated in a microtreatment chamber to perform sterilization. Simulation of microelectrode structures was carried out by COMSOL software to optimize microchip performance. A third-generation device, which consisted of a structure of a parallel plate was developed. The microchamber inactivated *E. coli,*

TABLE 13.1 Some Microorganisms Investigated for Inactivation by PEF

Genus	Species	Genus	Species
Salmonella	*enteridis* *dublin,* *thypimurium* *panama*	*Streptococcus*	*faecalis* *cremoris*
Enterobacter	*aerogenes* *sakazakii*	*Micrococcus*	*radiodurans* *lysodeikticus*
Gluconobacter	*oxydans*	*Pichia*	*fermentans*
Staphylococcus	*aureus*	*Zygosaccharomyces*	*bailii*
Saccharomyces	*cerevisiae* *senftenberg*	*Candida*	*humilis*
Listeria	*monocytogenes* *innocua*	*Sarcina*	*lutea*
Byssochlamys	*fulva conidiospores*	*Clostridium*	*weichii*
Cronobacter	*sakazakii*	*Neosartoria*	*fischeri ascospores*
Lactobacillus	*plantarum* *brevis* *casei* *fermentum* *lactis* *delbrueckii*	*Bacillus*	*cereus* *pumilus* *subtilis*
Yersinia	*enterocolitica*	*Pseudomonas*	*aeruginosa* *fluorescens*
Campylobacter	*jejuni*	*Leuconostoc*	*mesenteroides*
Escherichia	*coli*		

(Barbosa-Cánovas, 1999; Gurtler et al., 2010; McNamee et al., 2010; Bukow et al., 2013; Barba et al. 2015; Pillet et al., 2016; Ou et al., 2017; Timmermans et al., 2019)

Saccharomyces cerevisiae, and *Staphylococcus aureus* by 5.32, 5.42, and 4.77 $\log_{10}$ cycles, respectively.

13.3 FOOD PROCESSING BY PEF

1. Besides microbial destruction and enzyme inactivation, PEF also have been applied in food processing, such as macromolecules modification, extraction, drying, freezing, and thawing, fruit, and vegetable expressing, pesticide degradation (Liu et al., 2010) among others, such as starch modification (Hong et al., 2018). In addition, PEF also has found application in the manufacturing process of wine, beer, and rice wine, which has been widely described by Yang et al. (2016). By-products also have been processed by electroporation, for instance, Barba et al. (2015) summarized the PEF treatment applied to food by-products and wastes, as well as the use of PEF for drying, freezing, and rehydration purposes. Table 13.2 shows selected examples of PEF treatment in separation, aggregation, and extraction. Both batch and continuous systems have been tested. Proteins have been the subject of separation and aggregation, due to PEF causes protein polarization, hydrophobic amino acid (AA) exposition to solvent, and unfolding protein changing into aggregates at high electric strength field (Han et al., 2018).
2. In the case of extraction, it has been focused on juices, sugars, anthocyanins, proteins, phenolic compounds, oils, and so forth.
   Table 13.3 enlists some examples in which PEF have been used in freeze, drying, and degradation of compounds. Most of the processing has been carried out in batch mode. Potato has been chosen as a model product for processing, among others. Optimized methods for freeze-drying (Liu et al., 2017) seen are examples of recent improvements in this specific application.

13.3.1 SENSORY EFFECTS AND QUALITY ATTRIBUTES AFTER PEF PROCESSING

13.3.1.1 FRUIT JUICES

1. The impact of PEF treatment on bioactive compounds and antioxidant activity of orange juice was studied by Sánchez-Moreno et al. (2004)

TABLE 13.2 Examples of the Use of Pulsed Electric Fields in Separation, Aggregation, and Extraction

Reference	Purpose		Processing Parameters	Circuit Configuration Pulses	Chamber	Mode	Achievements
Li et al. (2007)	Separation	Soybean proteins	E = 30 kV/cm; t = 288 µs; pulse width = 2 µs; f = 500 Hz	Bipolar square	Six cofield flow tubular	Continuous	Denaturation of protein molecule
Wu et al. (2015)	Aggregation	Multiproteins	E = 25 kV/cm; t = 200, 400, 600, 800 µs; f = 100 Hz; pulse width = 2 µs	Square	Six cofield tubular	Continuous	No aggregates in ovalbumin and lysozyme solution after 800 µs.
Grimi et al. (2007)	Extraction	Carrot sugars	E = 0.25–1 kV/cm; N = 100; t = 100 ms; pulse width = 2 s	Monopolar near rectangular	Polypropylene frame, cylindrical cavity. Gauze electrode	Batch	Sugar-free concentrate rich in vitamins and carotenoids
Praporscic et al. (2007)		Grape juice	E = 0.25–1 kV/cm; N = 100; t = 100 µs	Monopolar near-rectangular	Polypropylene frame cylindrical. Gauze electrode, square holes	Batch	Yields increased from 49%–54% to 76%–78% at 45 min of pressing
Corrales et al. (2008)		Grape anthocyanins	E = 3 kV/cm; N = 30; f = 2 Hz	Exponential decay	Parallel plate	Batch	Total phenolic content 50% higher than control. Antioxidant activity increased four-fold.

TABLE 13.2 *(Continued)*

Reference	Purpose	Processing Parameters	Circuit Configuration Pulses	Chamber	Mode	Achievements
López et al. (2009a)	Sugar beet sucrose	$E = 1–7$kV/cm; 20, 40, 70 °C. Pulse width = 2–5 μs; $f = 1–10$ Hz	Square and exponential decay	Parallel-electrode	Batch	20 pulses at 7 kV/cm reduced 60% the extraction temperature from 70 to 40 °C in a 60 min process
López et al. (2009b)	Red beetroot betanine	$E = 0, 1, 3, 5, 7, 9$ kV/cm; $N = 5$; $t = 2$μs; $f = 1$Hz	Exponential decay	Parallel-electrode	Batch	Increase of rate extraction by 4.2-fold at 7 kV/cm. PEF at 7 kV/cm + pressing at 14 kg/cm^2 shortened the extraction time by 18-fold
Coustets et al. (2013)	Microalgae proteins	$E = 3, 6$ kV/cm; $t = 2$ ms; $N = 15$	Bipolar	–	Continuous	Treatment did not cause cell lysis. No effect of cell wall characteristics on protein electroextractability
Luengo et al. (2013)	Orange peel polyphenols	$E = 1–7$ kV/cm; $f = 1$ Hz; pulse width = 3 us; $N = 5–50$	Square	Cylindrical methacrylate tube closed with two cylinders,	Batch	Highest cell disintegration index at 20 pulses; 3 μs. After 30 min of pressurization (5 bars), total polyphenol extraction yield increased 20, 129, 153 and 159% at 1, 3, 5 and 7 kV/cm

TABLE 13.2 *(Continued)*

Reference	Purpose	Processing Parameters	Circuit Configuration Pulses	Chamber	Mode	Achievements
Baier et al. (2015)	Pea tissue	$E = 5$ kV/cm; $N = 667$; pulse width = 10–15 µs; $f = 2$ Hz	NR	Prismatic, polyoxymethylene	Batch	Enhancement of oligosaccharides liberation and efficiency of drying and rehydration. Improvement of mass transfer without affecting protein quality
Coustets et al. (2015)	Microalgae cytoplasmic proteins	$E = 3, 6$ kV/cm; $t = 2$ ms; $N = 9$	Bipolar	–	Continuous	Increase of porosity and leackage of cytoplasmic proteins from *Nannochloropsis salina, Chlorella vulgaris and Haematococcus pluvialis*. Chosen electrical treatment did not cause cell lysis
López-Giral et al. (2015)	Grape phenolic compounds	$E = 4.6$ kV/cm; $t = 20$ us; $f = 400$ Hz	Square bipolar	Colinear	Continuous	Significant differences in content of anthocyanins in *Tempanillo* variety. Major differences in gallic acid content, catechin, and epicatechin in *Grenache* variety. No difference in liberation of phenolic compounds in *Graciano* variety

TABLE 13.2 *(Continued)*

Reference	Purpose	Processing Parameters	Circuit Configuration Pulses	Chamber	Mode	Achievements
Medina-Meza and Barbosa-Cánovas (2015)	Plum and grape peel Anthocyanins and flavonols	$E = 25$ kV/cm; 6 μs pulse width; $f = 10$ Hz	Exponential decay rectangular	–	Continuous	PEF more successful at larger diameter of chamber and residence time and number of pulses greater. Increase of anhtocyanins and flavonols from grape peels but deleterious for ascorbic acid
Segovia et al. (2015)	Borage polyphenols	$E = 1$–7 kV/cm; $f = 1$ Hz; pulse width = 3 μs; $N = 5$–50	Square	Cylindrical methacrylate tube	Batch	Increase in polyphenols extraction and antioxidant activities. Decrease in the extraction time
Shorstkii et al. (2015)	Sunflower seeds oil	$E = 1, 3, 5, 6, 7$ kV/cm; $f = 0.5, 1.5, 5, 10, 15$ Hz; $t = 10, 20, 30$ 40, 50 μs	Square	Two electrodes (top and bottom)	Batch	Oil yield increased by 9.1% at 7.0 kV/cm, 1.5 Hz, 30 μs
Bobinaité et al. (2016)	Blueberry juice	$E =$ 1, 3, 5 kV/cm; $f =$ 10 Hz; pulse width = 20 μs	Monopolar square	Cylindrical	Batch	Increase in juice yield (+32%), total anthocyanins content (+55%), and DPPH-RSC (+40%) at 1 and 3 kV/cm

TABLE 13.2 *(Continued)*

Reference	Purpose	Processing Parameters	Circuit Configuration Pulses	Chamber	Mode	Achievements
Dellarosa et al. (2016)	Water from apple parenchyma tissue	E = 0.1, 0.25, 0.4 kV/cm; 100 μs pulse width; N = 20, 60	Near rectangular	Chamber	Batch	Water migration from vacuole to cytoplasm from 15% to 40% with 20 and 60 pulses. 250 and 400 V/cm removed the possibility to distinguish the cell compartments
Gjörek et al. (2016)	Sugar beet cossettes sugar	E = 0.52–1.52 kV/cm; t = 0.1–10 ms; f = 1, 10 Hz	Rectangular	Cylindrical, two rigid electrodes	Batch	Increase of 49.58% in juice yield
Leong et al. (2016)	Grapes anthocyanins	E = 1.4kV/cm, 50 Hz, t = 20.66 ms; N = 1033; pulse width = 20 μs; f = 50 Hz	Square bipolar	Two electrodes	Batch	Increase of maldivin, delphinidin, and petunindin liberation after 48 h

TABLE 13.3 Examples of the Use of Pulsed Electric Fields in Drying, Freeze Drying, Freezing, and Degradation of Compounds

Reference	Purpose		Processing Parameters	Circuit Configuration Pulses	Chamber	Mode	Achievements
Lebovka et al. (2007)	Drying	Potato tissue	E = 1.5 kV/cm; N = 1–30,000; t = 10^{-5}, 10^{-3} s; f = 1000 Hz	Near-rectangular	Two concentric polypropylene rings, wire gauze electrodes	Batch	Moisture diffusivity increased with increased degree of PEF induced damage. Decrease of drying temperature at 20 °C
Huang Xiaoli et al. (2009)		Potato	E = 0.1 kV/cm; f = 40 Hz	–	–	–	Time of microwave drying of potato reduced by 23%
Lamanauskas et al. (2015)		Actinidia fruit	E = 1–10 kV/cm; f = 20 Hz; t = 120 s, pulse width = 20 µs	Monopolar square	Vertical spacer in polycarbonate, two cylindrical electrodes	Batch	Highest drying efficiency at 5 kV/cm. Reduction of drying time 2 times. Ascorbic acid content and color not affected
Won et al. (2015)		Red pepper	E = 1.0–2.5 kV/cm; f = 100 Hz; t = 1, 2, 4 s, pulse width = 30 µs	–	–	–	Reduction of 34.7 and 20.4% in drying time after 2.5 kV/cm and 4 and 1 s. Color quality increased

TABLE 13.3 *(Continued)*

Reference	Purpose		Processing Parameters	Circuit Configuration Pulses	Chamber	Mode	Achievements
Jalté et al. (2009)	Freeze-drying	Potato	$E = \leq 0.4$ k V/cm; $N = 2$, $t = 100$ μs; $f = 0.5$ kHz	Bipolar near-rectangular	Polypropylene cylindrical tube, 26 mm diameter	Batch	Improvement in the rate of freeze-drying. More uniform shape, clearer color, less shrinkage and visually better quality of freeze-dried samples
Wu and Zhang, (2014)		Potato	$E = 1.5$ kV/cm; 120 μs, $N = 45$	–	–	–	Efficiency per productivity area increased by 32.28%, specific energy consumption decreased by 16.59%, drying time shortened by 31.47%, drying time rate improved by 14.31%
Dymek et al. (2015)	Freezing	Spinach baby leave	$E = 350$ V; $t = 100$ μs; $N = 25$	Exponential decay	Two parallel electrodes	Batch	Significant leakage of solutes and cytoplasmic constituents into extracellular spaces. Ice propagation rate of leaves impregnated with water decreased four-fold.

TABLE 13.3 *(Continued)*

Reference	Purpose		Processing Parameters	Circuit Configuration Pulses	Chamber	Mode	Achievements
Mok et al. (2015)		0.9% NaCl	$E = 1.78$ V/cm; $f =$ 1–20 kHz	Bipolar square	Parallel electrodes	Batch	PEF + static magnetic field synergistic process.
Wictor et al. (2015)		Apple tissue	$E = 0$, 1.85, 3, 5 kV/cm; $N = 0$, 10, 50, 100; pulse width = 0, 15, 20, 24 μs	Monopolar exponential	Parallel electrodes	Batch	Freezing time reduction by 3.5–17.2%. Phase transition stage during freezing up to 33% shorter than untreated. Thawing time reduced by 71.5%. Maximal mass loss at 5 kV/cm, N = 50.
[a]Chen et al. (2009)		Apple juice residual pesticide	$E = 8$, 12, 16, 20 kV/cm; $N = 6$, 9, 12, 19, 26, $f = 1$ Hz; 10 μs pulse width	Exponential decay	Round parallel-plate electrodes	Continuous	Degradation of methamidophos and chlorpyrifos. Chlorpyrifos more labile to PEF than methamidophos
Zhang et al. (2012)	Degradation	Organo-phosphorus pestidices from apple juice	$E = 20$ kV/cm; $t =$ 260 μs; $f =$ 1Hz; 10 μs pulse width	Exponential decay	Round parallel-plate electrodes	Continuous	Diazinon degraded 47.6% and dimethoate 34.7%

and total carotenoid and flavanone contents were not modified after processing. Individual carotenoids content expressed in μg/100 mL was 142.27 for β-cryptoxanthin, 49.47 for α-cryptoxanthin, 284.25 for zeaxanthin, 304.37 for lutein, 44.79 for β-carotene, and 26.92 for α-carotene. Vitamin A content was 12.84 retinol activity equivalents/100 mL. In another study, Sánchez-Moreno et al. (2005) reported the effects on the antioxidant capacity of orange juice in terms of the percentage of inhibition of DPPH* (2,2-diphenyl-1-picrylhydrazyl). Studies were carried out with monopolar and bipolar pulses. Electric field strength was also varied (15, 25 y 35 kV/cm) and treatment time was 100, 400, and 1000 μs. Results were compared to those obtained after thermal pasteurization (90 °C, 1 min) and it was concluded that PEF treatments did not affect the antioxidant activity of orange juice (197.79 mL/g of DPPH*) meanwhile heat-treated products showed lower amount (183.14 mL/g of DPPH*). Aguilar-Rosas et al. (2007) pasteurized apple juice extracted from golden delicious fruits by means of PEF (bipolar pulse, 4 μs, 35 kV/cm, 1.2 kHz). Minimal variability in pH was observed and no significant changes were detected in acidity compared to thermal treatment (90 °C, 30 s). PEF increased the retention phenolics and most volatile compounds responsible for color, and flavor of apple juice. In another research, Sampedro et al. (2009) detected a loss in volatile compounds of orange juice after PEF processing (13.7–8.3% at 25 °C, 5.8–21.0% at 45 °C and 11.6–30.5% at 65 °C).

2. A sensory quality evaluation of control (without treatment), PEF treated, heating sterilized, and PEF treated combined with freeze concentrated orange juice was reported by Zhao et al. (2008). Preference of different treated juices was ranked according to preference as follows: PEF treated > PEF combined with frozen concentrated processing > heating-treated > control. This study indicated that PEF treated orange juice had the highest level of acceptance compared with traditional thermal treatments
3. Due to the good performance of PEF treatment in flavor and nutrients retention, Chen et al. (2009b) integrated the PEF processing with freeze concentration to concentrate fruit juice. Sensory analysis results indicated that the juice produced by freeze concentration combined with PEF had higher quality than pasteurized juice according to palatability.

4. Shelf life and sensory attributes of orange juice after the combination of batch thermosonication (55 °C, 10 min) and continuous PEF (40 kV/cm, 150 μs) was studied by Walkling-Ribeiro et al. (2009). Color, odor, sweetness, acidity, flavor, and overall acceptability of different orange juices processed with combined thermosonication and PEF and control (high- temperature-short-time pasteurization 94 °C, 26 s) were evaluated. All sensory attributes were rated equivalent for both treatments. No significant change in the physical properties was detected after both treatments during 168 days of shelf life ($p \geq 0.05$). The color attributes of orange juice were observed to be initially better for thermosonic and PEF treated juice, but during long term storage conventional pasteurization proved to be more consistent with regard to color stability.
5. Watermelon juice also has been processed by PEF (Aguiló-Aguayo et al., 2010). The treatment conditions were 35 kV/cm, 1727 μs, bipolar pulses (4 μs, 188 Hz). PEF treatment induced a rise (20%) in the concentrations of hexanal, (E)-2-nonenal, nonanal, 6-methyl-5-hepten-2-one, and geranylacetone than thermal treatment at 90 °C for 60 s. Moreover, less reductions in the retention of volatiles than thermal treatment were found. PEF treated juices showed better flavor retention than heat-treated samples for at least 21 days of storage. The stability of main carotenoids and flavanones after PEF treatment during 40 days of refrigerated storage of orange juice at 40 °C was investigated by Plaza et al. (2011). No significant changes of flavanone content ($p < 0.05$) compared to fresh orange juice were observed.
6. Mosqueda-Melgar et al. (2012) combined PEF treatment with antimicrobials such as citric acid and cinnamon bark oil to improve the microbiological shelf life of strawberry, orange, apple, pear, and tomato juices. The extension of the microbiological shelf life of fruit juices treated by PEF with or without antimicrobial substances was observed in comparison with nonprocessed juices. Strawberry and orange PEF treated juices did not show microbial growth along 91 days of storage at 5 °C. Resident microbial populations in apple, pear, and tomato juices only were controlled during that time of PEF combined with antimicrobials. No significant changes in the sensory attributes in all studied fruit juices PEF treated were found, but this was not the case for PEF treatment combined with antimicrobials, as changes in aroma, taste, and sourness were perceived. Barba et

al. (2012) evaluated the quality changes of blueberry juice during refrigerated storage (during 56 days at 4 °C) after PEF processing at 36 kV/cm and 100 μs. After treatment, the juices showed a decrease lower than 5% in ascorbic acid content compared to the untreated one. At the end of refrigerated storage, unprocessed and PEF treated juices showed similar ascorbic acid losses of 50% in relation to untreated juice. The juices exhibited fluctuations in total phenolic values with a decrease after 7 days in refrigerated storage. The feasibility of pasteurizing raw pomegranate juice in a commercial scale PEF processing system was studied by Guo et al. (2014). Processing conditions were 35 and 38 kV/cm for 281 μs at 55 °C with a flow rate of 100 L/h. No significant p (0.05) differences in pH and °Brix values between the PEF-treated juice and the unprocessed juice were detected. Moreover, PEF treatment did not alter total phenolics contents and anthocyanins compared to unprocessed juice. PEF treatment had less impact on the color of pomegranate juice than thermal processing. PEF-treated juice had the same consumer satisfaction scores as the unprocessed juice, which were higher than thermally processed juice.

13.3.1.2 FRUIT PUREES AND SMOOTHIES

1. PEF effect on smoothies was investigated by Walkling-Ribeiro (2010). The author combined moderate heat (55 °C after 60 s) and PEF (34 kV/cm, 60 μs). The microbiological shelf life of smoothies stored at 4 °C after the combined treatment was extended (21 days) compared to mildly pasteurized (14 days) at 72 °C, 15 s. Combined treatment caused better stability of Brix and viscosity than mild pasteurization, but better color stability was reported by mild pasteurized smoothies. Sensory evaluation of all samples indicated the overall acceptability of the products ($p \geq 0.05$).
2. A process for strawberry puree pasteurization in which a pilot plant PEF system at field strengths of 24.9–36.6 kV/cm was developed (Geveke et al., 2015). Outlet temperatures of 45.0–57.5 °C were detected and a flow rate of 100 L/h. After PEF processing, populations of *E coli* in the strawberry puree were reduced by 6.5 log in BPW at 30 kV/cm and 57.5 °C and 7.3 log at 24 kV/cm and 52.5 °C. Taste and color of strawberry beverage initially made from PEF

processed puree was fresh and bright red. This color remained for the first 3 months-equivalent of storage and a slight drop in flavor was detected.

13.3.1.3 MILK BEVERAGES

1. A mixed orange juice and milk beverage fortified with water-soluble vitamins, such as biotin, folic acid, pantothenic acid, and riboflavin, was developed by Rivas et al. (2007). Angiotensin-l-converting enzyme (ACE) inhibitory peptides were also added. The beverage was subjected to PEF (15–40 kV/cm; 0–700 μs) and thermal processing, at 84 and 95 °C and heating time 15–120 s. The effects of treatments and storage at 4°C for 81 days were evaluated and it was confirmed the stability of the vitamins and the ACE inhibitory activity after PEF processing and during storage.
2. Flavored strawberry milk also has been studied (Bermúdez-Aguirre et al., 2010). The aim was to investigate the degradation of coloring agent Allura Red at 40 kV/cm, 48 pulses (2.5 μs), and 55 °C. After PEF processing, only minor changes were observed in color, Allura Red concentration and pH. During storage for 32 days at 4 °C, the concentration of Allura red changed, reaching a maximum value during the middle of storage, attributed to microbial growth, pH reduction, or interaction to proteins.
3. Another milk beverage, but in this case orange juice-based was reported by Zulueta et al. (2010). The effect of PEF (15, 25, 35, and 40 kV/cm) and time (40–700 μs) on the carotenoid and vitamin A profile was assessed. Results were compared to thermal pasteurization and they concluded that PEF treatment enhanced the concentration of carotenoids. An increase of carotenoids at 15 kV/cm were found, whereas thermal pasteurization caused a reduction in total carotenoid concentration. Morales-de la Peña et al. (2012) evaluated the effects of PEF on the free AA profile of a fruit juice–soy milk beverage. Treatment was carried out at 35 kV/cm with 4 μs bipolar pulses at 200 Hz for 800 and 1400 μs. Immediately after 800 μs, no significant changes on individual free AA content were observed, except for Val. Total free AA content of PEF 800 μs treated beverage was similar to the untreated (64.50–64.79 mg/100 mL), whereas 1400 μs PEF or heat-treated beverages presented lower values (60.25–61.82 mg/100

mL). After 1400 μs or heat treatments, Glu, Gly, Tyr, Val, Leu, Phe, Lys, and Ile content was reduced, whereas the concentration of Arg, Ala, and Met slightly increased.

13.3.1.4 OIL

1. Arbequina olive paste for olive oil extraction was processed by PEF (Abenoza et al., 2013). Treatment consisted of 0–2 kV/cm, different malaxation times (0, 15, and 30 min), and temperature 15 °C and 26 °C. The extraction yield was improved by 54% after PEF treatment at 2 kV/cm without malaxation. Acidity, peroxide value, K_{232}, and K_{270} were not affected by PEF treatments, and any bad flavor or taste in the oil was generated after PEF, according to sensory analysis.
2. In another study, Arroniz olive oil production was investigated (Puértolas and Martínez-de-Marañón, 2015). Processing conditions were 2 kV/cm and 11.25 kJ/kg. The extraction yield was increased by 13.3% compared to a control. Olive oil obtained by means of PEF showed total phenolic content, total phytosterols, and total tocopherols significantly higher than control (11.5%, 9.9%, and 15.0%, respectively). The use of PEF had no negative effects on chemical and sensory characteristics of olive oil, maintaining the highest quality according to EU legal standards.

13.3.1.5 PLANT TISSUES

1. Two methods for PEF application at 400 kV/cm on whole apples before cutting and PEF treatment of apple slices after cutting were explored (Grimi et al., 2011). After PEF treatment, juice yield increased from 44 g/100 g apple (untreated samples) to 58 g/100 g apple (whole apples) and 64 g/100 g apple (slices) after 30 min of pressing. Conductivity and pH of untreated and PEF treated samples showed no significant difference. Total soluble matter content of juice increased after PEF treatment. The use of PEF improved the juice clarity and polyphenols and antioxidant capacities of juice were increased after PEF pre-treatment. Juice clarity and content of antioxidants were higher for the whole treated apples compared to untreated samples and PEF-treated apple slices. PEF treated accelerated browning, being more noticeable in whole samples.

2. Wiktor et al. (2015) analyzed the impact of PEF on bioactive compound content and color of plant tissue of apple and carrot tissues. Treatments were carried out at 0, 1.85, 3, and 5 kV/cm and 0, 10, and 50 pulses. The total color change of carrot tissue after PEF treatment immediately and after 60 min was smaller (Delta *E*, 0 min = 1.64–5.51; Delta *E*, 60 min = 133–3.91) than apple samples (Delta *E*, 0 min = 0.48–7.20; Delta *E*, 60 min = 1.25–21.87). After PEF treatment at 1.85 kV/cm total carotenoid content was increased to up to 11.34%. At this voltage, the maximal increase of total polyphenolic content and antioxidant activity was observed for apple tissue treated by 10 pulses.
3. Recently, Liu et al. (2019) studied the effect of electroporation on bunching onion bulb tissues. It was found that PEF induced disruption at the cellular level, which was detected by carbohydrate leakage, however, no structural changes at organ level were observed. It was concluded that PEF can be used to manipulate fructan contents in plant-based foods such as onion bulbs.

13.3.1.6 MEAT

1. Meat quality attributes such as weight loss, color, cook loss, and texture of beef (*longissimus thoracis et lumborum*) after PEF treatment (1.4 kV/cm, 10 Hz, 20 μs, 300 and 600 pulses) were studied by Arroyo et al. (2015a). No detrimental effect on cook loss, storage loss, and color regardless of the length of aging before PEF (2, 10, 18, and 26 days) was found. A tendency on reducing toughness was observed after PEF treatment, but the process did not affect the tenderization by means of aging itself. Moreover, PEF treated meat was scored with similar odor and better texture than untreated beef. In addition, in another study, the risk of off-flavor development on PEF in turkey meat tenderization was investigated (Arroyo et al., 2015b). A batch chamber with increasing electric field strength up to 3 kV/cm was used. Lipid oxidation in all PEF-treated samples progressed at the same rate with storage as the untreated samples. PEF did not induce differences in weight loss, cook loss, lipid oxidation, texture, and color. After a sensory evaluation, panelists detected slight differences between the PEF-treated samples and the controls

in terms of texture and odor. Table 13.4 shows details of the sensory analysis carried out to different products after PEF processing.

13.4 CONSUMER ACCEPTANCE

1. The food industry is considered as rapidly changing and the market is delimited by mergers and acquisitions (Barbosa-Cánovas, 1999). PEF-processed fruit juices were introduced to the US market in 2005, and no dangerous chemical reactions have been detected, a fact that makes PEF process considered safe (Frewer et al., 2011). Nielsen et al. (2009) performed a qualitative study on consumer attitudes toward PEF for juice and baby food. Respondents were 97 adults (age, 20–71) in Slovenia, Hungary, Serbia, Slovakia, Norway, and Denmark. Several baby food buyers had a positive attitude toward the higher price of PEF products, but juice buyers saw a higher price as negative. As high price indicates high quality in most of the cases, expensive baby food is sometimes preferred by consumers and the high price of baby food processed by PEF was accepted. Studies about consumers' opinions toward disposition to pay for PEF processed orange juice was carried out, and they indicated that unprocessed fresh juice had the highest mean price for sale. The second highest mean price was for PEF juice and it was followed by the not from concentrate option, and the lowest was from concentrate. Another study with 120 consumers in Norway, Denmark, Hungary, and Slovakia showed that PEF products were perceived to have advantages but some consumers appeared unsure about the risks (Olsen et al., 2010). Frewer et al. (2011) also studied consumer acceptance regarding PEF technology and they reported that perceived risks were rare and not high related to safety and allergenicity. Moreover, they found that consumers had negative attitudes toward food producers, such as association with irradiation and microwave ovens, with fear of electricity or electrical impulses and their consequences. Results showed also that consumers had low familiarity with PEF processing, and they expressed concern about long-term consequences.
2. In general, East-European consumers were more concerned about electrical impulses than Northern-European consumers. Jaeger et al. (2015) studied the consumers' attitude toward pulsed electric

TABLE 13.4 Sensory Analysis of Products After PEF Processing

Products		Type Test	N	Panel	Age	References
Juice	Orange	Attributes (Color, odor, sweetness, acidity, flavor)	37 (16F, 21 M)	Untrained	18–65	Walkling-Ribeiro et al. (2009)
	Strawberry, orange, apple, pear, tomato	Attributes (Aroma, color, taste, sourness)	30	Untrained	NR	Mosqueda-Melgar et al. (2012)
	Pomegranate	Attributes (Overall appearance, overall flavor, pomegranate flavor, acceptability)	30	Untrained	NR	Guo et al. (2014)
Smoothie	Pineapple, banana, Apple, orange, coconut milk	Attributes (Color, odor, sweetness, acidity, flavor, overall acceptability)	35 (19F, 16 M)	Untrained	18–65	Walkling-Ribeiro, (2010)
Puree	Strawberry puréé	Comments on appearance and flavor	NR	NR	NR	Geveke et al. (2015)
Oil	Olive	Negative and positive attributes (Fruity, bitter, pungent, fusty-muddy sediment, musty-humid-earthy, winey-vinegary-acid-sour, metallic, rancid, other)	10	Trained	NR	Abenoza et al., (2013)
	Olive	Negative and positive attributes (Fruity, bitter, pungent, fusty, musty/humid, winey-vinegary/acid sour, metallic, rancid, heated, hay/wood, rough, greasy, vegetable water, brine, esparto, earthy, grubby, cucumber, wet wood, other	12	Trained	NR	Puértolas and Martínez-de-Marañón, (2015)

TABLE 13.4 *(Continued)*

Products		Type Test	N	Panel	Age	References
Meat	Beef	Attribute (Tenderness)	20 (Balanced gender)	Untrained	Balanced	Arroyo et al. (2015a)
	Turkey breast	Discrimination (Color, tenderness, odor) Acceptance (Tenderness)	40	Untrained	NR	Arroyo et al. (2015b)

fields by means of surveys to elder age people (45 ≤ years), young adults (25–40 years), health-conscious consumers and experts (13) in technology, legislation, food policy, consumer, and nutrition sciences. Representatives of professional and consumer organizations and decision makers also were included in the study. The authors included a new term "micropulse," because it was known that "pulsed electric fields" terminology was associated to fear of electricity for consumers. Distrust, uncertainty, and fear were associated with pulsed electric fields. Consumers also associated PEF with radiation, roentgen, and electricity. Moreover, 25% more of doubts were reported compared to micropulse, and mostly negative associations, such as complicated, expensive, and dangerous, were found regarding the characteristics of PEF. After using the new terminology "micropulse," more neutral associations were found than in the case of PEF. Associations (30%) were regarding microwave, heating, and vibration. Respondents (23%) had associations of inquisitiveness, accepting, and uncertainty. Consumers expressed that micropulse is the technology of the future, has a small effect, it is rapid, complicated and expensive. According to experts, the name of pulsed electric fields is inappropriate, and "micro pulse" is not informative enough. Based on results, experts proposed the term "microelectrical pulse" for this technology.

3. China also was subjected to the consumer's perception. A study was conducted in Hangzhou, Zhejiang Province, which included respondents (44 males and 56 females; ≥ 20 years) from six different companies. Consumers of healthy beverages revealed that the dominant consumers' perceptions toward PEF were associated to radiation, lack of information, uncertainty, electricity, environmental friendly, distrust, unsafety, fear of electricity, harmful side effects to health, microorganisms destruction, alteration of original state and flavor of the products, destruction of good and bad bacteria, and nutritional loss. Most of the Chinese respondents (96%) were willing to purchase the products if the PEF treated beverages price were to be 10% below the current market price. Ninety percent of the respondents were willing to purchase the PEF treated product if sold at a 10% premium compared to heat treatment once detailed information about the technology was provided (Lee et al., 2015).
4. Studies about the consumer's opinions regarding PEF in other countries are still lacking. More research is needed to know the perception

of people from different places to develop a market strategy to place PEF treated products in the global market.

13.5 ENVIRONMENTAL IMPACT

1. PEF technology avoids the use of natural gas and boilers, the lack of steam generators could reduce the wastewater, increasing the water and energy savings. The partial reduction of cooling system requirements represents 50% of the total electricity consumption. If high-temperature treatments combined with synergetic heat effects take place, the required energy for PEF operation can be reduced close to the amount of energy needed for thermal pasteurization. Input could be approximate 20 kJ/kg, which is the energy needed for thermal pasteurization (Toepfl et al., 2006). Because of these reasons, PEF processing could be an alternative for energy saving, causing a positive impact on the environment (Pereira and Vicente, 2010). Although PEF could be classified in the waste-free process because the lower cooling requirements, a study reveals that the equivalent CO_2 annual emission for PEF pasteurization was about 700,000 kg, 777 times higher than for thermal pasteurization (90,000 kg) due to the higher electricity consumption (Sampedro et al., 2014).

13.6 LEGISLATION FRAMEWORKS

1. The Food and Drug Administration (FDA) expressed the need to regulate the fruit and vegetable juice industry by way of HACCP in 1997. However, to establish full regulations about PEF technology, some research was needed, and more statements needed to be added (Barbosa-Cánovas, 1999). Three regulatory issues were described outside the code of federal regulations (CFR), title 114, including compliance with good manufacturing practices. They suggested a demonstration of a minimum 5 log cycle reduction of pathogenic microorganisms of usual concern and insurance that no harmful substances are present in the product as a result of the process (Barbosa-Cánovas, 1999). Years later, the FDA made public information in which no objection for PEF was declared (Vega-Mercado et al., 2004).

2. On its part, the European Union requires that PEF treatment fulfills the requirements of the Novel Food Regulation (NFR)14, specifically to the principle of substantial equivalence. This principle indicates that safety can be assumed if no additives are introduced and there is no significant difference of treated product when compared to the untreated one, in a novel process (Mastwijk and Bartels, 2004). Roodenburg et al. (2005) indicated that PEF treated orange juice presented dissolved metals corresponding to the stainless-steel electrodes of chambers. These identified metals were iron (13–15 $\mu g\ L^{-1}$), nickel (0.7 $\mu g\ L^{-1}$). However, they found that the metal concentrations did not exceed the legislation values for fruit juices and the EU Drinking Water Directive for human consumption.

13.7 REMARKS

1. Consumer's demand for minimally processed foods with an increased quality indicates that nonthermal technologies such as PEF will have more opportunities to be implemented in the food industry in the next years. Characteristics of PEF treated products can be compared to the obtained after thermal treatment, and the nutritional, component, and sensory analysis indicated that PEF processing avoids the loss of important bioactive compounds. Consumer surveys pointed out that information about PEF operation and safety needs to be spread to increase the confidence in this technology. More countries need to be considered for the consumer's opinion research.

KEYWORDS

- **electroporation**
- **nonthermal**
- **PEF**
- **food quality**
- **consumers' opinions**
- **sensorial attributes**

REFERENCES

Abenoza, M.; Benito, M.; Saldaña, G.; Álvarez, I.; Raso, J.; Sánchez-Gimeno, A.C. Effects of Pulsed Electric Field on Yield Extraction and Quality of Olive Oil. *Food Bioprocess. Tech.* 2013, 6, 1367–1373.

Aguilar-Rosas, S.F.; Ballinas-Casarrubias, M.L.; Nevarez-Moorillon, G.V.; Martin-Belloso, O.; Ortega-Rivas, E. Thermal and Pulsed Electric Fields Pasteurization of Apple Juice: Effects on Physicochemical Properties and Flavour Compounds. *J. Food Eng.* 2007, 83, 41–46.

Aguiló-Aguayo, I.; Soliva-Fortuny, R.; Martín-Belloso, O. Optimizing Critical High-intensity Pulsed Electric Fields Treatments for Reducing Pectolytic Activity and Viscosity Changes in Watermelon Juice. *Eur. Food Res. Technol.* 2010, 231, 509–517.

Álvarez, I.; Condón, S.; Raso, J. Microbial Inactivation by Pulsed Electric Fields. In *Pulsed Electric fields for the Food Industry*; Raso, J.; Heinz, V., Eds.; Springer US: New York, 2006.

Arroyo, C.; Lascorz, D.; O'Dowd, L.; Noci, F.; Arimi, J.; Lyng, J.G. Effect of pulsed Electric Field Treatments at Various Stages During Conditioning on Quality Attributes of Beef *longissimus thoracis et lumborum* Muscle. *Meat Sci.* 2015a, 99, 52–59.

Arroyo, C.; Eslami, S.; Brunton, N.P.; Arimi, J.M.; Noci, F.; Lyng, J.G. An Assessment of the Impact of Pulsed Electric Fields Processing Factors on Oxidation, Color, Texture, and Sensory Attributes of Turkey Breast Meat. *Poult. Sci.* 2015b, 94, 1088–1095.

Ayhan, Z.; Yeom, H.W.; Zhang, Q.H.; Min, D.B. Flavor, Color, and Vitamin C Retention of Pulsed Electric Field Processed Orange Juice in Different Packaging Materials. *J. Agr. Food Chem.* 2001a, 49, 669–674.

Ayhan, Z.; Streaker, C.B.; Howard-Zhang, Q. Design, Construction and Validation of a Sanitary Glove Box Packaging System for Product Shelf-life Studies. *J. Food Process. Pres.* 2001b, 25, 183–196.

Barba, F.J.; Jäger, H.; Meneses, N.; Esteve, M.J.; Frígola, A.; Knorr, D. Evaluation of Quality Changes of Blueberry Juice During Refrigerated Storage After High-pressure and Pulsed Electric Fields Processing. *Innov. Food Sci. Emerg.* 2012, 14, 18–24.

Barba, F.J.; Parniakov, O.; Pereira, S.A.; Wiktor, A.; Grimi, N.: Boussetta, N.; Saraiva, J.A.; Raso, J.; Martin-Belloso, O.; Witrowa-Rajchert, D.; Lebovka, N.; Vorobiev, E. Current Applications and New Opportunities for the Use of Pulsed Electric Fields in Food Science and Industry. *Food Res. Int.* 2015, 77, 773–798.

Barbosa-Cánovas, G. *Preservation of Foods with Pulsed Electric Fields*. Academic Press: San Diego, 1999.

Barbosa-Cánovas, G.; Altunakar, B. *Pulsed Electric Processing of Foods, An Overview*. In *Pulsed Electric Fields Technology for the Food Industry*; Raso, J.; Heinz, V., Eds.; Springer: New York, 2006; pp 3–26.

Berk, Z. *Food Process Engineering and Technology*, 2nd ed.; Academic Press: San Diego, London, 2013.

Bermúdez-Aguirre, D.; Yáñez, J.A.; Dunne, C.P.; Davies, N.M.; Barbosa-Cánovas, G.V. Study of Strawberry Flavored Milk Under Pulsed Electric Field Processing. *Food Res. Int.* 2010, 43, 2201–2207.

Bi, X.; Liu, F.; Rao, L.; Li, J.; Liu, B.; Liao, X.; Wu, J. Effects of electric field strength and pulse rise time on physicochemical and sensory properties of apple juice by pulsed electric field. *Innov. Food Sci. Emerg.* 2013, 17, 85–92.

Buckow, R.; Ng, S.; Toepfl, S. Pulsed Electric Field Processing of Orange Juice: A Review on Microbial, Enzymatic, Nutritional, and Sensory Quality and Stability. *Compr. Rev. Food Sci. F.* 2013, 12, 455–467.

Chen, F.; Zeng, L.Q.; Zhang, Y.Y.; Liao, X.; Ge, Y.; Hu, X.; Jiang, L. Degradation Behavior of Methamidophos and Chlorpyrifos in Apple Juice Treated with Pulsed Electric Fields. *Food Chem.* 2009a, 112, 956–961.

Chen, M.; Gao, M.; Gong, X.; Chen, J.; Chen, Y. Integration of Freeze Concentration and PEF in the Processing of Fruit Juices. *Agr. Eng.* 2009b, 25, 237–241.

de Carvalho, R.J; de Souza, G.T; Pagán, E.; García-Gonzalo, D.; Magnani, M.; Pagán, R. Nanoemulsions of *Mentha piperita* L. Essential Oil in Combination with Mild Heat, Pulsed Electric Fields (PEF) and High Hydrostatic Pressure (HHP) as an Alternative to Inactivate *Escherichia coli* O157: H7 in Fruit Juices. *Innov. Food Sci. Emerg.* 2018, 48, 219–227.

Donsi, F.; Ferrari, G.; Pataro, G. Applications of Pulsed Electric Field Treatments for the Enhancement of Mass Transfer from Vegetable Tissue. *Food Eng. Rev.* 2010, 2, 109–130.

Dunn, J. Pulsed Electric Field Processing: An Overview. *In Pulsed Electric Fields in Food Processing, Fundamentals Aspects and Applications*; Barbosa-Cánovas, G.; Zhang, Q.H., Eds.; Technomic Press, Lancaster, 2001; pp 1–30.

Frewer, L.J.; Bergmann, K.; Brennan, M.; Lion, R.; Meertens, R.; Rowe, G.; Siegrist, M.; Vereijken, C. Consumer Response to Novel Agri-food Technologies: Implication for Predicting Consumer Acceptance of Emerging Food Technologies. *Trends Food Sci. Tech.* 2011, 22, 442–456.

Geveke, D.J.; Aubuchon, I.; Zhang, H.Q.; Boyd, G.; Sites, J.E.; Bigley, A.B.W. Validation of a Pulsed Electric Field Process to Pasteurize Strawberry Purée. *J. Food Eng.* 2015, 166, 384–389.

Grimi, N.; Mamouni, F.; Lebovka, N.; Vorobiev, E.; Vaxelaire, J. Impact of Apple Processing Modes on Extracted Juice Quality: Pressing Assisted by Pulsed Electric Fields. *J. Food Eng.* 2011, 103, 52–61.

Gurtler, J.; Rivera, R.; Zhang, H.; Geveke, J. Selection of Surrogate Bacteria in Place of *Escherichia coli* O157:H7 and *Salmonella Typhimurium* for Pulsed Electric Field Treatment of Orange Juice. *Int. J. Food Microbiol.* 2010, 139, 1–8.

Guo, M.; Jin, T.Z.; Geveke, D.J.; Fan, X.; Sites, J.E.; Wang, L. Evaluation of Microbial Stability, Bioactive Compounds, Physicochemical Properties, and Consumer Acceptance of Pomegranate Juice Processed in a Commercial Scale Pulsed Electric Field System. *Food Bioprocess Tech.* 2014, 7, 2112–2120.

Hamilton, W.A.; Sale, A.J.H. Effects of High Electric Fields on Microorganisms: II. Mechanism of Action of the Lethal Effect. *BBA-Gen. Subjects.* 1967, 148, 789–800.

Han, Z.; Cai, M.; Cheng, J.; Sun, D. Effects of Electric Fields and Electromagnetic Wave on Food Protein Structure and Functionality: A Review. *Trends Food Sci.* Tech. 2018, 75, 1–9.

Holdsworth, S.D.; Simpson, R. *Thermal Processing of Packaged Foods*, 2nd ed.; Springer: New York, 2007.

Holsinger, V.H.; Rajkowski, K.T.; Stabel, J.R. Milk Pasteurization and Safety: A Brief History and Update. *Rev. Sci. Tech. Off. Int. Epiz.* 1997, 16, 441–451.

Hong, J.; Zeng, X.; Han, Z.; Brennan, C.S. Effect of Pulsed Electric Fields Treatment on the Nanostructure of Esterified Potato Starch and their Potential in Glycemic Digestibility. *Innov. Food Sci. Emerg.* 2018, 45, 438–446.

Huang, K.; Wang, J. Designs of Pulsed Electric Fields Treatment Chambers for Liquid Foods Pasteurization Process: A Review. *J. Food Eng.* 2009, 95, 227–239.
Huang, X.; Yang, W.; Wang, N. Pulsed Electric Field Pretreatment on Potato Microwave. Drying Characteristics. *Proc. Agri. Prod.* 2009, 3, 190–192.
Jaeger, H.; Meneses, N.; Knorr, D. Impact of PEF Treatment Inhomogeneity Such as Electric Field Distribution, Flow Characteristics and Temperature Effects on the Inactivation of *E. coli* and Milk Alkaline Phosphatase. *Innov. Food Sci. Emerg.* 2009, 10, 470–480.
Jaeger, H.; Knorr, D.; Szabó, E.; Hámori, J.; Bánáti, D. Impact of Terminology on Consumer Acceptance of Emerging Technologies Through the Example of PEF Technology. *Innov. Food Sci. Emerg.* 2015, 29, 87–93.
Khadre, M.A.; Yousef, A.E. Susceptibility of Human Rotavirus to Ozone, High Pressure, and Pulsed Electric Field. *J. Food Prot.* 2002, 65, 1441–1446.
Knorr, D.; Froehling, A.; Jaeger, H.; Reineke, K.; Schlueter, O.; Schoessler, K. Emerging Technologies in Food Processing. *Annu. Rev. Food Sci. Technol.* 2011, 2, 203–235.
Lee, P.Y.; Lusk, K.; Mirosa, M.; Oey, I. Effect of Information on Chinese Consumers' Perceptions and Purchase Intention for Beverages Processed by High Pressure Processing, Pulsed-Electric Field and Heat Treatment. *Food Qual. Prefer.* 2015, 40, 16–23.
Lee, S.J.; Bang, I.H.; Choi, H.; Min, S.C. Pasteurization of Mixed Mandarin and Hallabong tangor Juice Using Pulsed Electric Field Processing Combined with Heat. *Food Sci. Biotechnol.* 2018, 27, 669–675.
Li, J.; Wei, X.; Xu, X.; Wang, Y. In *Bacteria Inactivation by PEF with Coaxial Treatment Chamber and Tube-plate Treatment Chamber* Proceedings of the 9th International Conference on Properties and Applications of Dielectric Materials, Harbin, China, July 19–23, 2009; Zhongguo, D.G.J.S.X.H.; Zhongguo, D.J.G.C.X.H. Eds.; Piscataway, New Jersey, 2009. .
Li, Y.Q.; Chen, Z.X.; Mo, H.Z. Effects of Pulsed Electric Field on Physicochemical Properties of Soybean Protein Isolates. *LWT-Food Sci. Technol.* 2007, 40, 1167–1175.
Liu, T.; Burritt, D.J.; Oey I. Understanding the Effect of Pulsed Electric Fields on Multilayered Solid Plant Foods: Bunching Onions (*Allium fistulosum*) as a Model System. *Food Res. Int.* 2019, 120, 560–567.
Liu, Z.; Song, Y.; Guo, Y.; Wang, H.; Wu Z. Influence of Pulsed Electric Field Pretreatment on Vacuum Freeze-dried Apples and Process Parameter Optimization. *Adv. J. Food Sci. Technol.* 2017, 13, 224–235.
Liu, F.; Sun, J.; Li, J.; Liao, X. Review on Novel Application of Pulsed Electric Fields in Food Processing. *Food Ferment. Indus.* 2010, 36, 138–42.
López, N; Puértolas, E.; Condón, S.; Raso, J.; Álvarez, I. Enhancement of the Solid-Liquid Extraction of Sucrose from Sugar Beet (*Beta vulgaris*) by Pulsed Electric Fields. *LWT-Food Sci. Technol.* 2009a, 42, 1674–1680.
López, N.; Puértolas, E.; Condón, S.; Raso, J.; Álvarez, I. Enhancement of the Extraction of Betanine from Red Beetroot by Pulsed Electric Fields. *J. Food Eng.* 2009b, 90, 60–66.
Mahnič-Kalamiza, S.; Vorobiev E.; Miklavčič D. Electroporation in Food Processing and Biorefinery. *J. Membrane Biol.* 2014, 247, 1279–1304.
Mañas, P.; Pagán R. A Review Microbial Inactivation by New Technologies of Food Preservation. *J. Appl. Microbiol.* 2005, 98, 1387–1399.
Mastwijk, H.C.; Bartels, P.V. Pulsed Electric Field (PEF) Processing in the Fruit and Dairy Industries. *Int. Rev. Food Sci. Technol. IUFoST.* 2004, 106–108.

McNamee, C.; Noci, F.; Cronin, D.A.; Lyng, J.G.; Morgan, D.J.; Scannell, A.G.M. PEF based Hurdle Strategy to Control *Pichia fermentans*, *Listeria innocua* and *Escherichia coli* k12 in Orange Juice. *Int. J. Food Microbiol.* 2010, 138, 13–18.

Medina-Meza, I.G.; Barbosa-Cánovas, G.V. Assisted Extraction of Bioactive Compounds from Plum and Grape Peels by Ultrasonics and Pulsed Electric Fields. *J. Food Eng.* 2015, 166, 268–275.

Meza-Jiménez, M.L.; Pokhrel, P.; Robles de la Torre, R.R.; Barbosa-Cánovas, G.V.; Hernández-Sánchez, H. Effect of Pulsed Electric Fields on the Activity of Food-grade Papain in a Continuous System. *LWT-Food Sci. Technol.* 2019, 109, 336–341.

Morales-de la Peña, M.; Salvia-Trujillo, L.; Garde-Cerdán, T.; Rojas-Graü, M.A.; Martín-Belloso, O. High Intensity Pulsed Electric Fields or Thermal Treatments on the Amino Acid Profile of a Fruit Juice Soy-milk Beverage During Refrigeration Storage. *Innov. Food Sci. Emerg.* 2012, 16, 47–53.

Mosqueda-Melgar, J.; Raybaudi-Massilia, R.M.; Martín-Belloso, O. Microbiological Shelf Life and Sensory Evaluation of Fruit Juices Treated by High-intensity Pulsed Electric Fields and Antimicrobials. *Food Bioprod. Process.* 2012, 90, 205–214.

Nielsen, H.B.; Sonne, A.; Grunert, K.G.; Banati, D.; Pollák-Tóth, A.; Lakner, Z.; Olsen, N.V.; Žontar, T.P.; Peterman, M. Consumer Perception of the Use of High-pressure Processing and Pulsed Electric Field Technologies in Food Production. *Appetite.* 2009, 52, 115–126.

Olsen, N.V.; Grunert, K.G.; Sonne, A.M. Consumer Acceptance of High-pressure Processing and Pulsed-electric Field: A Review. *Trends Food Sci. Tech.* 2010, 21, 464–472.

Ortega-Rivas, E. Critical Issues Pertaining to Application of Pulsed Electric Fields in Microbial Control and Quality of Processed Fruit Juices. *Food Bioprocess. Tech.* 2011, 4, 631–645.

Ou, Q.; Nikolic-Jaric, M.; Gänzle M. Mechanisms of Inactivation of *Candida humilis* and *Saccharomyces cerevisiae* by Pulsed Electric Fields. *Bioelectrochemistry.* 2017, 115, 47–55.

Pagán, R.; Mañas, P. Fundamental Aspects of Microbial Membrane Electroporation. In *Pulsed Electric Fields Technology for the Food Industry;* Raso, J.; Heinz, V., Eds.; Springer: New York, 2006; pp 73–94.

Peng, J.; Tang, J.; Barrett, D.M.; Sablani, S.S.; Anderson, N.; Powers, J.R. Thermal Pasteurization of Ready-to-eat foods and Vegetables: Critical Factors for Process Design and Effects of Quality. *Crit. Rev. Food Sci.* 2017, 57, 2970–2995.

Pereira, R.N.; Vicente, A.A. Environmental Impact of Novel Thermal and Non-thermal Technologies in Food Processing. *Food Res. Int.* 2010, 43, 1936–1943.

Pillet, F.; Formosa-Dague, C.; Houda, B.; Dague, E.; Rolls, M-P. Cell Wall as a Target for Bacteria Inactivation by Pulsed Electric Fields. *Sci. Rep-UK.* 2016, 6, 19778.

Plaza, L.; Sánchez-Moreno, C.; De-Ancos, B.; Elez-Martínez, P.; Martín-Belloso, O.; Cano, M.P. Carotenoid and Flavanone Content During Refrigerated Storage of Orange Juice Processed by High-pressure, Pulsed Electric Fields and Low Pasteurization. *LWT- Food Sci. Technol.* 2011, 44, 834–839.

Praporscic, I.; Levobka, N.; Vorobiev, E.; Mietton-Peuchot, M. Pulsed Electric Field Enhanced Expression and Juice Quality of White Grapes. *Sep. Purif. Technol.* 2007, 52, 520–526.

Puértolas, E.; Martínez de Marañón, I. Olive Oil Pilot-production Assisted by Pulsed Electric Field: Impact on Extraction Yield, Chemical Parameters and Sensory Properties. *Food Chem.* 2015, 167, 497–502.

Rajkovik, A.; Smigic, N.; Devlieghere, F. Contemporary Strategies in Combating Microbial Contamination in Food Chain. *Int. J. Food Microbiol.* 2010, 141, S29–S42.

Raso, J.; Calderón, M.L.; Góngora, M.; Barbosa-Cánovas, G.; Swanson, B.G. Inactivation of Mold Ascospores and Conidiospores Suspended in Fruit Juices by Pulsed Electric Fields. *Lebens-Wisse Technol.* 1998, 31, 668–672.

Rivas, A.; Rodrigo, D.; Company, B.; Sampedro, F.; Rodrigo, M. Effects of Pulsed Electric Fields on Water-soluble Vitamins and ACE Inhibitory Peptides Added to a Mixed Orange Juice and Milk Beverage. *Food Chem.* 2007, 104, 1550–1559.

Rodriguez-Gonzalez, O. Hurdle Technologies: Microbial Inactivation by Pulsed Electric Fields During Milk Processing. PhD Dissertation, University of Guelph, Guelph, Ontario, 2010.

Roodenburg, B.; Morren, J.; Berg, H.E.; de Haan, S.W.H. Metal Release in a Stainless Steel Pulsed Electric Field (PEF) System Part II. The Treatment of Orange Juice; Related to Legislation and Treatment Chamber Lifetime. *Innov. Food Sci. Emerg.* 2005, 6, 337–345.

Sack, M.; Sigler, J.; Frenzel, S.; Eing, C.; Arnold, J.; Michelberger, Th.; Frey, W.; Attmann, L.; Stukenbrock, L.; Müller, G. Research on Industrial-scale Electroporation Devices Fostering the Extraction of Substances from Biological Tissue. *Food Eng. Rev.* 2010, 2, 147–156.

Sampedro, F.; Geveke, D.J.; Fan, X.; Zhang, H.Q. Effect of PEF, HHP and Thermal and Thermal Treatment on PME Inactivation and Volatile Compounds Concentration of an Orange Juice-Milk Based Beverage. *Innov. Food Sci. Emerg.* 2009, 10, 463–469.

Sampedro, F.; McAloon, A.; Yee, W.; Fan, X.; Geveke, D.J. Cost Analysis and Environmental Impact of Pulsed Electric Fields and High Pressure Processing in Comparison with Thermal Pasteurization. *Food Bioprocess Tech.* 2014, 7, 1928–1937.

Sánchez-Moreno, C.; Cano, P.; De-Ancos, B.; Plaza, L.; Olmedilla, B.; Granado, F.; Elez-Martínez, P.; Martín-Belloso, O.; Martín, A. Pulsed Electric Fields-processed Orange Juice Consupmption Increases Plasma Vitamin C and Decreases F2-isoprostanes in Healthy Humans. *J. Nutr. Biochem.* 2004, 15, 601–607.

Sánchez-Moreno, C.; Plaza, L.; Ellez-Martínez, P.; De-Ancos, B.; Martín-Belloso, O.; Cano, M.P. Impact of High Pressure and Pulsed Electric Fields on Bioactive Compounds and Antioxidant Activity of Orange Juice in Comparison with Traditional Thermal Processing. *J. Agr. Food Chem.* 2005, 53, 4403–4409.

Silva, F.V.M.; Gibbs, P.A. *Principles of Thermal Processing: Pasteurization*. In *Engineering Aspects of Thermal Food Processing*; Simpson, R., Ed.; CRC Press, Taylor and Francis Group, Boca Raton, 2009; pp 13–48.

Terefe, N.S.; Buckow, R.; Versteeg, C. Quality-related Enzymes in Plant-based Products: Effects of Novel Food Processing Technologies Part 2: Pulsed Electric Field Processing. *Crit. Rev. Food Sci. Nutr.* 2015, 55, 1–15.

World Health Organization. The International Pharmacopeia. Geneva: World Health Organization, Methods of Sterilization, 8th ed., 2018.

Timmermans, R.A.H.; Mastwijk, H.C.; Berendsen, L.B.J.M.; Nederhoff, A.L.; Matser, A.M.; Van Boekel, M.A.J.S.; Nierop Groot, M.N. Moderate Intensity Pulsed Electric Fields (PEF) as Alternative Mild Preservation Technology for Fruit Juice. *Int. J. Food Microbiol.* 2019, 298, 63–73.

Toepfl, S.; Mathys, A.; Heinz, V.; Knorr, D. Review: Potential of High Hydrostatic Pressure and Pulsed Electric Fields for Energy Efficient and Environmentally Friendly Food Processing. *Food Rev. Int.* 2006, 22, 405–423.

Van Boekel, M.A.J.S. Effect of Heating on Maillard Reactions in Milk. *Food Chem.* 1998, 62, 403–414.

Van Loey, A.; Verachtert, B.; Hendrickx, M. Effects of High Electric Field Pulses on Enzymes. *Trends Food Sci. Tech.* 2001, 12, 94–102.

Vega-Mercado, H.; Gongora-Nieto, M.M.; Barbosa-Canovas, G.V.; Swanson, B.G. Pulsed Electric Fields in Food Preservation. *Food Sci. Technol.-NY.- Marcel Dekker* – 167, 2004, 183.

Vega-Mercado, H.; Gongora-Nieto, M.; Barbosa-Canovas, G.V.; Swanson, B.G. Pulsed Electric Fields in Food Preservation. In *Handbook of Food Preservation* 2nd ed.; Rahman, M.S., Ed.; CRC Press: New York, 2007; pp 783–813.

Vorobiev, E.; Lebovka, N. Enhanced Extraction from Solid Foods and Biosuspensions by Pulsed Electrical Energy. *Food Eng. Rev.* 2010, 2, 95–108.

Walkling-Ribeiro, M.; Noci, F.; Cronin, D.A.; Lyng, J.G.; Morgan, D.J. Shelf Life and Sensory Evaluation of Orange Juice After Exposure to Thermosonication and Pulsed Electric Fields. *Food Bioprod. Process.* 2009, 87, 102–107.

Walkling-Ribeiro, M.; Noci, F.; Cronin, D.A.; Lyng, J.G.; Morgan, D.J. Shelf Life and Sensory Attributes of a Fruit Smoothie-type Beverage Processed with Moderate Heat and Pulsed Electric Fields. *LWT-Food Sci. Technol.* 2010, 43, 1067–1073.

Wang, Q.; Li, Y.; Sun, D.; Zhu Z. Enhancing Food Processing by Pulsed and High Voltage Electric Fields: Principles and Applications. Crit. Rev. Food Sci. 2018, 58, 2285–2298.

Wang, R.; Zou, J.; Liao, M.; Fang, T.; Yan, Z. Non-thermal Processing of Orange Juice by PEF in Continuous Chamber. *J. Jiangsu Univ.* 2009, 30, 169–173.

Wiktor, A.; Schulz, M.; Knorr, D.; Witrowa-Rajchert, D. Impact of Pulsed Electric Field on Kinetics of Immersion Freezing, Thawing, and on Mechanical Properties of Carrot. *Food Sci. Technol. Qual.* 2015, 21.

Wu, Y.; Zhang, D. Effect of Pulsed Electric Field on Freeze-Drying of Potato Tissue. *Int. J. Food Eng.* 2014, 10, 857–862.

Wu, L.; Zhao, W.; Yang, R.; Yan, W.; Sun, Q. Aggregation of Egg White Proteins with Pulsed Electric Fields and Thermal Processes. *J. Sci. Food Agr.* 2015, 96, 3334–3341.

Yang, N.; Huang, K.; Lyu, C.; Wang, J. Pulsed Electric Field Technology in the Manufacturing Processes of Wine, Beer, and Rice Wine: A Review. *Food Control.* 2016, 61, 28–38.

Yang, R.-J.; Li S.-Q.; Zhang, Q.H. Effects of Pulsed Electric Fields on the Activity of Enzymes in Aqueous Solution. *J. Food Sci.* 2004, 69, 241–248.

Yeom, H.W.; Streaker, C.B.; Zhang, Q.H.; Min, D.B. Effects of Pulsed Electric Fields on the Activities of Micro-organisms and Pectin Methyl Esterase in Orange Juice. *J. Food Sci.* 2000, 65, 1359–1363.

Yeom, H.W.; Zhang, Q.H.; Chism, G.W. Inactivation of Pectin Methyl Esterase in Orange Juice by Pulsed Electric Fields. *J. Food Sci.* 2002, 67, 2154–2159.

Yu, L.J. Application of Pulsed Electric Field Treated Milk of Cheese Processing: Coagulation Properties and Flavor Development. PhD Dissertation, McGill University, Montreal, Quebec, 2009.

Zhao, J.; Fang, T.; Chen, M.; Gao, M.; Lin, M.; Chen, J. Application of Fuzzy Mathematics on Sensory Evaluation of Different Processed Orange Juice. *J. Henan Univ. Technol.* (Natural Science Edition). 2008, 29, 72–75.

Zhong, L. Studies on orange PE (pectinesterase). Surveys and Reviews. *Southwest College of Food Science, Research Direction: The Theory of Modern Food Processing and Technology*. 2008, 9–11.

Zhu, N.; Wang, Y.; Zhu, Y.; Yang, L.; Yu, N.; Wei, Y.; Zhang, H.; Sun, A. Design of a Treatment Chamber for Low-voltage Pulsed Electric Field Sterilization. *Innov. Food Sci. Emerg.* 2017, 42, 180–189.

Zulueta, A.; Barba, F.; Esteve, M.; Frígola, A. Effects on the Carotenoid Pattern and Vitamin A of a Pulsed Electric Field-treated Orange Juice-milk Beverage and Behavior During Storage. *Eur. Food Res. Technol.* 2010, 231, 525–534.

CHAPTER 14

Security and Biodisponibility of Derivatives from Medicinal Plants in Food Consumption

MARIA DEL CARMEN RODRÍGUEZ SALAZAR[1], LUIS ENRIQUE COBOS PUC[1], HILDA AGUAYO MORALES[1], JOSÉ EZEQUIEL VIVEROS VALDEZ[2], CRYSTEL ALEYVICK SIERRA RIVERA[1], JUAN JOSÉ GAYTÁN ANDRADE[1], and SONIA YESENIA SILVA BELMARES[1*]

[1]*Research Group of Chemist-Pharmacist-Biologist, School of Chemistry, Autonomous University of Coahuila, Blvd. Venustiano Carranza, Col. República Oriente, Saltillo, Coahuila 25280, Mexico*

[2]*Department of Chemistry, School of Biological Sciences, Autonomous University of Nuevo León, Pedro de Alba S/N, between Av. Alfonso Reyes and Av. Fidel Velázquez, University City. San Nicolás de los Garza, Nuevo León 66455, Mexico*

[*]*Corresponding author. E-mail: yesenia_silva@uadec.edu .mx*

ABSTRACT

Consumption of plant derivatives (whole plant, extracts, and active compounds) is increased worldwide because they are part of nutritional supplements, nutraceuticals, or herbal products. The main reason to consume these types of products is to preserve health since plants contain compounds with biological properties to treat various diseases. Their consumption is associated with conditions associated with metabolic syndrome and aging. For this reason, many plant derivatives are used as antiobesogenic, antidiabetic, anticholesterolemic, and energetic. Additionally, some plants are used to treat infections or inflammatory diseases. Unfortunately, consumers and suppliers do not know the mechanisms of action and interactions of these products. However, the insufficiency of governmental regulation systems around the world increases the risk of adverse effects. Although there is a scientific breakthrough in the effectiveness and toxicity of a large number of medicinal plants, this information is not described on the label of the products that marketed. Therefore, regulatory agencies should consider their inclusion as well as the dosage. Pharmacokinetics and bioavailability studies of the active compounds from plants must be carried out to ensure their efficacy and safety.

14.1 INTRODUCTION

In recent years, the consumption of dietary supplements in the world is growing (AlTamimi, 2019). In this regard, some classes of these products contain extracts or compounds derived from medicinal plants that improve the health of the consumers (Huang et al., 2019). For instance, dietary supplements added with ginger prevent type II diabetes and have negligible side effects, so they are safe according to the Food and Drug Administration (Huang et al., 2019). Another example is found in foods supplemented with antioxidants, such as anthocyanins, carotenoids, flavonoids, and vitamins, which reduce the risk of macular degeneration (Deshpande, 2012; Rasmussen and Johnson, 2013). Actually, numerous foods containing derivatives from medicinal plants are commercialized, but in many cases, the collection, taxonomic identification, standardization of the products, pharmacological, and clinical assays are handled at free will by the manufacturer (Miousse et al., 2017).

Frequently, herbal remedies perceived as harmless, but some could cause liver and kidney damage (Donet et al., 2016). To illustrate this, in different countries, the consumption of food supplements is associated with hepatocellular lesions and high levels of transaminases (Medina-Caliz et al., 2018; Ettel et al., 2017). Biological activities of some supplements are well known, but the biological attributes of most of these products remain in the dark. Consequently, studies should be implemented to determine the adequate dose, efficacy, and bioavailability of the active ingredients of each supplement (Khoo et al., 2019). Moreover, if that was not enough, the majority of the consumers use concomitantly more than one supplement, so studies that validate their impact on health are required (AlTamimi, 2019). Moreover, the stability and bioavailability of the active compounds are fundamental since affects their functionality.

14.2 MAIN REASONS TO CONSUME FOOD AND NUTRACEUTICAL SUPPLEMENTS

Nowadays, some sectors of the human population are more interested in conserving their health, so they consume frequently food supplements derived from plants. Curiously, the consumption of these products increases the incidence of suffering metabolic diseases such as insulin resistance, type 2 diabetes, hyperlipidemia, hyperglycemia, hypertension, atherosclerosis, and nonalcoholic fatty liver disease, due to the rise in fat mass, caused by the storage of excess nutrients (Kim et al., 2018; Iñiguez et al., 2018). Consequently, people use supplements derived from plants as antiobesogenic to maintain a healthy life. Also, these supplements are used to prevent the aging process or as energetic.

Consumption phytosterols and curcumin as dietary supplements reduce low-density lipoproteins (LDL) cholesterol (Ferguson et al., 2019). For this reason, phytosterols represent an alternative for the prevention of cardiovascular diseases in low or medium risk populations (Vilahur et al., 2019). However, the consumption of dietary flavonoids containing cocoa, apple, tea, citrus fruits, and berries decreases cardiovascular and cerebrovascular risk (Rees et al., 2018). It has been reported that tea consumption improves type 2 diabetes. For instance, vine tea reduces the serum levels of glucose and lipids, inhibits gluconeogenesis, and reduces the synthesis of fatty acids (Xiang et al., 2019).

Foods that contain polyphenols and are consumed in the diet, such as green tea, ameliorate the metabolic syndrome by reducing body mass index and waist circumference and augmenting lipid metabolism. Therefore, the products with a high content of polyphenols could represent an effective nutritional strategy to improve the health of patients suffering from metabolic syndrome (Amiot et al., 2016). Berberine is a plant extract that contains an isoquinoline alkaloid marketed as a dietary supplement in the United States and used for the treatment of type 2 diabetes mellitus and dyslipidemia. However, the quality of the product is variable and presents inconsistencies in safety and efficacy (Funk et al., 2018). Several nutraceuticals obtained from plants contain antioxidants, fiber, and other phytochemicals by which, they have antiobesogenic and antidiabetic effects through the modulation of various cellular and physiological pathways. These effects include reduced appetite, modulation of lipid absorption and metabolism, increased insulin sensitivity, thermogenesis, and changes in intestinal microbiota (Martel et al., 2017).

The interest in slowing down or reversing the aging process has fostered the growth of the antiaging industry. In this regard, many people consume food supplements to prevent premature aging (Bradford et al., 2018). Resveratrol is a compound incorporated in food supplements because it has anti-ageing properties. However, adequate doses for clinical studies are unknown since its bioavailability data is not available (Tou, 2015).

In the past years, human activities and necessities have augmented considerably resulting in high levels of stress. Therefore, the consumption of food supplements with energizing properties usually ingested as beverages is growing. Energy drinks are an example but lack objective bases for their use, which is why their functionality and safety should be evaluated just like other food supplements (Sather et al., 2018). The root of Rhodiola is used in traditional Chinese medicine to increase the resistance of an organism to stress and promote longevity by improving oxidative stress. However, more scientific research is required to validate its functionality (Jafari et al., 2007).

Plant extracts or active compounds from plants are incorporated frequently into foods as coadjuvants for health or to preserve food. For this, some studies have evaluated the functionality and appropriate doses of several plant products that are part of food supplements (Gutierrez et al., 2016) which is essential to avoid harmful treatments (Bradford et al., 2018). However, a large number of plants lack studies that guarantee their safe use.

14.3 IMPORTANCE OF THE MEDICAL CONDITION DURING THE CONSUMPTION OF FOOD SUPPLEMENTS

Food supplements are marketed as products to improve different health conditions. However, dietary supplements are not suitable for all people (Petroni et al., 2019; Konishi et al., 2019; Touillaud et al., 2019). Therefore, their proper use depends on the health condition of each person and some hereditary factors. Additionally, to ensure the safety of the consumption of supplements containing plant derivatives (complete plant, extracts, and active compounds), it is necessary to understand their interaction with other supplements, foods or medications (Petroni et al., 2019; Konishi et al., 2019; Touillaud et al., 2019).

Consumption of soy-derived supplements is a classic example because people often ingest them to avoid sarcopenia and mitigate the symptoms of menopause (Petroni et al., 2019). Additionally, these supplements contain proteins and isoflavones that improve the symptoms of diabetes (Konishi et al., 2019). However, they contain steroidal compounds that promote the development of breast cancer in women with a family history in which their use is controversial. However, this group of women should avoid the consumption of soy derivatives during the premenopausal or postmenopausal stage (Touillaud et al., 2019). Although phytosterols are a dietary alternative to prevent cardiovascular diseases and have been shown that reduce the oxidative stress and inflammation, not all effects observed experimentally have resulted in clinical benefits (Vilahur et al., 2019).

It is to draw attention that there exist few studies of the pharmacokinetics, bioavailability, interaction with foods and medications of plant supplements. In one study, it was shown that dietary pattern and obesity affect the absorption and metabolism of polyphenolic compounds in humans (Novotny et al., 2017). For instance, lutein is transferred from the small intestine to the lymphatic system where it accumulates, instead of circulating through the bloodstream. Therefore, to improve the absorption of these components is a challenge in this field (Sato, 2019). Pterostilbene is a natural analog of resveratrol consumed as a dietary supplement, which potentiates the effect of clonazepam and increases the activity of carbamazepine but has no pharmacokinetic interaction with carbamazepine, oxcarbazepine, and valproate (Nieoczym et al., 2019) suggesting a pharmacodynamic interaction.

In different regions of the world, nutraceutical products are consumed as infusions, capsules, tablets, tinctures, and functional foods prepared from plants used in traditional medicine. Frequently, the collection, taxonomic identification, extraction techniques, standardization, and homogenization of the products and pharmacological and clinical assays, are handled at free will of the manufacturers (Miousse et al., 2017). Although many people consume them, their usefulness, and safety in most cases are unknown. Some dietary supplements could cause interactions with foods or medications such as insulin. Therefore, valuable information on healthy foods should be disseminated in a simple language for consumers (Saito, 2018).

Food safety is a concern in many countries since there are numerous adverse events associated with the consumption of healthy foods (Toda and Uneyama, 2018). These products may contain undeclared ingredients, toxic compounds, natural toxins for the plants or high levels of heavy metals added illegally and intentionally that could cause adverse health effects.

14.4 SANITARY REGULATION TO INCORPORATE NATURAL PRODUCTS IN FOOD

In recent years, health regulations regarding food, plants, and food supplements have grown considerably. Currently, there exist pharmacopeias, compendiums, and guides that describe the tests to demonstrate that the raw materials and finished products meet the established specifications. These specifications focus on the identity, purity, quality, packaging, and labeling of various products [(Diario Oficial de la Federación, 1999; Regulation (EU) No 1169/2011, 2011; NOM-051-SCFI/SSA1-2010, 2010)].

Currently, there is a tendency to formulate food, beverages, and food supplements with plant extracts. Although many plants are used in traditional medicine, some food products contain higher amounts than those used usually. Since years, various plant derivatives that had not been added formerly have been incorporated into foods. Therefore, government regulators offices in different countries have an interest in ensuring that plants and the quantity used are safe (Diario Oficial de la Federación, 2003; Diario Oficial de la Federación, 1999).

In the United States, herbal substances incorporated into food must previously recognize as safe (Generally Recognized as Safe) (Food and Drug Administration, 2014). In Mexico, the Federal Commission for Protection against Health Risks (COFEPRIS) is the competent authority in matters of regulation. Therefore, it authorizes the use of plants in food and food supplements provided that they do not promote any therapeutic effect and comply with the requirements, norms, and regulations established. Otherwise, these formulations must meet other criteria and register as herbal remedies or herbal medicine (Diario Oficial de la Federación, 2003).

The herbalist pharmacopeia includes plants with known toxicity as well as those prohibited or allowed in food (Diario Oficial de la Federación, 1999; Secretaría de Salud, 2013). Plants used as raw materials to formulate food and food supplements must undergo treatments to reduce microbial flora as well as physical or chemical residues that can damage health (Food and Drug Administration, 2014). Simultaneously, these products must comply with the standards established by Good Hygiene Practices to avoid contamination (NOM-251-SSA1-2009, 2008). Therefore, the label of the commercialized products must not present confusing or misleading information about its composition, origin, or effects. The labeling must not show preventive or therapeutic indications either (Diario Oficial de la Federación, 1999; Regulation [EU] No 1169/2011, 2011; Diario Oficial de la Federación, 2003).

14.5 SAFETY WHEN USING DIETARY SUPPLEMENTS FORMULATED WITH PLANT DERIVATIVES

At present, the consumption of food products with nutraceutical properties has increased with the idea of maintaining a healthier life. Additionally, in recent years, some research on the effectiveness and toxicity, support the use of plant derivatives as functional ingredients in food. As a result, vegetable extracts have been incorporated in some dietary supplements. However, studies on interactions with foods or medications are limited, as are studies on their functionality, bioavailability, and safety (Hussain et al., 2015). Nevertheless, some derivatives of medicinal plants are used for culinary purposes and are incorporated in the preparation of foods to improve their appearance, taste, and smell.

As a general rule, herbal remedies are perceived as harmless. However, some damage the liver and kidney (Donet et al., 2016). Additionally, in different countries, the consumption of food supplements is associated with hepatocellular lesions and high levels of transaminases (Medina-Caliz et al., 2018; Ettel et al., 2017). On the other hand, dietary supplements are marketed as a panacea to reduce the weight of morbidly obese people. Nevertheless, the associated harmful side effects are worrisome and not recognized by consumers. *Garcinia cambogia* and products that contain are associated with fulminant liver failure (Lunsford et al., 2016; Gavrić et al., 2018).

However, most of the information consulted by consumers of herbal products comes from unreliable sources, which favors the consumption of these products lacking scientific validity. Although some remedies eliminated for lack of effectiveness, others still used, so they continue to cause damage to organs such as liver and kidney, if not appropriately administered (Dourakis et al., 2002). Sometimes small elevations appear in the liver enzymes while in others a fulminating liver failure that requires a transplant. Although regulation by the Health Office is part of the solution, it is necessary to implement awareness in both public and professional education on the proper use of herbal preparations (Pak et al., 2004).

It is important to note that not all plant extracts represent a problem, some are endorsed by methodologically rigorous studies, such as the antidepressant properties of *Hypericum perforatum*, the sedative effect of different valerian species, linden (*Ternstroemia pringlei*); the mechanism of action of extracts of mullein (*Gnaphalium liebmannii*), used in Mexico for the treatment of respiratory diseases (Balderas et al., 2008; Sánchez-Mendoza et al., 2007; Salazar-Leal et al., 2006). *Akebia quinata* contains components that can be used to treat acute hepatotoxicity (Lee et al., 2017). The immunomodulatory effect of extracts of *Allium sativum*, *Camellia sinensis*, *Zingiber officinale*, *Echinacea purpurea*, *Nigella sativa*, *Glycyrrhiza glabra*, and *H. perforatum* has also been evaluated since some of their compounds eliminate free radicals and have anti-inflammatory properties. However, the researchers propose to study the interaction of the components for their safe use (Sultan et al., 2014). Another study found that *Artemisia capillaris* extract can prevent weight gain in obese rats and improve lipid metabolism (Lim et al., 2013). In contrast, functionality and toxicity are related to the chemical composition of each plant. Therefore, many types of research focus on identifying their active compounds through different analytical techniques.

In the field of nutrition, exploring the links between food and health is one of the main areas of research, given that, the information on foods added with plant extracts is scarce. The evaluation of the functionality of the dietary supplement colostrum-NONI is one of the best known and was shown to improve some intestinal alterations (Côté et al., 2010). Table 14.1 shows the plants that are consumed most frequently as food supplements, nutraceuticals, or condiments, as well as their active compounds, properties, and evaluation of toxicity.

14.6 ASPECTS TO CONSIDER FOR THE APPROPRIATE ADMINISTRATION OF PLANT DERIVATIVES

As was described, the consumption of herbal products has increased, since they contain plant ingredients with preventive and curative properties of diseases known for many years. Therefore, the food industry often adds plant derivatives in food supplements. The most frequent use of plant derivatives is as a preservative since some have antimicrobial properties (Kinsella et al., 2018). However, there is no standardized regulation to guarantee the safety of the use of natural supplements similar to the one that regulates allopathic medicines. Currently, the systems of regulation depend on each country, which creates confusion in the use of food supplements (Mehmood et al., 2019; Santini et al., 2018).

Therefore, the main challenge of herbal products is to understand their mechanism of action to prevent and treat diseases without causing harm to the consumer. Accordingly, the dose is the primary aspect that must be regulated, since adjusting the dose ensures the efficacy of the herbal product and

TABLE 14.1 Relationship Between Active Principles from Plants Used as Food Supplements with Their Biological Actions and Toxicity

Plant	Active Compounds	Properties	Toxicity	References
Rosmarinus officinalis L.	– Carnosol, carnosic acid, and rosmarinic acid (extract). – Alpha-pinene, (−)-bornyl acetate, camphor, and eucalyptol (essential oil).	– Flavoring and condiment – Food additive: antimicrobial and antioxidant agent. – Anti-inflammatory, headache, abdominal pain, antispasmodic, arthritis, gout affections.	– Hypersensitivity (contact dermatitis and asthma). The frequency is not known.	(Ribeiro-Santos et al., 2015; Bernatoniene et al., 2016)
Zingiber officinale	– Essential oil – Sesquiterpene hydrocarbons and relatively low monoterpene hydrocarbons.	– Prevent oxidative damage of food – Free radical scavenging, antioxidant, and antiperoxidative effects – Antimicrobial – Antitumor – Anti-inflammatory – Cholesterol-lowering properties – Treatment for nausea	– Antiplatelet effect	(Pai Jakribettu et al., 2016; Munda et al., 2018; Marx et al., 2015)
Panax ginseng	– Saponins (ginsenosides)	– Immunoregulatory–Cardiovascular homeostatic – Antitumoral – Antioxidant – Combat respiratory infections – Diabetes mellitus (decrease glucose) – Reduce hyperlipidemia and hypertension. – Treatment of dementia	– Risk of bleeding for concomitant use with anticoagulant or antiplatelet drugs	(Fung et al., 2017; Ralla et al., 2017; Husain et al., 2018)

TABLE 14.1 *(Continued)*

Plant	Active Compounds	Properties	Toxicity	References
Garcinia gummi-gutta	– Alkaloids, tannins, phenolic flavonoids, carbohydrates, and proteins. – Low content of steroids, terpenoids, phlobatannin, and cardiac glycosides, hydroxyl citric acid	– Antifungal, antibacterial, antihelmintic – Decrease glucose and lipids – Weight loss – Increase erythrocytes, leucocytes, thrombocytes, hemoglobin. – Anticancer, – Anti-inflammatory – Antioxidant – Hepatoprotective – Antiulcer	Safe to use	(Jacob, 2015; Semwal et al., 2015)
Curcuma longa	– Flavonoids, tannins, anthocyanin, phenolic compounds, oil, organic acids, and inorganic compounds. – Curcumin	– Hepatobiliary diseases – Functional gastrointestinal disorders – Joint pain – Anti-inflammatory – Antioxidant – Antifungal – Antihypertensive – Neuroprotective	– Antiplatelet activity has been shown by in vitro and in animal experiments but there had been no clinical studies to validate this effect.	(Fung et al., 2017; Ayati et al., 2019)
Amphipterygium adstringens	– Anacardic acid	– Anti-inflammatory – Antioxidant – Anticáncer – Antiulcer activity – Hypocholesterolemic, antifungal and antiprotozoal	– Safe to use	(Rodriguez et al., 2015)

TABLE 14.1 *(Continued)*

Plant	Active Compounds	Properties	Toxicity	References
Aloe vera	– Fat-soluble vitamins, minerals, enzymes, – Simple/complex polysaccharides, – Phenolic compounds – Organic acids	– Cosmetic, medic and nutraceutical applications	– The foliar pulp at 100 mg/kg decreases body weight, vital organs, and erythrocyte count. Additionally, it causes spermatogenic dysfunction in humans. – Aqueous extract from leaf pulp at 500 mg/kg does not exhibit acute toxicity. However, at higher doses, decreases the activity of the central nervous system in mice	(Radha and Laxmipriya, 2015)
Vaccinium myrtillus	– Polyphenols: flavonoids, anthocyanins, phenolic acids, tannins, ellagitannins	– Antioxidant – Antibacterial – Reduce the risk of cardiovascular disease, cancer, inflammation, obesity, and diabetes. – Treatment of cardiovascular disease and in ophthalmology	– In rats, consumption of bilberries as a food supplement should not pose a risk of interacting with co-administered drugs based on their metabolism.	(Korus et al., 2015; Prokop et al., 2019)
Vaccinium macrocarpon	– Proanthocyanidins: Poly-flavan-3-ols	– Prevent urinary tract infections – Suppress the growth of cancer cells in vitro	– Interaction with warfarin increased INR (international normalized ratio) in patients.	(Wu et al., 2018; Edwards et al., 2015; Paeng et al., 2007)

TABLE 14.1 *(Continued)*

Plant	Active Compounds	Properties	Toxicity	References
Punica granatum	– Polyphenolic compounds including punicalagin isomers, ellagic – Acid derivatives oleanolic acid, ursolic acid, and gallic acid – Anthocyanins (delphinidin, cyanidin, and pelargonidin 3-glucosides and 3,5-diglucosides)	– Treatment of diarrhea, dysentery, acidosis, helminthiasis, bleeding, and respiratory disorders – Anticancer activity – Antihyperglycemic and antihyperlipidemic activity – Antioxidant	– Interactions of the bioactive compounds from pomegranate with solute carrier transporters of the liver and kidney which will result in altered pharmacokinetics of drugs.	(Modaeinama et al., 2015; El-Hadary and Ramadan, 2019; Li et al., 2014)
Lycium barbarum	– Polysaccharides – Carotenoids – Flavonoids	– Antioxidant – Immunomodulatory – Neuroprotective properties – Hematopoietic properties – Treatment of diabetes mellitus and hypertension – Infertility – Abdominal pain – Dry cough – Fatigue – Headache	– Interaction with warfarin Roots: – Vomiting appears in dogs receiving 120 g/kg orally or 30 g/kg intraperitoneally. In rabbits, drowsiness was observed at doses of 80 g/kg (orally). – An increase in the weight of heart, liver, and lungs was found in the course of the assessment of subacute toxicity in rats with high doses (5 and 10 g/kg) orally over a test period of 14 days. – Rise of blood urea nitrogen (BUN) and a decrease of creatinine were also noted as well as a raised count of white blood cells. – No reports of adverse effects due to Goji fruits	(Potterat, 2010)

TABLE 14.1 *(Continued)*

Plant	Active Compounds	Properties	Toxicity	References
Litchi chinensis	– **Leaves:** (–)Epicatechin, procyanidin A2, and procyanidin B2 – **Fruit:** 5-hydroxymethyl-2-furfurolaldehyde (5-HMF), benzyl alcohol, hydrobenzoin, and (+) – catechi – **Seed:** Leucocyanidin, cyanidin glycoside and malvidin glycoside, and saponins – **Pericarp:** 5-2-(2-hydroxy-5-(methoxy carbonyl) phenoxy) benzoic acid, bis-(8-epicatochinyl) methane, butylated hydroxy toluene, epicatechin, dehydrodiepicatechin A, methyl shikimate, ethyl shikimate, isolariciresinol, kaempferol, methyl 3,4-dihydroxy benzoate, proanthocyanidin A1, A2, rutin, and stigmasterol	– Antioxidant, – Hepatoprotective, – Anti-inflammatory, – Analgesic, – Anti-lipase – Cardiovascular activity – Cytotoxic in cancer cells – Antimicrobial and antiviral activity	– No toxic – Lychee may lower blood sugar levels. Caution is advised when taking insulin or drugs for diabetes: consumption by mouth should be monitored closely. – Lychee may increase the risk of bleeding when taken with medications including aspirin, anticoagulants such as warfarin or heparin, antiplatelet drugs such as clopidogrel, and nonsteroidal anti-inflammatory drugs such as ibuprofen or naproxen. – It also interacts with anticancer agents, anti-inflammatory agents, antivirals, cardiovascular agents, cholesterol- or lipid-lowering agents, immune-modulating agents, or pain relievers.	(Kilari and Putta, 2016)

TABLE 14.1 *(Continued)*

Plant	Active Compounds	Properties	Toxicity	References
Camellia sinensis	– Polyphenolic compounds known as catechins including epigallocatechin gallate, catechin, epicatechin, gallocatechin, gallocatechin gallate, epigallocatechin, and epicatechin gallate	– Hepatoprotective – Cardioprotective – Neuroprotective – Anticancer – Antiobesity – Antidiabetic – Antibacterial – Antiviral – Antioxidant	– Showed toxic effects and animal morbidity and mortality in rats – The increasing use of Green Tea by consumers proposes potentials for interactions with conventional medications.	(Bedrood et al., 2018)
Verbascum thapsus	– Polysaccharides: galactose, arabinose – Phenolic acids – Flovonoeids – Triterpene saponins, verbasco saponin – Iridoid glycoside – Tannen: acobin, katapol and related compounds	– Antimicrobial – Expectorant and demulcent: treat various respiratory problems such as bronchitis, dry coughs, whooping cough, tuberculosis, asthma and – Hoarseness. – Mildly diuretic – Mild sedative – Anti-inflammatory	– Toxicity; there are no data on genotoxicity, carcinogenicity, reproductive	(Jamshidi-Kia et al., 2018, Prakash et al., 2016)
Chamaemelum nobile	– Volatile Oil, – Sesquiterpenes, – Hydroperoxides, – Flavonoids, – Catechins, – Coumarins,	– Gastric debility with flatus – Diaphoretic – Emetic – To relieve colds – Dysmenorrhea to decrease pain and facilitate the flow,	– Safe to use	(Al-Snafi, 2016)

TABLE 14.1 *(Continued)*

Plant	Active Compounds	Properties	Toxicity	References
	– Polyacetylenes – Phenolic Acids – Triterpenes – Steroids – Polysaccharides	– Antiemetic, – Antispasmodic, – Sedative – Antibacterial – Anti-inflammatory – Antioxidant – Cytotoxic effect in tumor cell lines – Nervous effect Antiasthmatic effects		
Garcinia mangostana	– Prenylated and oxygenated xanthones	– Antioxidant – Anticancer and cytotoxic – Anti-inflammatory – Antiallergy – Analgesic – Antibacterial, antifungal, antiviral – Antiobesity – Antihelmintic and antiparasitic – Treatment of Alzheimer's disease (AD)	– Safe to use	(Ibrahim et al., 2016; Pedraza-Chaverri et al., 2008)
Moringa oleifera	– Phenolic acids: gallic acid predominating followed by ellagic and caffeic acids – Alkaloids – Glucosinolates – Isothiocyanates – Thiocarbamates.	– Treating stomach pain and ulcers, – Poor vision – Joint pain and – For aiding digestion – Antimicrobial activity – Antidiabetic	– Safe to use	(Leone et al., 2016; Stohs and Hartman, 2015)

TABLE 14.1 *(Continued)*

Plant	Active Compounds	Properties	Toxicity	References
Salvia officinalis	– Alkaloids, – Carbohydrate – Fatty acids – Glycosidic derivatives (e.g., cardiac glycosides, flavonoid glycosides, saponins) – Phenolic compounds (e.g., coumarins, flavonoids, tannins) – Polyacetylenes – Steroids – Terpenoids (e.g., monoterpenoids, diterpenoids, triterpenoids, sesquiterpenoids)	– Anticancer and antimutagenic effect – Antioxidant – Anti-inflammatory – Antinociceptive – Antiseptic – Cognitive- and memory-enhancing effects – Antiglycemic – Antiobesity	– In the case of prolonged use or following overdose some unwanted effects such as vomiting, salivation, tachycardia, vertigo, hot flushes, allergic reactions, tongue swallowing, cyanosis, and even convulsion may occur. – Toxic effects on the fetus and newborn	(Ghorbani and Esmaeilizadeh, 2017; Lopresti, 2017)

prevents adverse effects on the consumer (Izzo et al., 2016). Consequently, herbal products must undergo pharmacokinetic studies (ADME) similar to those of conventional medicines.

Additionally, it is necessary to evaluate the bioavailability of the active compound of herbal product that reaches the blood circulation. Since factors as chemical instability during digestion, poor solubility in liquids, slow absorption of the gastrointestinal tract, and metabolism of the first step, affect the adequate circulation of the active compound (Mehmood et al., 2019). However, the pharmacokinetic studies of herbal products are complicated due to their phytochemical profile. Since the phytochemical composition of each plant varies according to the factors to which it is exposed, an example is the environment. Additionally, the composition of each plant can be affected by the extraction conditions of each plant structure (leaf, flower, root, stem), making it difficult to establish the pharmacokinetics and pharmacodynamics of natural products (Bonati, 1991).

Another aspect to consider is the interaction of herbal products with conventional drugs. Since the interaction can decrease or increase the bioavailability of the drug and cause adverse effects or affect drug efficacy. This aspect should be taken into account by patients with diabetes, hypertension, and other cardiovascular diseases since they depend on the drug to improve their condition (Mehmood et al., 2019; Izzo et al., 2016; Santini et al., 2017).

Derivatives from plants like any drug can give rise to toxic effects with a high-risk factor that depends on the gender, age, nutritional status, genetic factors, and diseases of each consumer. Currently, there are very few studies about the adverse effects and drug–herbal product interactions, so the information available is theoretical or comes from toxicological profiles in vitro. However, weaker or null effects have been detected in vivo studies than in vitro models, so an adequate dosage should be established (Mehmood et al., 2019; Izzo et al., 2016; Gupta et al., 2018).

14.7 BIOAVAILABILITY OF ACTIVE PRINCIPLES OF PLANTS IN FORMULATED FOODS

At present, studies on the stability and bioavailability of the active ingredients in food supplements are limited. Consequently, it is of great interest to investigate the phytochemicals incorporated in dietary supplements. Because they have proven benefits in human health, and some contribute to the sensory properties of formulated products providing flavor, color, and stability. For this reason, its application in the food industry is increasing, since research guarantees its beneficial effects and results in the safety of its use to open new markets and develop new functional foods and nutraceuticals (Cory et al., 2018).

Now it is known that not all phytochemical compounds added in food supplements are released in sufficient quantities to provide a therapeutic effect, so they are not bioavailable or bioaccessible. Since bioavailability is defined as the concentration of the bioactive compound that reaches the bloodstream, absorption corresponds to the introduction of the phytochemical compound in the cells of the intestine or until it arrives at the therapeutic target and produces the physiological response (Oracz et al., 2019). Consequently, it is necessary to know the number of phytonutrients consumed, as well as those absorbed by the body, to understand the effect of digestion and absorption on the bioactivity of phytonutrients.

For several years, there has been a wide variety of studies to evaluate the functionality of the phytochemical compounds that propose their incorporation in dietary supplements and polyphenols, sulfur compounds, phytosterols, terpenes, and carotenoids are some examples of proposed compounds since these groups of compounds have beneficial properties for the health of those who consume them. Most of the studies focused on evaluating biological activity using in vitro models (Koss-Mikołajczyk et al., 2019; Viveros-Valdez et al., 2016; Xiao and Bai, 2019). However, these models have a variety of limitations such as the conditions used in the reaction, use synthetic substrates, high concentration of initiators, among others. Therefore, the results of some investigations not extrapolated to physiological conditions. On the other hand, the speed, reproducibility, and low cost of this type of method are the main reason for its routine use.

Currently, ex vivo methods are presented as a breakthrough in emulating complex physiological conditions (Carranza-Torres et al., 2019). These models are carried out as a preliminary stage to models in vivo studies. However, the most unequivocal evidence on the therapeutic effectiveness of phytochemicals is the ability to improve the health status of people, as well as changes in specific biomarkers (Collins, 2005). In these studies, phytochemicals or their metabolites are detected in plasma and urine, to demonstrate their bioavailability, absorption and, urinary excretion (Prior, 2003; Manach et al., 2005), and even its incorporation into LDL (Nigdikar et al., 1998).

Accordingly, to evaluate the bioavailability of dietary phytochemicals, the first step is to quantify the compounds ingested with the diet. Therefore, its great structural diversity, as well as the influence of the agronomic factors of the processing and storage, varies their content, making it difficult to monitor them. For its detection, chromatographic techniques coupled to different detectors are usually used, such as diode array or tandem mass spectrometry (Burgos-Edwards et al., 2018).

Evaluate bioavailability encompasses a large number of stages, which ends when evaluating the biological effect of dietary phytochemicals (Rein et al., 2013).

Therefore, a large number of variables must be considered, such as:

- Digestion and the metabolic process by the microflora.
- Intestinal absorption.
- Circulation through the blood and lymphatic stream.
- Liver metabolism.
- The process of recruitment in therapeutic targets
- The period of tissue accumulation.
- Excretion (biliary or urinary).

The chemical structure of the compound influences digestion and absorption through the intestinal tract as well as transport, metabolism, and biological effect. Moreover, the food matrix plays a vital role in the bioavailability of the compounds, because they could interact with proteins, lipids, and polysaccharides. As well, gastric and intestinal pH, intestinal fermentation, biliary excretion, and transit time in the digestive tract can affect the bioavailability and absorption of phytochemicals, because the process of abetting phytochemicals begins in the stomach and ends when it reaches the small intestine.

During the absorption, the compounds can be hydrolyzed or deconjugated, and once inside the enterocyte, they conjugated by methylation reactions, glucuronidation, sulfation, or their combinations. Therefore, only the aglycones and some glycosides are absorbed in the small intestine (Piskula et al., 1999; Passamonti et al., 2005; Crespy et al., 2002). The compounds that are not absorbed, reach the large intestine, where they metabolized by the bacteria present in this organ. Then they reach the colon with a different chemical structure, for example, as glucuronides (Winter and Bokkenheuser, 1987). Additionally, enzymes from the microbiota can hydrolyze the glycosides into aglycones. Therefore, interindividual varieties and the influence of the composition of the microbiota should be taken into account (Bento-Silva et al., 2019). Therefore, identifying and quantifying microbial metabolites is an important field of research, since some could have physiological effects or function as biomarkers of phytochemical intake (Clifford et al., 2000).

14.8 MODELS OF DIGESTION AND BIOAVAILABILITY IN VITRO

Some models that simulate physiological conditions of the gastrointestinal tract are used to evaluate the release, viability, and resistance of phytochemicals in different formulations. These models are classified into conventional and dynamic models (Gbassi and Vandamme, 2012). The conventional model simulates the conditions of the stomach or intestine separately. This model consists of a reactor with agitation and physiological temperature, in which the conditions of digestion are simulated by dividing it into three phases, which are gastric, enteric and final enteric (Juárez-Roldán and Jiménez-Munguía, 2013).

During the first phase, an isotonic medium and a pH of 1–3 similar to that of the gastric fluid is used, and frequently pepsin and lipase added to this fluid. For the enteric phase, a medium containing sodium salts, glucose, bile, and yeast extract is added. This medium adjusts to a pH from 6 to 8, and in some cases, pancreatic enzymes are added. In the final enteric phase, the same conditions of the enteric phase are used. However, the pH is adjusted from 6.7 to 7.5, and sometimes pectin, glucose, starch, yeast extract, and sterilized fecal water are added (Zhao et al., 2012).

The dynamic models are simulators of the human gastrointestinal tract and consist of a reactor for each part of the digestive tract (stomach, small intestine, ascending colon, transverse colon, and descending colon) (Afkhami et al., 2007). There are two outstanding models, in which temperature and pH are controlled. The first model consists of taking an aliquot of the gastric phase and adding it to the enteric phase or modifying the pH of the container containing the gastric fluid. In the second model, the reactors maintained with agitation and constant temperature, and flow is generated using a peristaltic pump. The simulator of the microbial ecosystem of the human intestine called SHIME (Simulator of the Human Intestinal Microbial Ecosystem) and the gastrointestinal model TIM by its acronym TNO GastroIntestinal Model are examples of these models (Juárez-Roldán and Jiménez-Munguía, 2013). The TIM was the first model adapted to control the dynamic parameters of the gastric lumen. This model has four compartments, and each one has two glass units with flexible walls inside. These compartments connected through peristaltic valves (Cordonnier et al., 2015). Gastric secretion contains electrolytes, pepsin, and lipase and duodenal secretion consist of electrolytes, bile, and pancreatin. The secretion rates are programmable over time and two different systems eliminate digestion products.

The water-soluble products are eliminated by dialysis through membranes connected to the compartment of the jejunum and ileum. The lipophilic products removed through a filter that allows the passage of micelles and retains the drops of fat. Therefore, this model aims to study the availability of a compound for absorption through the intestinal wall (bioavailability). While the SHIME model simulates the passage of chyme (to which the encapsulated probiotics are added) by the gastrointestinal rectum, gastric, pancreatic, and intestinal enzymes are added in a controlled manner in this model.

Additionally, bile is added, and the pH, temperature, composition of the food, anaerobic environment, time of transit through the gastrointestinal tract, and the addition of human microbiota are controlled (Yoo and Chen, 2006). The SHIME consists of six reactors the first reactor simulates the conditions of the stomach, and the second those of the duodenum. Therefore, both simulate the conditions of food digestion. The third reactor simulates the conditions of the jejunum at a pH of 6.5–7. Whereas, the remaining three reactors simulate the colon microbiota since they are inoculated with a fecal suspension (Wu et al., 2018).

Accordingly, the biological effect of dietary phytochemicals depends on the bioavailability and stability that is determined in part by their resistance to gastrointestinal digestion. However, there is still a long way to go, as some of these phytochemicals can be difficult to trace because the microbiota can biotransform the compounds. Therefore, an essential parameter to guarantee bioactivity is to measure the biomarkers of stress, and not only the presence or absence of phytochemicals in blood, urine, or feces. In contrast, it is necessary to know the mechanisms of bioavailability of the active compounds of plants, to avoid possible interactions with food or drugs and to avoid adverse reactions.

14.9 FINAL COMMENTS

As described, to ensure the safety and bioavailability of the active compounds of plants, mechanisms of action and absorption must be investigated. Also, research on the interaction with medicines and foods is scarce which represents an area of opportunity in the field. However, the study of the appropriate doses should deepen, and the information generated should be informed to the consumer through the label or an informative brochure to create awareness in the user about their consumption and avoid adverse effects. Consequently, regulatory agencies should consider their inclusion.

Accordingly, it is necessary to continue with investigations that deepen the studies on the safe use of the incorporation of plants in food products. Therefore, the most efficient cooperation between researchers, private industry, and regulatory systems is necessary, since this allows the marketing of safer food.

KEYWORDS

- **security**
- **biodisponibility**
- **derivatives from plants**

REFERENCES

Afkhami, F.; Ouyang, W.; Chen, H.; Lawuyi, B.; Lim, T.; Prakash, S. Impact of Orally Administered Microcapsules on Gastrointestinal Microbial Flora: *In-Vitro* Investigation Using Computer Controlled Dynamic Human Gastrointestinal Model. *Artif Cell Blood Sub.* 2007, *35*, 359–375.

Al-Snafi, A.E. Medical Importance of *Anthemis nobilis* (*Chamaemelum nobile*)—A Review. *Asian J Pharm Sci Technol.* 2016, *6*, 89–95.

AlTamimi, J.Z. Awareness of the Consumption of Dietary Supplements Among Students in a University in Saudi Arabia. *J Nutr Metab.* 2019, *2019*, 4641768.

Amiot, M.J.; Riva, C.; Vinet, A. Effects of Dietary Polyphenols on Metabolic Syndrome Features in Humans: A Systematic Review. *Obes Rev.* 2016, *17*, 573–586.

Ayati, Z.; Ramezani, M.; Amiri, M. S.; Moghadam, A. T.; Rahimi, H.; Abdollahzade, A.; Emami, S. A.; Sahebkar, A. Ethnobotany, Phytochemistry and Traditional Uses of *Curcuma spp.* and Pharmacological Profile of Two Important Species (*C. longa and C. zedoaria*): A Review. *Curr Pharm Des.* 2019, *25*, 871–925.

Balderas, J.L.; Reza, V.; Ugalde, M.; Guzmán, L.; Serrano, M.I.; Aguilar, A.; Navarrete, A. Pharmacodynamic Interaction of the Sedative Effects of *Ternstroemia pringlei* (Rose) Standl. with Six Central Nervous System Depressant Drugs in Mice. *J Ethnopharmacol.* 2008, *119*, 47–52.

Bedrood, Z.; Rameshrad, M.; Hosseinzadeh, H. Toxicological Effects of *Camellia sinensis* (Green Tea): A Review. *Phytother Res.* 2018, *32*, 1163–1180.

Bento-Silva, A.; Koistinen, V.M.; Mena, P.; Bronze, M.R.; Hanhineva, K.; Sahlstrøm, S.; Kitrytė, V.; Moco, S.; Aura, A.M. Factors Affecting Intake, Metabolism and Health Benefits of Phenolic Acids: Do We Understand Individual Variability? *Eur J Nutr.* [Online early access]. https://doi.org/10.1007/s00394-019-01987-6. May 21, 2019. https://link.springer.com/article/10.1007%2Fs00394-019-01987-6

Bernatoniene, J.; Cizauskaite, U.; Ivanauskas, L.; Jakstas, V.; Kalveniene, Z.; Kopustinskiene, D. M. Novel Approaches to Optimize Extraction Processes of Ursolic, Oleanolic and Rosmarinic Acids from *Rosmarinus Officinalis* Leaves. *Ind Crops Prod.* 2016, *84*, 72–79.

Bonati, A. Formulation of Plants Extracts into Dosage Forms. In *The Medicinal Plant Industry*; CRC Press: New York, 1991; Chapter 9; pp 107–114.

Bradford, S.; Ramsetty, A.; Bragg, S.; Bain, J. Endocrine Conditions in Older Adults: Anti-Aging Therapies. *FP Essent.* 2018, *474*, 33–38.

Burgos-Edwards, A.; Jiménez-Aspee, F.; Theoduloz, C.; Schmeda-Hirschmann, G. Colonic Fermentation of Polyphenols from Chilean Currants *(Ribes spp.*) and its Effect on Antioxidant Capacity and Metabolic Syndrome-associated Enzymes. *Food Chem.* 2018, *30*, 144–155.

Carranza-Torres, I.E.; Viveros-Valdez, E.; Guzmán-Delgado, N.E.; García-Davis, S.; Morán-Martínez, J.; Betancourt-Martínez, N.D.; Balderas-Rentería, I.; Carranza-Rosales, P. Protective Effects of Phenolic Acids on Mercury-induced DNA Damage in Precision-cut Kidney Slices. *Iran J Basic Med Sci.* 2019, *22*, 367–375.

Clifford, M.N.; Copeland, E.L.; Bloxsidge, J.P.; Mitchell, L.A. Hippuric Acid as a Major Excretion Product Associated with Black Tea Consumption. *Xenobiotica.* 2000, *30*, 317–326.

Collins, A.R. Assays for Oxidative Stress and Antioxidant Status: Applications to Research into the Biological Effectiveness of Polyphenols. *Am J Clin Nutr.* 2005, *81*, 261S–267S.

Cordonnier, C.; Thévenot, J.; Etienne-Mesmin, L.; Denis, S.; Alric, M.; Livrelli, V.; Blanquet-Diot, S. Dynamic *In Vitro* Models of the Human Gastrointestinal Tract as Relevant Tools to Assess the Survival of Probiotic Strains and Their Interactions with Gut Microbiota. *Microorganisms.* 2015, *3*, 725–745.

Cory, H.; Passarelli, S.; Szeto, J.; Tamez, M.; Mattei, J. The Role of Polyphenols in Human Health and Food Systems: A Mini-Review. *Front Nutr.* 2018, *5*, 87.

Côté, J.; Caillet, S.; Doyon, G.; Sylvain, J.F.; Lacroix, M. Bioactive Compounds in Cranberries and Their Biological Properties. *Crit Rev Food Sci Nutr.* 2010, 50, 666–79.

Crespy, V.; Morand, C.; Besson, C.; Manach, C.; Demigne, C.; Remesy, C. Quercetin, but not Its Glycosides, Is Absorbed from the Rat Stomach. *J. Agric. Food Chem.* 2002, *50*(3): 618–621.

Deshpande, S. Role of Anti-oxidants in Prevention of Age-related Macular Degeneration. *J. Med. Nutr. Nutraceut.* 2012, *1*, 83–86.

Diario Oficial de la Federación, Acuerdo por el que se determinan las plantas prohibidas o permitidas para tés, infusiones y aceites vegetales comestibles. México, 15 de diciembre de 1999. http://www.salud.gob.mx/unidades/cdi/nom/compi/a1512993.html (accessed June 24, 2019).

Diario Oficial de la Federación, Ley General de Salud, México, 30 de abril de 2003. Art. 200 Bis, 212, 215 y 216. http://www.salud.gob.mx/unidades/cdi/legis/lgs/LEY_GENERAL_DE_SALUD.pdf (accessed June 24, 2019).

Diario Oficial de la Federación, Reglamento de Control Sanitario de Productos y Servicios, México, 09 de agosto de 1999. Art. 168, 170, 173. http://www.salud.gob.mx/unidades/cdi/nom/compi/rcsps.html (accessed June 24, 2019).

Donet, J.A.; Sornmayura, K.; Gulau, M.; Schiff, E. Hepatotoxicity by Herbs: A Practical Review of a Neglected Disease. *Rev Gastroenterol Peru.* 2016, *36*, 350–353.

Dourakis, S.P.; Papanikolaou, I.S.; Tzemanakis, E.N.; Hadziyannis, S.J. Acute Hepatitis Associated with Herb (*Teucrium capitatum* L.) Administration. *Eur J Gastroenterol Hepatol.* 2002, *14*, 693–695

Edwards, S. E.; Rocha, I. D.; Williamson, E. M.; Heinrich, M. Bilberry; Blueberry. In *Phytopharmacy*; John Wiley & Sons, Ltd: UK, 2015; Chapter 10; pp 47–49.

El-Hadary, A. E.; Ramadan, M. F. Phenolic Profiles, Antihyperglycemic, Antihyperlipidemic, and Antioxidant Properties of Pomegranate (*Punica granatum*) peel extract. *J Food Biochem.* [Online early access]. doi.org/10.1111/jfbc.12803. Published Online: February 11, 2019.

Ettel, M.; Gonzalez, G.A.; Gera, S.; Eze, O.; Sigal, S.; Park, J.S.; Xu, R. Frequency and Pathological Characteristics of Drug-induced Liver Injury in a Tertiary Medical Center. *Hum Pathol.* 2017, *68*, 92–98.

Ferguson, J.J.A.; Wolska, A.; Remaley, A.T.; Stojanovski, E.; MacDonald-Wicks, L.; Garg, M.L. Bread Enriched with Phytosterols with or Without Curcumin Modulates Lipoprotein Profiles in Hypercholesterolaemic Individuals. A Randomised Controlled Trial. *Food Funct.* 2019, *10*, 2515–2527.

Food and Drug Administration, Guidance for Industry, Considerations Regarding Substances Added to Foods, Including Beverages and Dietary Supplements, United States American, January 2014. https://www.fda.gov/regulatory-information/search-fda-guidance-documents/guidance-industry-considerations-regarding-substances-added-foods-including-beverages-and-dietary-0 (accessed June 24, 2019).

Fung, F. Y.; Wong, W. H.; Ang, S. K.; Koh, H. L.; Kun, M. C.; Lee, L. H.; Li, X.; Ng, H. J.; Tan, C. W.; Zhao, Y.; et al. A Randomized, Double-blind, Placebo-controlled Study on the Anti-haemostatic Effects of *Curcuma longa*, *Angelica sinensis* and *Panax ginseng*. *Phytomedicine.* 2017, *32*, 88–96.

Funk, R.S.; Singh, R.K.; Winefield, R.D.; Kandel, S.E.; Ruisinger, J.F.; Moriarty, P.M.; Backes, J.M. Variability in Potency Among Commercial Preparations of Berberine. *J Diet Suppl.* 2018, *15*, 343–351.

Gavrić, A.; Ribnikar, M.; Šmid, L.; Luzar, B.; Štabuc, B. Fat Burner-induced Acute Liver Injury: Case Series of Four Patients. *Nutrition.* 2018, *47*,110–114.

Gbassi, G.; Vandamme, T. Probiotic Encapsulation Technology: From Microencapsulation to Release into the Gut. *Pharmaceutics.* 2012, *4*, 149–163.

Ghorbani, A.; Esmaeilizadeh, M. Pharmacological Properties of *Salvia officinalis* and its Components. *J Tradit Complement Med.* 2017, *7*, 433–440.

Gupta, R. C.; Srivastava, A.; Lall, R. Toxicity Potential of Nutraceuticals. *Methods Mol Biol.* 2018, *1800*, 367–394.

Gutierrez, J.L.; Bowden, R.G.; Willoughby, D.S. *Cassia cinnamon* Supplementation Reduces Peak Blood Glucose Responses but Does not Improve Insulin Resistance and Sensitivity in Young, Sedentary, Obese Women. *J Diet Suppl.* 2016, *13*, 461–71.

Huang, F.Y.; Deng, T.; Meng, L.X.; Ma, X.L. Dietary Ginger as a Traditional Therapy for Blood Sugar Control in Patients with Type 2 Diabetes Mellitus: A Systematic Review and Meta-analysis. *Medicine (Baltimore).* 2019, *98*, e15054.

Husain, I.; Ahmad, R.; Chandra, A.; Raza, S. T.; Shukla, Y.; Mahdi, F. Phytochemical Characterization and Biological Activity Evaluation of Ethanolic Extract of *Cinnamomum zeylanicum*. *J Ethnopharmacol*. 2018, *219*, 110–116.

Hussain, S.A.; Panjagari, N.R.; Singh, R.R., Patil, G.R. Potential Herbs and Herbal Nutraceuticals: Food Applications and Their Interactions with Food Components. *Crit Rev Food Sci Nutr*. 2015, *55*, 94–122.

Ibrahim, M. Y.; Hashim, N. M.; Mariod, A. A.; Mohan, S.; Abdulla, M. A.; Abdelwahab, S. I.; Arbab, I. A. α-Mangostin from *Garcinia mangostana* Linn: An Updated Review of its Pharmacological Properties. *Arab J Chem*. 2016, *9*, 317–329.

Iñiguez, M.; Pérez-Matute, P.; Villanueva-Millán, M.J.; Recio-Fernández, E.; Roncero-Ramos, I.; Pérez-Clavijo, M.; Oteo, J.A. *Agaricus bisporus* Supplementation Reduces High-fat Diet-induced Body Weight Gain and Fatty Liver Development. *J Physiol Biochem*. 2018, *74*, 635–646.

Izzo, A. A.; Hoon-Kim, S.; Radhakrishnan, R.; Williamson, E. M. A Critical Approach to Evaluating Clinical Efficacy, Adverse Events and Drug Interactions of Herbal Remedies. *Phytother Res* 2016, *30*, 691–700.

Jacob K. M. Evaluation of Antibacterial and Antioxidant Activity of *Garcinia gummigutta*. *Int J Drug Dev Res*. 2015, *7*, 62–64.

Jafari, M.; Felgner, J.S.; Bussel, I.I.; Hutchili, T.; Khodayari, B.; Rose, M.R.; Vince-Cruz, C.; Mueller, L.D. Rhodiola: A Promising Anti-aging Chinese Herb. *Rejuvenation Res*. 2007, *10*, 587–602.

Jamshidi-Kia, F.; Lorigooini, Z.; Asgari, S.; Saeidi, K. Iranian Species of Verbascum: A Review of Botany, Phytochemistry, and Pharmacological Effects. *Toxin Rev*. 2018, *0*, 1–8.

Juárez-Roldán, A.R.; Jiménez-Munguía, M.T. Condiciones gastrointestinales modelo utilizadas para evaluar probióticos encapsulados. *Temas selectos de ingeniería de alimentos*. 2013, *7* -(2), 15–24.

Khoo, H.E.; Ng, H.S.; Yap, W.S.; Goh, H.J.H.; Yim, H.S. Nutrients for Prevention of Macular Degeneration and Eye-related Diseases. *Antioxidants (Basel)*. 2019, 8, pii: E85.

Kilari, E. K.; Putta, S. Biological and Phytopharmacological Descriptions of *Litchi chinensis*. *Pharmacogn Rev*. 2016, *10*, 60–65.

Kim, J.K.; Jeong, H.W.; Kim, A.Y.; Hong, Y.D.; Lee, J.H.; Choi, J.K.; Hwang, J.S. Green satsuma mandarin Orange (*Citrus unshiu*) Extract Reduces Adiposity and Induces Uncoupling Protein Expression in Skeletal Muscle of Obese Mice. *Food Sci Biotechnol*. 2018, *28*, 873–879.

Kinsella, L. J.; Grossberg, G. T.; Prakash, N. Drug-dietary Interactions: Over-the-counter Medications, Herbs, and Dietary Supplements. In *Clinical Psychopharmacology for Neurologists: A Practical Guide*; Springer International Publishing: Cham, 2018; Chapter 12; pp 213–224.

Konishi, K.; Wada, K.; Yamakawa, M.; Goto, Y.; Mizuta, F.; Koda, S.; Uji, T.; Tsuji, M.; Nagata, C. Dietary Soy Intake is Inversely Associated with Risk of type 2 Diabetes in Japanese Women but not in Men. *J Nutr*. 2019, 11, pii: nxz047.

Korus, A.; Jaworska, G.; Bernaś, E.; Juszczak, L. Characteristics of Physico-chemical Properties of Bilberry (*Vaccinium myrtillus* L.) Jams with Added Herbs. *J Food Sci Technol* 2015, *52*, 2815–2823.

Koss-Mikołajczyk, I.; Kusznierewicz, B.; Wiczkowski, W.; Płatosz, N.; Bartoszek, A. Phytochemical Composition and Biological Activities of Differently Pigmented Cabbage (*Brassica oleracea* var. capitata) and Cauliflower (*Brassica oleracea* var. botrytis) varieties.

J Sci Food Agric. [Online early access]. https://doi.org/10.1002/jsfa.9811. Published Online: May 17, 2019. https://onlinelibrary.wiley.com/doi/pdf/10.1002/jsfa.9811 (accessed June 24, 2019).

Lee, S.H.; Song, Y.S.; Jeong, Y.; Ko, K.S. Antioxidative and Anti-inflammatory Activities of *Akebia quinata* Extracts in an In Vitro Model of Acute Alcohol-induced Hepatotoxicity. *J Med Food.* 2017, *20*, 912–922.

Leone, A.; Spada, A.; Battezzati, A.; Schiraldi, A.; Aristil, J.; Bertoli, S. *Moringa oleifera* Seeds and Oil: Characteristics and Uses for Human Health. *Int J Mol Sci.* 2016, *17*, 1–14.

Li, Z.; Wang, K.; Zheng, J.; Cheung, F. S. G.; Chan, T.; Zhu, L.; Zhou, F. Interactions of the Active Components of *Punica granatum* (Pomegranate) with the Essential Renal and Hepatic Human Solute Carrier Transporters. *Pharm Biol.* 2014, *52*, 1510–1517.

Lim, D.W.; Kim, Y.T.; Jang, Y.J.; Kim, Y.E.; Han, D. Anti-obesity Effect of *Artemisia capillaris* Extracts in High-fat Diet-induced Obese Rats. *Molecules.* 2013, *18*, 9241–52.

Lopresti, A. L. Salvia (Sage): A Review of its Potential Cognitive-enhancing and Protective Effects. *Drugs R D.* 2017, *17*, 53–64.

Lunsford, K.E.; Bodzin, A.S.; Reino, D.C.; Wang, H.L.; Busuttil, R.W. Dangerous Dietary Supplements: Garcinia Cambogia-associated Hepatic Failure Requiring Transplantation. *World J Gastroenterol.* 2016, *22*, 10071–10076.

Manach, C.; Williamson, G.; Morand, C.; Scalbert, A.; Rémésy, C. Bioavailability and Bioefficacy of Polyphenols in Humans. I. Review of 97 Bioavailability Studies. *Am J Clin Nutr.* 2005, *81*, 230S–242S.

Martel, J.; Ojcius, D.M.; Chang, C.J.; Lin, C.S.; Lu, C.C.; Ko, Y.F.; Tseng, S.F.; Lai, H.C.; Young, J.D. Anti-obesogenic and Antidiabetic Effects of Plants and Mushrooms. *Nat Rev Endocrinol.* 2017, *13*, 149–160.

Marx, W.; McKavanagh, D.; McCarthy, A. L.; Bird, R.; Ried, K.; Chan, A.; Isenring, L. The effect of Ginger (*Zingiber officinale*) on Platelet Aggregation: A Systematic Literature Review. *PLoS One.* 2015, *10*, 1–13.

Medina-Caliz, I.; Garcia-Cortes, M.; González-Jiménez, A.; Cabello, M.R.; Robles-Díaz, M.; Sanabria-Cabrera, J.; Sanjuan-Jimenéz, R.; Ortega-Alonso, A.; García-Muñoz, B.; Moreno, I.; Jiménez-Pérez, M.; Fernández, M.C.; Ginés, P.; Prieto, M.; Conde, I.; Hallal, H.; Soriano, G.; Roman, E.; Castiella, A.; Blanco-Reina, E.; Montes, M.R.; Quiros-Cano, M.; Martin-Reyes, F.; Lucena, M.I.; Andrade, R.J. Herbal and Dietary Supplement-induced Liver Injuries in the Spanish DILI Registry. *Clin Gastroenterol Hepatol.* 2018, *S1542-3565*, 30010–30017.

Mehmood, Z.; Khan, M. S.; Qais, F. A.; Samreen; Ahmad, I. In *New Look to Phytomedicine*; Academic Press: UK, 2019; Chapter 18; pp 503–520.

Miousse, I.R.; Skinner, C.M.; Lin, H.; Ewing, L.E.; Kosanke, S.D.; Williams, D.K.; Avula, B.; Khan, I.A.; ElSohly, M.A.; Gurley, B.J.; Koturbash, I. Safety Assessment of the Dietary Supplement OxyELITE™ Pro (New Formula) in Inbred and Outbred Mouse Strains. Food *Chem Toxicol.* 2017, *109*, 194–209.

Modaeinama, S.; Abasi, M.; Abbasi, M. M.; Jahanban-Esfahlan, R. Anti Tumoral Properties of *Punica granatum* (Pomegranate) Peel Extract on Different Human Cancer Cells. *Asian Pac J Cancer Prev.* 2015, *16*, 5697–5701.

Munda, S.; Dutta, S.; Haldar, S.; Lal, M. Chemical Analysis and Therapeutic Uses of Ginger (*Zingiber officinale* Rosc.) Essential Oil: A Review. *J Essential Oil Bear Plants.* 2018, *21*, 994–1002.

Nieoczym, D.; Socała K; Jedziniak, P.; Wyska, E.; Wlaź, P. Effect of Pterostilbene, A Natural Analog of Resveratrol, on the Activity of Some Antiepileptic Drugs in the Acute Seizure Tests in Mice. Neurotox Res. [Online early access]. https://doi.org/10.1007/s12640-019-00021-1. Published Online: March 15, 2019. https://link.springer.com/article/10.1007%2Fs12640-019-00021-1 (accessed June 24, 2019).

Nigdikar, S.V.; Williams, N.R.; Griffin, B.A.; Howard, A.N. Consumption of Red Wine Polyphenols Reduces the Susceptibility of Low-density Lipoproteins to Oxidation *In Vivo*. *Am J Clin Nutr*. 1998, *68*, 258–265.

NOM-051-SCFI/SSA1-2010, Diario Oficial de la Federación. Especificaciones generales de etiquetado para alimentos y bebidas no alcohólicas preenvasados-Información comercial y sanitaria, México, 05 de abril de 2010. http://www.economia-noms.gob.mx/normas/noms/2010/051scfissa1mod.pdf (accessed June 24, 2019).

NOM-251-SSA1-2009, Diario Oficial de la Federación. Prácticas de higiene para el proceso de alimentos, bebidas o suplementos alimenticios, México, 10 de octubre de 2008. http://www.dof.gob.mx/normasOficiales/3980/salud/salud.htm (accessed June 24, 2019).

Novotny, J.A.; Chen, T.Y.; Terekhov, A.I.; Gebauer, S.K.; Baer, D.J.; Ho, L.; Pasinetti, G.M.; Ferruzzi, M.G. The Effect of Obesity and Repeated Exposure on Pharmacokinetic Response to Grape Polyphenols in Humans. *Mol Nutr Food Res*. [Online early access]. https://doi.org/10.1002/mnfr.201700043. Published Online: November, 2017. https://www.ncbi.nlm.nih.gov/pmc/articles/PMC5668187/pdf/nihms905375.pdf (accessed June 24, 2019).

Oracz, J.; Nebesny, E.; Zyzelewicz, D.; Budryn, G.; Luzak, B. Bioavailability and Metabolism of Selected Cocoa Bioactive Compounds: A Comprehensive Review. *Crit Rev Food Sci Nutr*. 2019, 24, 1–39.

Paeng, C.H.; Sprague, M.; Jackevicius, A. Interaction Between Warfarin and Cranberry Juice. *Clin Therapeut*. 2007, *29*, 1730–1735.

Pai Jakribettu, R.; Boloor, R.; P. Bhat, H.; Thaliath, A.; Haniadka, R.; P. Rai, M.; George, T.; Baliga, S. Ginger (*Zingiber officinale* Rosc.) Oils. In *Essential Oils in Food Preservation, Flavor and Safety*; Academic Press: 2016; Chapter 50; pp 447–454.

Pak, E.; Esrason, K.T.; Wu, V.H. Hepatotoxicity of Herbal Remedies: An Emerging Dilemma. *Prog Transplant*. 2004, *14*, 91–96.

Passamonti, S.; Vrhovsek, U.; Vanzo, A.; Mattivi, F. Fast Access of Some Grape Pigments to the Brain. *J. Agric. Food Chem.* 2005, *53*, 7029–7034.

Pedraza-Chaverri, J.; Cárdenas-Rodríguez, N.; Orozco-Ibarra, M.; Pérez-Rojas, J. M. Medicinal Properties of Mangosteen (*Garcinia mangostana). Food Chem Toxicol*. 2008, *46*, 3227–3239.

Petroni, M.L.; Caletti, M.T.; Dalle Grave, R.; Bazzocchi, A.; Gómez, M.P.A.; Marchesini, G. Prevention and Treatment of Sarcopenic Obesity in Women. *Nutrients*. 2019, *11*, 1–25.

Piskula, M.K.; Yamakoshi, J.; Iwai, Y. Daidzein and Genistein But Not Their Glucosides are Absorbed from the Rat Stomach. *FEBS Letters*. 1999, *44*, 287–291.

Potterat, O. Goji (*Lycium barbarum and L. chinense*): Phytochemistry, Pharmacology and Safety in the Perspective of Traditional Uses and Recent Popularity. *Planta Med.* 2010, *76*, 7–19.

Prakash, V.; Rana, S.; Sagar, A. Studies on Antibacterial Activity of *Verbascum thapsus. J Med Plants Stud.* 2016, *4*, 101–103.

Prior, R.L. Fruits and Vegetables in the Prevention of Cellular Oxidative Damage. *Am J Clin Nutr*. 2003, *78*, 570S–578S.

Prokop, J.; Lněničková, K.; Cibiček, N.; Kosina, P.; Tománková, V.; Jourová, L.; Láníčková, T.; Skálová, L.; Szotáková, B.; Anzenbacher, P.; et al. Effect of Bilberry Extract (*Vaccinium myrtillus* L.) on Drug-metabolizing Enzymes in Rats. *Food Chem Toxicol* 2019, *129*, 382–390.

Radha, M. H.; Laxmipriya, N. P. Evaluation of Biological Properties and Clinical Effectiveness of *Aloe vera*: A Systematic Review. *J Trad Compl Med.* 2015, *5*, 21–26.

Ralla, T.; Herz, E.; Salminen, H.; Edelmann, M.; Dawid, C.; Weiss, J. Emulsifying Properties of Natural Extracts from *Panax ginseng* L. *Food Biophys.* 2017, *12*, 479–490.

Rasmussen, H.M.; Johnson, E.J. Nutrients for the Aging Eye. *Clin. Interv. Aging.* 2013, 8,741–748.

Rees, A.; Dodd, G.F.; Spencer, J.P.E. The Effects of Flavonoids on Cardiovascular Health: A Review of Human Intervention Trials and Implications for Cerebrovascular Function. *Nutrients.* 2018, 10, pii: E1852.

Regulation (EU) No 1169/2011, Official Journal of the European Union, of the European Parliament and of the Council, on the Provision of Food Information to Consumers, Strasbourg, France, 25 de octubre 2011. https://eur-lex.europa.eu/LexUriServ/LexUriServ.do?uri = OJ:L:2011:304:0018:0063:EN:PDF (accessed June 24, 2019).

Rein, M.J.; Renouf, M.; Cruz-Hernandez, C.; Actis-Goretta, L.; Thakkar, S.K; da Silva Pinto, M. Bioavailability of Bioactive Food Compounds: A Challenging Journey to Bioefficacy. *Br J Clin Pharmacol.* 2013, *75*, 588–602.

Ribeiro-Santos, R.; Carvalho-Costa, D.; Cavaleiro, C.; Costa, H. S.; Albuquerque, T. G.; Castilho, M. C.; Ramos, F.; Melo, N. R.; Sanches-Silva, A. A Novel Insight on An Ancient Aromatic Plant: The Rosemary (*Rosmarinus officinalis* L.). *Trends Food Sci Technol.* 2015, *45*, 355–368.

Rodriguez, A.; T A Peixoto, I.; Verde-Star, M.; De la Torre, S.; Avilés Arnaut, H.; Ruiz, A. *In vitro* Antimicrobial and Antiproliferative Activity of *Amphipterygium adstringens. Evid Based Complementary Altern Med.* 2015, *2015*, 1–7.

Saito, Y. Current Status of Health Foods Including Their Interactions with Drugs and Adverse Events. *Yakugaku Zasshi.* 2018, *138*, 1511–1516.

Salazar-Leal, M.E.; Flores, M.S.; Sepúlveda-Saavedra, J.; Romero-Díaz, V.J.; Becerra-Verdin, E.M.; Tamez-Rodríguez, V.A.; Martínez, H.R.; Piñeyro-López, A.; Bermúdez, M.V. An Experimental Model of Peripheral Neuropathy Induced in rats by *Karwinskia humboldtiana* (buckthorn) Fruit. *J Peripher Nerv Syst.* 2006, *11*, 253–61.

Sánchez-Mendoza, M.E.; Torres, G.; Arrieta, J.; Aguilar, A.; Castillo-Henkel, C.; Navarrete, A. Mechanisms of Relaxant Action of a Crude Hexane Extract of *Gnaphalium liebmannii* in Guinea Pig Tracheal Smooth Muscle. *J Ethnopharmacol.* 2007,*111*, 142–147.

Santini, A.; Cammarata, S. M.; Capone, G.; Ianaro, A.; Tenore, G. C.; Pani, L.; Novellino, E. Nutraceuticals: Opening the Debate for a Regulatory Framework. *Br J Clin Pharmacol.* 2018, *84*, 659–672.

Santini, A.; Tenore, G. C.; Novellino, E. Nutraceuticals: A Paradigm of Proactive Medicine. *Eur J Pharm Sci.* 2017, *96*, 53–61.

Sather, T.E.; Woolsey, C.L.; Delorey, D.R.; Williams, R.D. J. Energy Drink and Nutritional Supplement Beliefs Among Naval Aviation Candidates. *Aerosp Med Hum Perform.* 2018, *89*, 731–736.

Sato, Y. Study of Formulation Development Based on the Pharmacokinetic Properties of Functional Food Components. *Yakugaku Zasshi.* 2019, *139*, 341–347.

Secretaría de Salud, Farmacopea Herbolaria de los Estados Unidos Mexicanos, Segunda Edición, México, 2013.

Semwal, R. B.; Semwal, D. K.; Vermaak, I.; Viljoen, A. A Comprehensive Scientific Overview of *Garcinia cambogia. Fitoterapia*. 2015, *102*, 134–148.

Stohs, S. J.; Hartman, M. J. Review of the Safety and Efficacy of *Moringa oleifera. Phytother Res*. 2015, *29*, 796–804.

Sultan, M.T.; Butt, M.S.; Qayyum, M.M.; Suleria, H.A. Immunity: Plants as Effective Mediators. *Crit Rev Food Sci Nutr*. 2014, *54*, 1298–308.

Toda, M.; Uneyama, C. Current Problems Associated with Overseas Health Products. *Yakugaku Zasshi*. 2018, *138*, 1531–1536.

Tou, J.C. Resveratrol Supplementation Affects Bone Acquisition and Osteoporosis: Pre-clinical Evidence Toward Translational Diet Therapy. *Biochim Biophys Acta*. 2015, *1852*, 186–194.

Touillaud, M.; Gelot, A.; Mesrine, S.; Bennetau-Pelissero, C.; Clavel-Chapelon, F.; Arveux, P.; Bonnet, F.; Gunter, M.; Boutron-Ruault, M.C.; Fournier, A. Use of Dietary Supplements Containing Soy Isoflavones and Breast Cancer Risk Among Women Aged >50 y: A Prospective Study. *Am J Clin Nutr*. 2019, *109*, 597–605.

Vilahur G, Ben-Aicha S, Diaz E, Badimon L, Padro T. Phytosterols and inflammation. *Curr Med Chem*. [Online early access]. https://doi.org/10.2174/0929867325666180622151438. Published Online: June 22, 2018. http://www.eurekaselect.com/163194/article (accessed June 24, 2019).

Viveros-Valdez, E.; Jaramillo-Mora, C.; Oranday-Cardenas, A.; Mordn-Martinez, J.; Carranza-Rosales, P. Antioxidant, Cytotoxic and Alpha-glucosidase Inhibition Activities from the Mexican Berry "Anacahuita" (*Cordia boissieri*). *Arch Latinoam Nutr*. 2016, *66*, 211–218.

Winter, J.; Bokkenheuser, V.D. Bacterial Metabolism of Natural and Synthetic Sex Hormones Undergoing Enterohepatic Circulation. *J Steroid Biochem*. 1987, *27*, 1145–50.

Wu, T.; Grootaert, C.; Pitart, J.; Vidovic, N.K.; Kamiloglu, S.; Possemiers, S.; Glibetic, M.; Smagghe, G.; Raes, K.; Van de Wiel, T.; Van Camp, J. Aronia (*Aronia melanocarpa*) Polyphenols Modulate the Microbial Community in a Simulator of the Human Intestinal Microbial Ecosystem (SHIME) and Decrease Secretion of Proinflammatory Markers in a caco-2/Endothelial Cell Coculture Model. *Mol Nutr Food Res*. 2018, *62*, e1800607.

Wu, X.; Song, M.; Cai, X.; Neto, C.; Tata, A.; Han, Y.; Wang, Q.; Tang, Z.; Xiao, H. Chemopreventive Effects of Whole Cranberry (*Vaccinium macrocarpon*) on Colitis-associated Colon Tumorigenesis. *Mol Nutr Food Res*. 2018, *62*, 1–21.

Xiang, J.; Lv, Q.; Yi, F.; Song, Y.; Le, L.; Jiang, B.; Xu, L.; Xiao, P. Dietary Supplementation of Vine Tea Ameliorates Glucose and Lipid Metabolic Disorder Via Akt Signaling Pathway in Diabetic Rats. *Molecules*. 2019, 24, pii: E1866.

Xiao, J.; Bai, W. Bioactive Phytochemicals. *Crit Rev Food Sci Nutr*. 2019, *59*, 827–829.

Yoo, J.; Chen, X. GIT Physicochemical Modeling a Critical Review. *Int J Food Eng*. 2006, *2*, 1–10.

Zhao, Q.; Mutukumira, A.; Lee, S.; Maddox, I.; Shu, Q. Functional Properties of Free and Encapsulated *Lactobacillus reuteri* DPC16 during and After Passage Through a Simulated Gastrointestinal Tract. *World J Microbiol Biotechnol*. 2012, *28*, 61–70.

CHAPTER 15

Dough Viscoelasticity of the Bread-Making Process Using Dynamic Oscillation Method: A Review

JESÚS ENRIQUE GERARDO-RODRÍGUEZ[1], BENJAMÍN RAMÍREZ-WONG[1*], PATRICIA ISABEL TORRES-CHÁVEZ[1], ANA IRENE LEDESMA-OSUNA[1], CONCEPCIÓN LORENIA MEDINA-RODRÍGUEZ[1], BEATRIZ MONTAÑO-LEYVA[1], and MARÍA IRENE SILVAS-GARCÍA[2]

[1]*Departamento de Investigación y Posgrado en Alimentos, Universidad de Sonora, Hermosillo, Sonora 83000, Mexico*

[2]*Universidad Estatal de Sonora, Hermosillo, Sonora, Mexico*

[*]*Corresponding author. E-mail: bramirez@guaymas.uson.mx*

ABSTRACT

Bread quality is an important factor to consider before buying it. The bread makers elaborate on them, and they need parameters of quality to obtain the best product. To evaluate the bread quality, there are some methods used, such as bread volume measurement and texture profile analysis. However, there are rheological studies carried out in dough which give an idea of the quality through viscoelasticity evaluation. Oscillatory dynamic studies are used to define the changes in the chemical components in baking. The objective of this review is to report the viscoelastic effects experienced at each stage of the process due to changes in gluten and starch of the dough using the dynamic oscillation method. For example, the elastic moduli (G') decrease when the protein is damaged in the mixing, fermentation, and baking. On the other hand, tan δ decreases at high temperatures in the oven due to the starch gelatinization and water evaporation, forming a rigid and porous structure characteristic of the bread.

15.1 INTRODUCTION

In baking, viscoelasticity is a behavior used to determine the use of dough in different products. Elastic flour is reflected in high-volume bread because this parameter is related to good-quality proteins. The transient stress response of the dough up to the application of constant stress is often a viscous response. On the contrary, the dough as viscous material shows a high level of retarded elasticity with a long terminal relaxation time (Lefebvre, 2006).

Viscoelasticity is a behavior of materials when they are subjected to stress (solid–liquid ratio). All foods have this behavior and its magnitude depends on their components. There are different methods used to obtain the viscoelastic parameters and other associated measurements to determine the characteristics of a product from its behavior to compression stress.

In the oscillatory dynamic method, sinusoidal parallel stress (or strain) is applied through oscillating rotating plates. The viscoelastic material responds to the strain. The elastic component represents the storage moduli (G') and the viscous component is measured as loss moduli (G''). The ratio between both modules (G''/G') is equal to tan δ or phase angle (Ponce-García, 2013). In a stress/strain plot, amplitude versus time, the phase angle determines the viscous or elastic ratio. While the angle is closer to 90° it is a more viscous behavior, and when it is closer to 0°, it is more elastic (Dogan and Kokini, 2007). Farris (1984) developed the impulse viscoelasticity technique in polymer sciences to investigate the properties of materials in the aging process; however, it has not been used in food rheology. The aim of this chapter is to study the viscoelastic behavior of each stage of the bread-making process (fresh bread, frozen dough, and part-baked bread) through rheological investigations using the dynamic oscillatory method.

15.2 VISCOELASTICITY OF PRODUCTS FROM DIFFERENT STEPS

15.2.1 WHEAT KERNEL

The characteristics of the different wheat varieties depend on genetic conditions. This may be due to different proportions of protein, starch, and moisture. Depending on the wheat variety it is used to obtain different products such as bread, tortillas, cookies, and pasta. It is important to measure the viscoelastic properties of the wheat kernel, as it will be reflected in dough and bread quality. Wheat and other agricultural products are considered viscoelastic materials. The creep recovery test is used to determine viscoelastic properties in products made from cereals such as wheat (Edwards et al., 2001), wheat sourdough (Clarke et al., 2004), and biscuit dough (Pedersen et al., 2004). Example of the potential of viscoelasticity evaluation in wheat is the study reported by Hernandez et al. (2012) in which a correlation between the recovery strain and the volume of the bread and measurement of the quality of several varieties of wheat is obtained (Van Bockstaele et al., 2008). In hard wheat cultivars (high content of high molecular weight glutenins) lower delay times have been found significantly in comparison to soft wheat varieties, indicating that gluten shows a faster recovery after creep (Tronsmo et al., 2003). However, no relationship with the volume of bread was reported. On the other hand, high retardation times have been obtained and thus demonstrate a slow recovery. On the contrary, studies by Kawai et al. (2006) found an inverse relationship with time delay and bread volume.

Due to the nature of the wheat, it is not possible to use the dynamic oscillatory method. Ponce-García et al. (2008) determined viscoelasticity using the uniaxial compression test. The results showed that the yield point had similar strength in all samples of different wheat varieties (hard, soft, and durum). On the other hand, the trend of the force–strain plots allowed to establish that the elastic work behavior (W_e) and the moduli of elasticity of kernels were higher and different depending of the wheat variety (soft < hard < durum).

In other studies, Gorji et al. (2010) measured the resistance of kernel fracture in terms of strength and energy. The force required for rupture decreased with moisture. When the grain was subjected to horizontal compression, it was more elastic. Therefore, with less moisture, higher stress.

15.2.2 DOUGH

The rheological analyzes to determine flour quality are carried out in water–flour dough. The fundamental measurements (dynamic oscillation) and empirical (farinograph) measurements are used to characterize the properties of different varieties of wheat flour (Sandeep and Narpinder, 2013).

Rheological studies provide important information on how the composition of wheat flour affects machinability and baking operations (Dogan, 2002). On the other hand, authors such as Singh et al. (2011) have performed farinographic and mixographic studies that detail information about water absorption, development time, dough stability, and extensibility/tenacity ratio. Empirical measurements, such as the farinograph, offer useful information but require large strain. The above can define the destination of the product to be elaborated. Oscillatory dynamic measurements involve small strain to study the fundamental structures and properties of wheat dough (Song and Zheng, 2007). Some rheological studies have been carried out by Dobraszczyk and Morgenstern (2003) to analyze dough viscoelastic properties by different methods, which vary in sample size, geometry, and strain size applied.

Figure 15.1 shows the behavior of the elastic moduli in water–flour dough of formulated dough and part-baked dough carry out by a dynamic oscillatory method using a rheometer. These tests are indicators of the molecular structures of gluten and starch and predict their behavior in bread making, using either descriptive or empirical techniques (Collar and Bollaín, 2004).

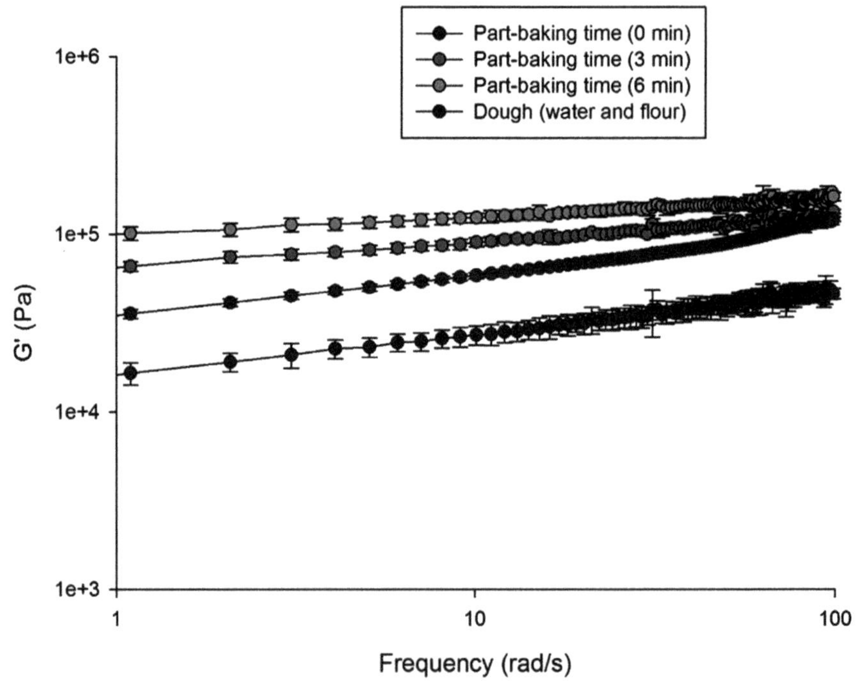

FIGURE 15.1 Effect of part-baking time on G' of dough and part-baked bread using the frequency sweep test.

Previously, Petrofsky and Hoseney (1995) demonstrated the existence of interactions between gluten and starch in the dough at 25 °C and Addo et al.

(2001) suggest that starch is trapped initially within granules interlaces between the gluten network and reinforces it. In a study by dynamic oscillation, when the dough is subject to less shear stress it shows high values of the modules due to the presence of high proportions of starch compared to gluten (Khatkar and Schofield, 2002). A viscoelastic evaluation carried out by Dreese and Hoseney (1990) in dough showed that the values of tan δ increased with high frequencies. A similar change in G', G'' and tan δ was observed in previous studies by other authors.

It has been reported that dynamic modulus (G' and G'') is affected by proteins of the flour as well as the water content in the dough. Studies by Letang et al. (1999) have recorded a decrease of G' and G'' and an increase in water content. Autio et al. (2001) reported low G' and G'' values for hard flours, whereas Janssen et al. (1996) reported the opposite.

15.2.3 DOUGH VISCOELASTICITY IN MIXING STEP

During the mixing of flour with water and other ingredients, the viscoelastic properties are developed due to the formation of three-dimensional matrices through protein–protein interactions of gluten and protein–carbohydrate of starch (Addo et al., 2001).

When mixing and hydration in the dough is insufficient, the result is the formation of a discontinuous protein network structure with an inadequate number of hydrated flour particles (Kim et al., 2008). The presence of these rigid flour particles under mixing and dough formation results in weak gluten, which is responsible for a solid behavior when viscoelastic analyses with a low strain are performed. This results in high values of G' and low values of tan δ. On the other hand, when the dough is formed with flour of hard wheat shows viscoelastic values lower than optimum mixed but higher than overmixed dough. This indicates that overmixed causes greater damage in hard dough due to the high breaking of the gluten network, resulting in low G' and G'' values (Sandeep et al., 2013).

The changes in the chemical components of dough and the formation of the gluten network during the processing impact on the quality of the final product. That is why many dough studies are performed in relation to mixing in the correct way (Dobraszczyk and Morgenstern, 2003). Extensional biaxial strain and dough tension show a significant relationship according to studies reported by Collar and Bollaín (2004). This provides additional information to the small strain performed by a dynamic oscillation test.

Figure 15.2 shows the viscoelastic behavior in the temperature sweep test (tan δ) of water–flour dough formulated dough and part-baked dough. This is important for baking because the dough is exposed to large strains and leave the linear region where viscoelasticity is measured (Lefebvre, 2006). Applying high stress (Khatkar and Schofield, 2002) or strain could be useful to predict the dough potential in baking.

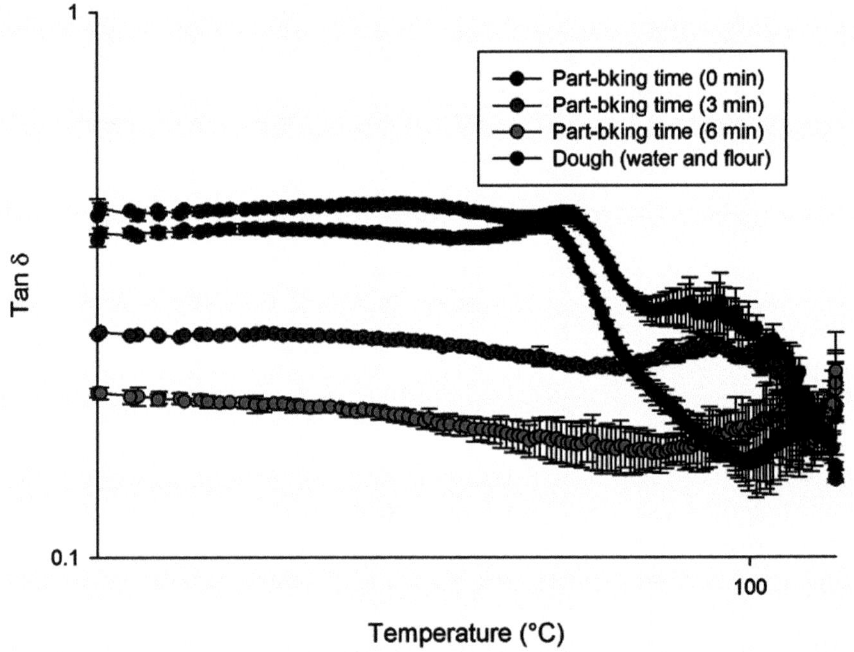

FIGURE 15.2 Effect of part-baking time on tan δ of dough and part-baked bread using the temperature sweep test.

Dough mixing time related to viscoelastic behavior has been studied for some researchers (Chin and Campbell, 2005; Zheng et al., 2000). Generally, it is accepted that mixing characteristics, such as torque versus time plots with a microscale mixer, are strongly related to the final dough properties (Dobraszczyk and Morgenstern, 2003). Water absorption, protein content, and protein quality have a great influence on the mixing properties and their final consistency.

The mixing process is a complex combination of strain by rotational and elongational stresses (Jongen et al., 2003). The polymers respond to the stress by three main processes: polymer disarray (it is a special form of bond rupture), reorientation of the chains, and rupture of the bonds (Singh and MacRitchie, 2001). The extensional stress of the mixture causes gluten dispersion in the dough (Peighambardoust et al., 2006). Applying large strain helps to understand the effects of the formation of disulfide bonds and their importance in the expansion of gas bubbles in the dough. Extension properties show large variations and similar properties in linear viscoelasticity. This gives an important contribution to the arrangements of dough strength and differences in extensional behavior are due to the arrangement of high molecular weight glutenins (Edwards et al., 2001).

As expected, the storage (G') and loss (G'') modules decrease as water content increases. Several studies show that the magnitude of this effect is large (Zheng et al., 2000), and depends on flour composition. It has been observed that the sensitivity of G' and G'' to moisture content decreases when protein content is higher. In the bread-making process, other ingredients such as yeast, salt, and additives are included to make dough highly thixotropic (it takes some time to stabilize) and complicating its rheological behavior.

15.2.4 FERMENTED DOUGH

During fermentation, extensional strain affects the rheological properties and bread texture. Extension in the fermentation stage was studied by several researchers using empirical, extensographic (Tlapale-Valdivia et al., 2010), and mixographic methods (Sapirstein et al., 2007). These techniques are also used to know rheological properties during mixing to decrease fermentation time.

There are several studies on rheological properties of fermented dough, some of which have been carried out by Esselink et al. (2003) and Newberry et al. (2002), in which yeast has kept active during measurements. Rheological properties of fermented dough are difficult to determine due to continuous changes in the system (Oliver and Brock, 1997; Wehrle and Arendt, 1998). The oscillatory measurements are based mainly on the stress while the dough strain during fermentation is dominated by its extensional flow (Dobraszczyk and Morgenstern, 2003). This suggests the benefit of using the viscoelastic momentum in dough experiments.

Wehrle and Arendt (1998) suggested that higher viscous behavior during fermentation should be related to the amount of gas bubbles, which interfere with the elastic network of dough. The same trend was observed using an acoustic technique (Skaf et al., 2009). It is apparent that these rheological changes are related to the addition of yeast. Baker's products that depend on yeast (CO_2 and acid production) could affect the action of enzymes and the solubility of proteins due to changes in pH as well as the dough geometry due to the growth of air bubbles. As a result, bonds involved in the gluten network of dough should be affected, altering its rheological properties.

The effect of yeast on the dough has been studied in order to produce bread in dynamic oscillation 24 h after its activity (sponge method). Lee et al. (2004) investigated the changes in rheological properties of dough during fermentation using ultrasound and extensional tests where G' and G'' and the extensional viscosity of the fermented dough were analyzed. The authors reported a loss of the elastic behavior of fermented dough.

Newberry et al. (2002) studied methods of inactivation of yeast using different freeze-thaw cycles. They concluded that a fast freezing rate followed by two stages of thawing had less yeast damage. Most authors recommend the addition of double yeast in the preparation of frozen dough (Phimolsiripol, 2009). During freezing, the integrity of the yeast membrane is subject to high osmotic pressure. Due to the above, a high amount of phospholipids in the membrane is necessary to prevent rupture (Codon et al., 2003). Hohmann (1997) suggested that exposure to hyperosmotic stress results in rapid dehydration of the cells, which limits CO_2 production; for this reason, is necessary to increase fermentation time.

15.2.5 PART-BAKED BREAD

After the fermentation step, the next stage of bread making is part-baking or full-baking. The dough is exposed to high temperatures and is where most of the chemical changes occur; this causes changes in viscoelasticity.

When determinations are made by oscillatory methods in the part-baked dough, the values of the modules decrease because heating operation causes gelatinization of the starch and all interactions in the system decrease (Dogan, 2002). This could be due to the amylase enzyme activity of the damaged starch, which is in the early stage of baking (low temperatures) and release the absorbed water, reducing G' and G''. The temperatures at which G' and G'' increase have been related to increased viscosity due to the escape of amylose from the starch granules and forming a gel (Addo et al., 2001; Kasapis et al., 2000).

Generally, during gelatinization water penetrates the starch granule and it is swollen. Upon reaching a certain temperature, the absorbed water causes a strong destabilizing effect and leads to disorganization and disruption of the crystalline zones (amylopectin branches). The gelatinization of the starch is possible to be observed as a loss of birefringence, which takes place in a temperature range of 10 °C, depending on the type of starch. When performing a temperature sweep test using the oscillatory dynamic method at constant deformation, it is possible to observe the tan δ peak after 60 °C, gelatinization and viscosity increase followed by an abrupt fall toward more elastic zones (Figure 15.3).

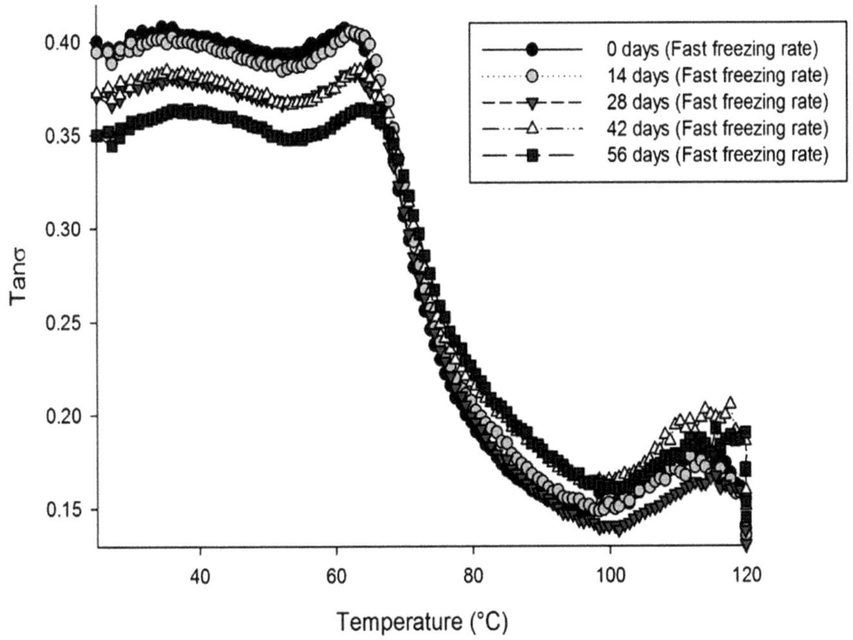

FIGURE 15.3 Effect of frozen storage time on tan δ of dough and part-baked bread using the temperature sweep test.

The temperature corresponding to the minimum values of *G′* (critical temperature) is in a range between 60 and 70 °C depending on the different wheat varieties. The decrease in *G′* values indicates the weakening of gluten and could be attributed to a partial denaturation of the proteins. An increase in the gelatinization temperature causes an increase in *G′*. The rheological behavior of gluten shows a viscous behavior before 60 °C and at higher baking temperatures it changes toward an elastic behavior. This indicates that gluten proteins are responsible for viscous behavior (gliadins) and probably entangled with a complex polymer protein structure (glutenins) with increasing temperature (Singh and MacRitchie, 2011). López-Da-Silva et al. (2007) observed a similar effect in G' at high temperatures that could be attributed to gluten-induced entanglement interactions during the formation of the structural network. It has been found that glutenin fraction in gluten is more sensitive to warming than gliadin fraction. When the temperature is above 75 °C, glutenins tend to denature due to the disulfide-sulfhydryl exchange reaction (Angioloni and Dalla Rosa, 2007).

15.2.6 FROZEN DOUGH

The baking industry has used advantages of the application of freezing technologies in the food, and has developed a special interest to know the needs of the customers increasing the useful life of their products (Giannou et al., 2003). The baking market offer added value and this has increased in recent decades due to consumer demands for convenience in obtaining a quality product (Bhattacharya et al., 2003).

Some advantages offered by the freezing process is to extend the useful life and best organization in bread making reducing nocturnal working hours. However, due to the use of the freezing process, there are quality implications that involve deficiencies in the products compared to traditional methods. Such reductions can be attributed to decreasing hydrophobic interactions between proteins, resulting in precipitation due to water redistribution. These results suggest that the protein folded beta conformation and alpha-helix are sensitive to freezing. The folded beta formation is partially evolved under the effect of freezing. This can be interpreted as forming a range of new folded beta-sheets together with a shift to bound structures by less strong hydrogen bonds (Georget et al., 2006) resulting in an increase in viscosity. Therefore, the decrease of the quality is related to the proteins and is the reason why being reduced in amount would reduce the elasticity. Additionally, to the protein changes, the quality of frozen bread is highly affected by the formulation of the dough (Rouille et al., 2000).

A way to measure the damage to dough components and understanding the loss of quality is through rheological tests. The measurement of viscoelasticity is related to proteins and their interactions. The viscoelastic changes that occur with a slow freezing rate are explained by the formation of large crystals and their mechanical action on the gluten network. On the other hand, a fast freezing rate forms small crystals with less dough damage. Results of studies suggest that G', G'', and tan δ of frozen dough were affected by different freezing treatments except for liquid nitrogen. This may be because the gluten network is weakened by traditional freezing treatments (Angioloni et al., 2008).

On the other hand, during freezing yeast dies and releases reducing agents such as glutathione (Collins and Haley, 1992). Glutathione weakens dough by reducing the disulfide bonds of gluten proteins, which is a determining factor in the gluten viscoelasticity. Frozen dough is used by international companies to reduce production costs and provide standard products at any time (Rosell and Gomez, 2007). Frozen dough is often exposed to a lower specific volume when making bread. This phenomenon is responsible for an increase of fermentation time compared to a fresh dough (Anon et al., 2004).

Esselink et al. (2003) studied the effect of frozen dough storage using spectroscopy and microscopy analysis. Due to the sensitivity of the gluten network and its rheological properties to temperature, freezing affected the textural properties of dough (Gélinas et al., 2004). When the dough is thaw, it is possible to perform an oscillatory test that allows to determinate damage through the behavior of the elastic and viscous modules. Generally, a decrease in the elastic modulus (G') is indicative of gluten weakening due to the disruption of terminal cysteine disulfide bonds of glutenins. Figure 15.4 shows the effect of the freezing rate and the storage time of frozen dough.

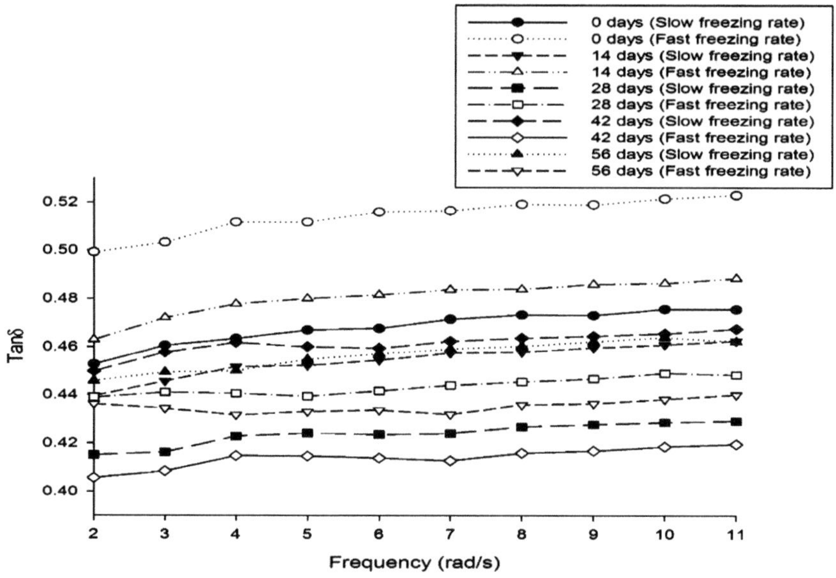

FIGURE 15.4 Effect of storage time and freezing rate on tan δ of dough and part-baked bread using the frequency sweep test.

The decrease in bread volume made with frozen dough is related to freezing storage, which reduces the fermentative capacity of the yeast and loses in gluten network integrity. This behavior affects dough machinability creating a problem in the baking industry (reducing product use life) because dough quality is reduced (Selomulyo and Zhou, 2007). The resistance of gluten to CO_2 decreases with breakage of disulfide bonds, resulting in poor gas retention and loss of volume during and after baking. This is a result of the redistribution of water in the protein matrix during the freezing step (Giannou and Tzia, 2007). On the other hand, Gormley et al. (2002) showed that temperature fluctuations in storage altered freezing and decreased the dough quality due to recrystallization. In addition, gasification power depends on the number of yeast cells, physiological state, and the amount of fermentable sugar (Teunissen et al., 2002).

Through some studies (Inoue and Bushuk, 1992; Havet et al., 2000; Giannou et al., 2003) it has been considered that quality attributes of bread made with frozen dough can be improved using additives and more yeast in the formulation by reducing changes in the dough structure.

15.2.7 BREAD

The firmness of the bread is often used as a measurement of quality of the crumb, which has been successfully determined with a texture analyzer using double compression mode (Baik and Chinachoti, 2000; Carson and Sun, 2001). Results obtained with this instrument are limited to empirical correlations because this test does not provide fundamental rheological data.

Measurement of firmness and volume of the bread are the most usual methods to evaluate bread quality. Viscoelastic tests by oscillatory strain are not very useful to evaluate bread quality due to the highly porous nature of the crumb, which leads to results in variable data. Multivariate analyzes of analytical data of the rheological profile of dough and bread have been used to evaluate the potential relation between functional properties and the final products. In addition, several studies have shown the effect of temperature above the glass transition (in the case of frozen dough), which accelerates starch retrogradation; therefore increases bread hardness (Charoenrein and Preechathammawong, 2010).

In the whole bread-making process, mainly in the freezing step of dough or part-baked bread, there are starch changes that cause water redistribution in the dough matrix and change the protein behavior (depolymerization).

The above also affects the rheological properties due to retrogradation that suffers when bread. This causes a faster moisture loss (syneresis) and a rearrangement of the amylopectin chains, causing a firmer crumb with less volume, a defect known as the aging of bread.

15.3 CONCLUSIONS

Viscoelastic measurement using oscillatory dynamic strain test has been shown to be a useful tool to know the integrity and changes of chemical components caused on each step of the bread-making process. In bread making it is useful to be aware of the effect of added ingredients and make decisions about their use. Measurement of viscoelastic behavior is useful to know in order to obtain parameters of the bread-making process such as the optimal fermentation and mixing time and the damages caused by freezing, and storage steps to proteins and starch. The dough is the material that suffers most of the physicochemical changes; this allows knowing in what step and temperature would occur these changes and how it affects baking speed and time.

KEYWORDS

- **bread-making process**
- **viscoelasticity**
- **water–flour dough**
- **fermentation**
- **wheat kernel**
- **baking**

REFERENCES

Addo, K., Xiong, Y. L., & Blanchard, S. P. 2001. Relationship of Polymeric Proteins and Empirical Dough Rheology with Dynamic Rheology of Dough and Gluten from Different Wheat Varieties. *Food Hydrocolloids,* 33, 342–348.

Angioloni, A., Balestra, F., Pinnavaia, G. G., Rosa, M.D. 2008. Small and Large Deformation Tests for the Evaluation of Frozen Dough Viscoelastic Behavior. *Journal of Food Engineering*, 87 (4), 527–531.

Angioloni, A., Rosa D.M. 2007. Effects of Cysteine and Mixing Conditions on White/Whole Dough Rheological Properties. *Journal of Food Engineering*, 80, 18–23.

Anon, M. C., Le-Bail, A., Leon, A. E. 2004. Effect of Freezing on Dough Ingredients. In: Hui, Y.H. et al., Eds.; *Handbook of Frozen Foods*. Marcel Dekker, New York, pp 571–580.

Autio, K., Flander, L., Kinnunen, A., Heinonen, R. 2001. Bread Quality Relationship with Rheological Measurements of Wheat Flour Dough. *Cereal Chemistry*, 78, 654–657.

Baik, M. Y., Chinachoti. 2000. Moisture Redistribution and Phase Transition During Bread Staling. *Cereal Chemistry*, 77, 484–488.

Bhattacharya, N., Black, E. L., Christensen, T. E., Larson, C. R. 2003. Assessing the Relative Informativeness and Permanence of Pro Forma Earnings and GAAP Operating Earnings. *Journal of Accounting and Economics*, (36), 285–319.

Carson, L., Sun, X. S. 2001. Creep-recovery of Bread and Correlation to Sensory Measurements of Textural Attributes. *Cereal Chemistry*, 78, 101–104.

Charoenrein, S., Preechathammawong, N. 2010. Undercooling Associated with Slow Freezing and its Influence on the Microstructure and Properties of Rice Starch Gels. *Journal of Food Engineering*, 100 (2), 310–314.

Chin, N. L., Campbell, G. M. 2005. Dough Aeration and Rheology: Part 2. Effects of Flour Type, Mixing Speed and Total Work Input on Aeration and Rheology of Bread Dough. *Journal of Science and Food Agriculture*, 85 (13), 2194–2202.

Clarke, C. I., Schober, T. J., Dockery, P., O'Sullivan, K., Arendt, E. K. 2004. Wheat Sourdough Fermentation: Effects of Time and Acidification on Fundamental Rheological Properties. *Cereal Chemistry*, 81 (3), 409–417.

Codon, A. C., Rincon, A. M., Moreno-Mateos, M. A., Delgado-Jarana, J., Rey, M., Limon, C., Rosado, I. V., Cubero, B., Penate, X., Castrejon, F., Benitez, T. 2003. New Saccharomyces Cerevisiae Baker's Yeast Displaying Enhanced Resistance to Freezing. *Journal of Agricultural and Food Chemistry*, 51 (2), 483–491.

Collar, C., Bollaín, C. 2004. Impact of Microbial Transglutaminase on the Viscoelastic Profile of Formulated Bread Doughs. *European Food Research and Technology*, 218 (2), 139–146.

Collins, B., Haley, S., 1992. Frozen Bread Doughs: Effect of Ascorbic Acid Addition and Dough Mixing Temperature on Loaf Properties. *Chorleywood Digest*, 114, 21–23.

Dobraszczyk, B. J., Morgenstern, M. P. 2003. Rheology and the bread Making Process. *Journal of Cereal Science*, 38 (3), 229–245.

Dogan, I. S. 2002. Dynamic Rheological Properties of Dough as Affected by Amylases from Various Sources. *Nahrung*, 46, 399–403.

Dogan, H., Kokin, J. L. 2007. Rheological Properties of Food. In Heldman, R. D. & Lund, B. D., Ed.; *Handbook of Food Engineering*; USA: CRC Press.

Dreese, P. C., Hoseney, R. C. 1990. The Effect of Water-extracted Solubles from Gluten on its Baking and Rheological Properties. *Cereal Chemistry,* 67, 400–404.

Edwards, N. M., Peressini, D., Dexter, J. E., Mulvaney, S. J. 2001. Viscoelastic Properties of Durum Wheat and Common Wheat Dough of Different Strengths. *Rheologica Acta* 40, 142–153.

Esselink, E., Van Aalst, H., Maliepaard, M., Henderson, T. M. H., Hoekstra, N. L. L., Van Duynhoven, J. 2003. Impact of Industrial Dough Processing on Structure: A Rheology, Nuclear Magnetic Resonance, and Electron Microscopy Study. *Cereal Chemistry*, 80 (4), 419–423.

Farris, R. J. 1984. An Impulse Approach to Linear Viscoelasticity. *Journal of Rheology,* 28 (4), 347–354.

Gélinas, P., McKinnon, C. M. 2004. Effect of Flour Heating on Dough Rheology. *Lebensmittel-Wissenschaft und-Technologie*, 37 (1), 129–131.
Georget, Dominique, M. R., Belton, Peter, S. 2006. Effects of Temperature and Water Content on the Secondary Structure of Wheat Gluten Studied by FTIR Spectroscopy. *American Chemical Society*, Washington, DC, USA.
Giannou, V., Kessoglou V., Tzia C. 2003. Quality and Safety Characteristics of Bread Made from Frozen Dough. *Trends in Food Science and Technology*, 14, 99–108.
Giannou, V. and Tzia, C. 2007. Frozen Dough Bread: Quality and Textural Behavior during Prolonged Storage e Prediction of Final Product Characteristics. *Journal of Food Engineering*, 79, 929–934.
Gorji, A., Rajabipour, A., Tavakoli, H. 2010. Fracture Resistance of Wheat Grain as a Function of Moisture Content, Loading Rate and Grain Orientation. *Australian Journal of Crop Science*, 4, 448–452.
Gormley, R., Walshe, T., Hussey, K., Butler, F. 2002. The Effect of Fluctuating vs. Constant Frozen Storage Temperature Regimes on Some Quality Parameters of Selected Food Products. *LWT—Food Science and Technology*, 35 (2), 190–200.
Janssen, A. M., van Vliet, T., Vereijken, J. M. 1996. Fundamental and Empirical Rheological Behaviour of Wheat Flour Doughs and Comparison with Bread Making Performance. *Journal of Cereal Science*, 23, 43–54.
Jongen, T. R. G., Bruschke, M. V., Dekker, J. G. 2003. Analysis of Dough Kneaders Using Numerical Flow Simulations. *Cereal Chemistry*, 80, 383–389.
Havet, M., Mankai, M., Le Bail, A. 2000. Influence of the Freezing Condition on the Baking Performances of French Frozen Dough. *Journal of Food Engineering,* 45 (3), 139–145.
Hernandez-Estrada, Z.J., Figueroa, J.D.C., Rayas-Duarte, P., Peña, R.J. 2012. Viscoelastic Characterization of Glutenins in Wheat Kernels Measured by Creep Tests. *Journal of Food Engineering*, 113, 19–26.
Hohmann, S. 1997. Shaping Up: The Response of Yeast to Osmotic Stress. In Hohmann, S., Mager, W.H., Eds.; *Yeast Stress Responses*. Springer, New York. pp 101–146.
Inoue, Y., Bushuk, W. 1992. Studies on Frozen Doughs II. Flour Quality Requirements for Bread Production from Frozen Dough. *Cereal Chemistry*, 69 (4), 423– 428.
Kasapis, S., Sablani, S. S., Biliaderis, C. G. 2000. Dynamic Oscillation Measurements of Starch Networks at Temperatures above 100 °C. *Carbohydrate Research*, 329, 179–187.
Kawai, T., Lal, A., Yang, X., Galban, S., Mazan-Mamczarz, K., Gorospe, M. 2006. Translational Control of Cytochrome c by RNA-binding Proteins TIA-1 and HuR. *Molecular and Cellular Biology*, 26, 3295–3307.
Khatkar, B. S., Schofield, J. D. 2002. Dynamic Rheology of Wheat Flour Dough. I. Non-Linear Viscoelastic Baheviour. *Journal of Science Food and Agricultural*, 82, 827–829.
Kim, Y. R., Cornillon, P., Campanella, O. H., Stroshine, S., Lee, S., Shim, J. Y. 2008. Small and Large Deformation Rheology for Hard Wheat Flour Dough as Influenced by Mixing and Resting. *Journal of Food Science*, 73, 1–8.
Lee, S. Y., Pyrak-Nolte, L. J., Campanella, O. 2004. Determination of Ultrasonic Based Rheological Properties of Dough During Fermentation. *J. Texture Stud.*, 35 (1), 33–51.
Lefebvre, J. 2006. An Outline of Dough Non-linear Viscoelastic Behavior in Shear. *Rheologica Acta*, 45, 525–538.
Letang, C., Piau, M., Verdier, C. 1999. Characterization of Wheat Flour Water Doughs. I. Rheometry and Microstructure. *Journal of Food Engineering*, 41, 121–132.

López-Da-Silva, J. A., Santos, D. M. J., Freitas, A., Brites, C., Gil, A. M. 2007. Rheological and Nuclear Magnetic Resonance (NMR) Study of the Hydration and Heating of Undeveloped Wheat Doughs. *Journal of Agricultural and Food Chemistry*, 55, 5636–5644.

Newberry, M. P., Phan-Thien, N., Larroque, O. R., Tanner, R. I., Larsen, N. G. 2002. Dynamic and Elongation Rheology of Yeasted Bread Doughs. *Cereal Chemistry*, 79, 874–879.

Oliver, G., Brock, C. J., 1997. A Rheological Study of Mechanical Dough Development and Long Fermentation Processes for Cream–cracker Dough Production. *J. Sci. Food Agric.*, 74, 294–300.

Pedersen, L., Kaack, K., Bergsøe, M. N., Adler-Nissen, J. 2004. Rheological Properties of Biscuit Dough from Different Cultivars, and Relationship to Baking Characteristics. *Journal of Cereal Science*, 39, 37–46.

Peighambardoust, S. H., van der Goot, A. J., van Vliet, T., Hamer, R. J., Boom, R. M. 2006. Microstructure Formation and Rheological Behavior of Dough Under Simple Shear Flow. *Journal of Cereal Science*, 43, 183–197.

Petrofsky, K. E, Hoseney R. C. 1995. Rheological Properties of Dough Made with Starch and Gluten from Several Cereal Sources. *Cereal Chemistry*, 72, 53–58.

Phimolsiripol, Y. 2009. Shelf Life Determination of Frozen Bread Dough Stored Under Fluctuating Temperature Conditions. *Kasetsart Journal—Natural Science*, 43 (1), 187–197.

Ponce-García, N., Figueroa-Cárdenas, J. D., López-Huape, G. A., Martinez, H. E., Martinez-Peniche, R. 2008. Study of Viscoelastic Properties of Wheat Kernels Using Compression Load Method. *Cereal Chemistry*, 85, 667–672.

Ponce-García, N., Ramírez-Wong, B., Torres-Chávez, P. I. Figueroa-Cárdenas, J. D., Serna-Saldivar, S. O. Cortez-Rocha, M. O. 2013. Effect of Moisture Content on the Viscoelastic Properties of Individual Wheat Kernels Evaluated by the Uniaxial Compression Test Under Small Strain. *Cereal Chemistry*, 90, 558–563.

Rosell, C. M., Gomez, M. 2007. Frozen Dough and Partially Baked Bread: An Update. *Food Reviews International*, 23 (3), 303–319.

Rouille, J., Le Bail, A., Courcoux, P. 2000. Influence of Formulation and Mixing Conditions on Breadmaking Qualities of French Frozen Dough. *Journal of Food Engineering*, 43 (4), 197–203.

Sandeep, S., Narpinder, S. 2013. Relationship of Polymeric Proteins and Empirical Dough Rheology with Dynamic Rheology of Dough and Gluten from Different Wheat Varieties. *Food Hydrocolloids*, 33 342–348.

Sapirstein, H. D., David, P., Preston, K. R., Dexter, J. E. 2007. Durum Wheat Bread Making Quality: Effects of Gluten Strength, Protein Composition, Semolina Particle Size and Fermentation Time. *Journal of Cereal Science,* 45, 150–161.

Selomulyo, V. O., Zhou, W. 2007. Frozen Bread Dough: Effects of Freezing Storage and Dough Improvers. *Journal of Cereal Science,* 45, 1–17.

Singh, H., MacRitchie, F. 2001. Application of Polymer Science to Properties of Gluten. *Journal of Cereal Science*, 33, 231–243.

Singh, S., Singh, N., MacRitchie, F. 2011. Relationship of Polymeric Proteins with Pasting, Gel Dynamic- and Dough Empirical-rheology in Different Indian Wheat Varieties. *Food Hydrocolloids*, 25, 19–24.

Skaf, A., Nassar, G., Lefebvre, F., Nongaillard, B. 2009. A New Acoustic Technique to Monitor Bread Dough During the Fermentation Phase. *Journal of Food Engineering,* 93, 365–378.

Song, Y., Zheng, Q. 2007. Dynamic Rheological Properties of Wheat Flour Dough and Proteins. *Trends in Food Science and Technology*, 18, 132–138.

Teunissen, A., Dumortier, .F, Gorwa, M. F., Bauer, J., Tanghe, A., Loiez, A., Smet, P., Van Dijck, P. Thevelein, J. M. 2002. Isolation and Characterization of a Freeze-tolerant Diploid Derivative of an Industrial Baker's Yeast Strain and its Use in Frozen Doughs. *Applied and Environmental Microbiology*, 68 (10), 4780–4787.

Tlapale-Valdivia, A. D., Chanona-Pérez, J., Mora-Escobedo, R., Farrera-Rebollo, R. R., Gutiérrez-López, G. F., Calderón-Domínguez, G. 2010. Dough and Crumb Grain Changes During Mixing and Fermentation and Their Relation with Extension Properties and Bread Quality of Yeasted Sweet Dough. *International Journal of Food Science and Technology*, 45, 530–539.

Tronsmo, K. M., Fægestad, E. M., Schofield, J. D., Magnus, E. M. 2003. Wheat Protein Quality in Relation to Baking Performance Evaluated by the Chorleywood Bread Process and a Hearth Bread Baking Test. *Journal of Cereal Science*, 38 205–215.

Van Bockstaele, F., De Leyn, I., Eeckhout, M., Dewettinck, K. 2008. Rheological Properties of Wheat Flour Dough and the Relationship with Bread Volume. I. Creep-recovery Measurements. *Cereal Chemistry*, 85 (6), 753–761.

Wehrle, K., Arendt, E. K. 1998. Rheological Changes in Wheat Sourdough During Controlled and Spontaneous Fermentation. *Cereal Chemistry*, 75 (6), 882–886.

Zheng, H., Morgenstern, M. P., Campanella, O. H., Larsen, N. G. 2000. Rheological Properties of Dough During Mechanical Dough Development. *Journal of Cereal Science*, 32, 293–306.

CHAPTER 16

Physicochemical Characteristics and Gelling Properties of Arabinoxylans Recovered from Maize Wastewater: Effect of Lime Soaking Time During Nixtamalization

GUILLERMO NIÑO-MEDINA[1], ELIZABETH CARVAJAL-MILLAN[2*], BENJAMÍN RAMÍREZ-WONG[3], JORGE MÁRQUEZ-ESCALANTE[2], and AGUSTIN RASCÓN-CHU[2]

[1]*Laboratorio de Química y Bioquímica, Facultad de Agronomía, Universidad Autónoma de Nuevo León, Francisco Villa S/N, Col. Ex-Hacienda El Canadá, C.P. 66050, General Escobedo, Nuevo León, México*

[2]*Centro de Investigación en Alimentación y Desarrollo, CIAD, A. C. Carretera Gustavo Enrique Astiazarán Rosas No. 46, Col. La Victoria, Hermosillo, Sonora 83304, Mexico*

[3]*Departamento de Investigación y Posgrado en Alimentos, Universidad de Sonora, Hermosillo, Sonora 83000, México*

[*]*Corresponding author. E-mail: ecarvajal@ciad.mx*

ABSTRACT

In Mexico, alkali cooking called "nixtamalization" is extensively used to improve maize texture and nutritional value. From maize nixtamalization, "nixtamal" is obtained, which is good to produce soft dough and "nejayote," which is the wastewater generated from this process. During maize nixtamalization, the arabinoxylans (AXs) present in the maize cell wall are partially hydrolyzed generating low molecular weight AX. A nixtamalization process with prolonged lime soaking could result in more effective polysaccharide hydrolysis; however, to the best of our knowledge, the characteristics of AX generated by different maize lime soaking time have not been investigated elsewhere. In the present study, AX recovered from nejayote after 6 (AX6) or 12 h (AX12) of maize lime soaking presented a molecular weight of 74 and 57 kDa and ferulic acid content of 0.14 and 0.10 μg/mg polysaccharide, respectively. The intrinsic viscosity and arabinose to xylose ratios of AX6 and AX12 were 206 and 161 mL/g and 0.55 and 0.53, respectively. Both AX formed covalent gels induced by laccase. The storage modulus (G') AX6 and AX12 at 8% (w/v) were 4 and 2 Pa, respectively. These results indicate that an increase in lime soaking from 6 to 12 h during maize nixtamalization reduces the AX molecular size and ferulic acid content which forms, therefore, less elastic gels.

16.1 INTRODUCTION

In Mexico, alkali cooking called "nixtamalization" (from the Nahuatl nixtli = ashes and tamalli = dough) is extensively used to improve maize texture and nutritional value. From maize nixtamalization Nixtamal is obtained, which is ground to produce soft dough named masa and nejayote, which is the wastewater generated from this process. Maize nixtamalization is important in Mexico because half of the total volume of consumed food is maize, which provides approximately 50% of the energy intake; this proportion being even greater for lower income groups. In fact, the nixtamalization process is commonly used in the production of tortillas and other related maize-based food products commonly consumed by Mexicans. During maize nixtamalization, the AXs present in the maize cell wall are partially hydrolyzed generating low molecular weight gelling AX (<60 kDa) (Niño-Medina et al., 2009). AXs consist of a linear chain of D-xylopyranoses bound by β-(1–4) bonds which may be substituted by α-L-arabinofuranoses units (Fincher and Stone, 1974). AXs are from the few polysaccharides containing ferulic acid residues (4-hydroxy-3-methoxycinnamic acid) esterified to C5 on some arabinose residues (Smith and Hartley, 1983). AXs are dietary fiber (Van Laere et al., 2000), namely they are nonhydrolyzable in the stomach and small intestine but rather fermented by the colonic microbiota. In recent years, there has been an increase in AX interest due to numerous studies reporting its prebiotic properties (Broekaert et al., 2011; Marquez-Escalante et al., 2018). In rats fed with bread enriched with AX, the levels of short-chain fatty acids in the colon were increased, especially propionic acid

(Lopez et al., 1999). It has been reported that the intestinal microbiota and the metabolites that it generates have a positive impact on other processes of the organisms (Broekaert et al., 2011). In this sense, AX has been called promoters of good health, because they present cholesterol lowering, antioxidant, antitumoral, and antiobesogenic properties (Adam et al., 2002; Ou et al., 2007; Cao et al., 2011; Neyrinck et al., 2011; Mendez-Encinas et al., 2018). However, to the best of our knowledge, the effect of lime soaking time during nixtamalization on the physicochemical characteristics and gelling properties of AX has not been investigated elsewhere. It is possible that long lime soaking may generate low molecular weight AX which could present prebiotic properties while short lime soaking could generate AX exhibiting better gelling properties.

16.2 METHODOLOGY

16.2.1 MATERIALS

Nejayote was kindly provided by two tortilla making in Northern Mexico. Laccase (benzenediol:oxygen oxidoreductase, E.C.1.10.3.2) from *Trametes versicolor* and other chemical products were purchased from Sigma Co. (St. Louis, MO, USA).

16.2.2 AX EXTRACTION

AX from nejayote were extracted as described before (Carvajal-Millan et al., 2010). Maize wastewater was generated after 6 (AX6) or 12 (AX12) h of lime soaking during nixtamalization.

16.2.3 NEUTRAL SUGARS

Neutral sugar content in AX was determined after the hydrolysis with 2 N trifluoroacetic acid at 120 °C for 2 h. The reaction was stopped on ice and the extracts were evaporated under air at 40 °C and rinsed twice with 200 μl of water. The evaporated extract was solubilized in 500 μl of water. Mannitol was used as an internal standard. Samples were filtered through 0.2 μm (Whatman) and analyzed by high-performance liquid chromatography (HPLC) using a Supercogel Pb column (300 × 7.8 mm; Supelco, Inc.,

Bellefonte, PA, USA) eluted with 5 mM H_2SO_4 (filtered 0.2 μm, Whatman) at 0.6 mL/min and 50 °C. A Varian 9012 HPLC with Varian 9040 refractive index detector (Varian, St. Helens, Australia) and a Star Chromatography Workstation system control version 5.50 were used (Carvajal-Millan et al., 2007).

16.2.4 PHENOLIC ACIDS CONTENT

Ferulic acid (FA), dimers, and trimer of FA content in AX6 and AX12 were quantified by reverse-phase HPLC after de-esterification step as described by Vansteenkiste et al. (2004) and Niño-Medina et al. (2009). An Alltima, C18 column (250 × 4.6 mm) (Alltech Associates, Inc., Deerfield, IL, USA) and a photodiode array detector Waters 996 (Millipore Co., Milford, MA, USA) were used. Detection was followed by UV absorbance at 320 nm. The measurements were performed in triplicate.

16.2.5 PROTEIN

The protein amount in AX was determined according to the Bradford method (Bradford, 1976).

16.2.6 ASH

Ash content was determined as described by the AOAC methods (Association of Official Analytical Chemists, 2002).

16.2.7 MOLECULAR WEIGHT

The molecular weight distribution of AX6 and AX12 was determined by size exclusion-HPLC at 38 °C using a TSKgel (Polymer Laboratories, Shropshire, UK) G500 PMWX column (7.8 × 300 mm). 20 μl of AX1 and AX2 solutions (0.5% w/v in 0.1 M $LiNO_3$) filtered through 0.2 μm (Whatman) were injected, and a Water 2414 refractive index detector was used for detection. Isocratic elution was performed at 0.6 mL/min with 0.1 M $LiNO_3$ filtered through 0.2 μm. Molecular weights were estimated after universal calibration with pullulans as standards (P50–P800) (Martínez-López et al., 2013).

16.2.8 INTRINSIC VISCOSITY

Specific viscosity (η_{sp}) was measured by registering AX6 and AX12 solutions flow time in an Ubbelohde capillary viscometer at 25 ± 0.1 °C, immersed in a temperature-controlled bath. The intrinsic viscosity ([η]) was estimated from relative viscosity (η_{rel}), of AX solutions by extrapolation of Kraemer and Mead and Fouss curves to "zero" concentration (Carvajal-Millan et al., 2005).

16.2.9 GELLING

AX1 and AX2 solutions at 8% (w/v) were prepared in 0.1 M sodium acetate buffer pH 5.5. Laccase (1.675 nkat per mg polysaccharide) was used as a cross-linking agent. AX1 and AX2 gels were allowed to develop for 4 h at 25 °C (Carvajal-Millan et al., 2005).

16.2.10 RHEOLOGY

Small amplitude oscillatory shear was used to follow the gelation process of AX6 and AX12 solutions at 8% (w/v). Solutions were mixed with laccase (1.675 nkat per mg AX) and immediately placed on the parallel plate geometry (4.0 cm in diameter) of a strain-controlled rheometer (Discovery HR-2 rheometer; TA Instruments, New Castle, DE, USA). Exposed edges were covered with silicone oil to prevent evaporation. The dynamic rheological parameters used to evaluate the gel network were the storage modulus (G'), loss modulus (G''), crossover point ($G' > G''$), and tan delta (tan δ, G''/G'). AX1 and AX2 gelation were monitored at 0.25 Hz and 5% strain. At the end of the network formation, a frequency sweep (0.01–10 Hz) was carried out. Rheological measurements were performed in duplicate (Niño-Medina et al., 2009).

16.3 RESULTS AND DISCUSSION

The compositions of the AX6 and AX12 are shown in Table 16.1. Arabinose and xylose were the main monosaccharides detected. Residues of other neutral sugars, proteins, and ashes were also found. The FA content was 0.14 and 0.10 μg/mg of polysaccharide for AX6 and AX12, respectively, which is lower than that reported by Niño-Medina et al. (2009) in the first reported study on AX extracted from nejayote (0.23 μg FA/mg AX). In that study, the nejayote was recovered after only 4 h of maize lime soaking during the nixtamalization process. The longer resting time of the nejayote used in the present study (6 and 12 h) could explain the lower FA content in the AX6 and AX12 as under alkaline conditions this phenolic acid can be de-esterified from AX.

TABLE 16.1 Composition of Arabinoxylans Extracted from Nejayote After 6 and 12 h of Lime Soaking

	AX6	AX12
Arabinose (g/100 g)	27.0 ± 0.80^a	28.0 ± 1.10^a
Xylose (g/100 g)	49.0 ± 1.90^a	52.0 ± 1.00^a
Protein (g/100 g)	4.5 ± 0.20^a	3.9 ± 0.50^a
Ash (g/100 g)	5.10 ± 0.21^a	5.4 ± 0.32^a
Ferulic acid (μg/mg)	0.14 ± 0.01^a	0.10 ± 0.01^b

Note: Different letters indicate significant differences within the same row ($P < 0.05$). Values are the means + S.E.M. of triplicate experiments.

The A/X ratio, the intrinsic viscosity ([η]), and the molecular weight (Mw) of AX6 and AX12 are reported in Table 16.2. The A/X ratio was 0.55 and 0.53 for AX6 and AX12, respectively, indicating a moderately branched structure. Other authors have reported A/X values greater than 0.60 in AX extracted from maize husks and nejayote (Singh et al., 2000, Carvajal-Millan et al., 2007, Niño-Medina et al., 2009). [η] and Mw values decreased significantly by increasing the resting time of the nejayote from 6 to 12 h. This loss of polysaccharide chain size could also be attributed to alkaline hydrolysis of the glycosidic bond between the monosaccharides component of AX due to the high pH of nejayote. In the AX from nejayote reported by Niño-Medina et al. (2009), the values of [η] and Mw were 183 and 60 kDa, respectively, which are below the values found for AX6 (74 kDa) and above the values corresponding to AX12 (57 kDa). These variations could be related to the artisan character of the nixtamalization process at tortilla-making small industries; together with previous reports that the size and viscosity of AX can still vary in the same source (Izydorczyk and Biliaderis, 1995, Niño-Medina et al., 2010).

TABLE 16.2 Physicochemical Characteristics of Arabinoxylans Extracted from Nejayote After 6 and 12 h of Lime Soaking

	AX6	AX12
A/X	0.550 + 0.001[a]	0.530 + 0.001[b]
[η] mL/g	206 + 15[a]	161 + 9[b]
Mw (kDa)	74 + 6[a]	57 + 4[b]

Note: Different letters indicate significant differences within the same row ($P < 0.05$). Values are the means + S.E.M. of triplicate experiments.

Solutions of AX6 and AX12 at 8% (w/v) formed gels in the presence of laccase. The AX gelation profiles followed characteristic kinetics with an initial increase in G' and G'' followed by a plateau region for both gels (Figure 16.1). This behavior reflects an initial formation of covalent linkages between FA of adjacent AX molecules producing a three-dimensional network (Niño-Medina et al., 2010). At the end of gelation, G' and G'' were 8 and 3 Pa for AX6, respectively and 4 and 1 Pa for AX12, respectively. Similar kinetics of gelation has been previously reported for maize bran AX gels (Martínez-López et al., 2013; Paz-Samaniego et al., 2016; Niño-Medina et al., 2009). AX6 gel presented a higher G' value in comparison to AX12 gel, which can be attributed to its higher FA content in relation to that registered in AX12. Niño-Medina et al. (2011) reported nejayote AX gels (4% and 8%, w/v) with G' of 2 and 4 Pa, respectively, and crossover points of 150 min, which are in the range found in the present work. Such similarities might have its origin in the structural and/or conformational characteristics of these macromolecules. The tan δ (G''/G') value decreased during gelation indicating the development of an elastic covalent system (Ross-Murphy, 1984).

The amount of di-FA and tri-FA in the gels is reported in Table 16.3. At the end of the gelation, 80% of the FA initially present in the AX solutions was oxidized and only 16% was recovered as di-FA and tri-FA. This behavior has been reported by other authors (Lapierre et al., 2001; Carvajal-Millan et al., 2007) and has been attributed to the formation of FA structures that have not yet been identified so they cannot be quantified.

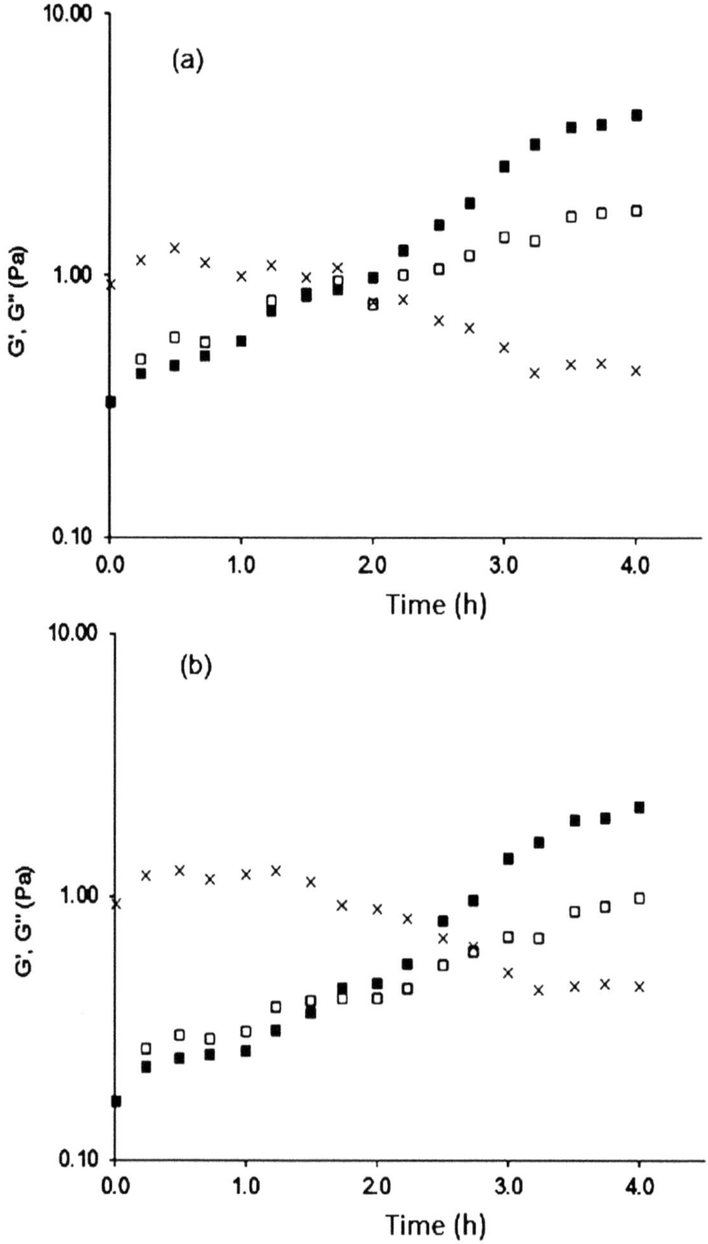

FIGURE 16.1 Monitoring the storage (G'■), loss modulus (G''□), and tan delta (x) of AX6 (a) and AX12 (b) solutions (8% w/v) during gelation by laccase at 0.25 Hz, 5% strain, and 25 °C.

TABLE 16.3 Di-FA and Tri-FA Content in Gels at 8% (w/v) in Arabinoxylans Extracted from Nejayote After 6 and 12 h of Lime Soaking

	AX6	AX12
di-FA (μg/mg AX)	0.025 + 0.001[a]	0.020 + 0.001[b]
tri-FA (μg/mg AX)	0.010 + 0.001[a]	traces

Note: Different letters indicate significant differences within the same row ($P < 0.05$). Values are the means + SEM of triplicate experiments.

16.4 CONCLUSIONS

An increase in maize lime soaking time from 6 to 12 h, during nixtamalization process, can modify the physicochemical characteristics and gelling properties of AX recovered from nejayote mainly by reducing the molecule Mw, [η] and FA content. The lower FA content in AX12 form gels presenting minor elasticity values. These results indicate that it is possible to generate tailored maize wastewater AX by modifying nixtamalization conditions. The study of maize wastewater AX recovered from different nixtamalization conditions might provide valuable information to support strategies aimed to design AX and AX gels with specific characteristics and functional properties.

KEYWORDS

- **ferulated arabinoxylans**
- **nejayote**
- **alkaline hydrolysis**
- **gelation**

REFERENCES

Adam, A. V., Crespy, V., Levrat-Verny, M. A., et al. *Journal of Nutrition* 2002, 132, 1962.

Association of Official Analytical Chemists. (2002). In *Official Methods of Analysis of AOAC International*. Arlington: AOAC International.

Bradford, M. (1976). A Rapid and Sensitive Method for the Quantification of Microgram Quantities of Protein Utilizing the Principle of Protein-dye Binding. *Analytical Biochemistry* 72, 248–254.

Broekaert W. F., Courtin C. M., Verbeke K. et al., (2011). Prebiotic and Other Health-Related Effects of Cereal-Derived Arabinoxylans, Arabinoxylan-Oligosaccharides, and Xylooligosaccharides. *Critical Reviews in Food Science and Nutrition*, 51(2), 178–194.

Cao, L. X., Liu, X. T., Qian, T. G. et al. 2011. *Int J Biol Macromol* 2011, 48,160.

Carvajal-Millan, E., Guigliarelli, B., Belle, V., Rouau, X., Micard, V. (2005). Storage Stability of Arabinoxylan Gels. *Carbohydrate Polymers*, 59, 181–188.

Carvajal-Millan, E., Rascón-Chu, A., Márquez-Escalante, J. (2010). Patente mexicana 278,768. Método para la obtención de goma de maíz a partir del líquido residual de la nixtamalización del grano de maíz.

Carvajal-Millan, E., Rascón-Chu, A., Márquez-Escalante, J. Ponce de León, N., Micard, V., Gardea, A. (2007). Maize Bran Gum: Characterization and Functional Properties. *Carbohydrate Polymers*, 69, 280–285.

Fincher, G. B., Stone, B. A. (1974). A Water-soluble Arabinogalactan-peptide from Wheat Endosperm. *Australian Journal of Biological Science* 27, 117–132.

Izydorczyk, M. S., Biliaderis C. G. (1995) Cereal Arabinoxylans: Advances in Structure and Physicochemical Properties. *Carbohydrate Polymers*, 28, 33–48.

Lapierre, C. Pollet, B., Ralet, M.C., Saulnier, L. (2001). The Phenolic Fraction of Maize Bran: Evidence for Lignin-heteroxylan Association. *Phytochemistry*, 57, 765–772.

Lopez, H. W., Levrat, M. A., Guy, C., et al. *J Nutr Biochem* 1999, 10, 500.

Marquez-Escalante, J., Carvajal-Millan, E., López-Franco, Y. L., et al. Ciencia UAT 2018, 13, 146.

Martínez-López, A. L., Carvajal-Millan, E., Rascón-Chu, A., et al. *CyTA—Journal of Food* 2013, 11, 22.

Mendez-Encinas, M. A., Carvajal-Millan, E., Rascon-Chu, A., Astiazaran-Garcia, H.F., Valencia-Rivera, D.E. *Oxidative Medicine and Cellular Longevity* 2018, DOI: 10.1155/2018/2314759.

Neyrinck, A. M., Possemiers, S., Druart, C. et al. *PLoS ONE* 2011, 6, e20944.

Niño-Medina, G., Carvajal-Millan, E., Rascon-Chu, A, Márquez-Escalante, J. A., Guerrero V., Salas-Muñoz, E. (2010). Feruloylated Arabinoxylans and Arabinoxylan Gels: Structure, Sources and Applications. *Phytochemistry Reviews*, 9, 111–120.

Niño-Medina, G., Carvajal-Millan, E., Rascón-Chu, A., Lizardi, J., Márquez-Escalante, J., Gardea, A., Martínez-López, A. L., Guerrero, V. (2009). Maize Processing Waste Water Arabinoxylans: Gelling Capability and Cross-linking Content. *Food Chemistry*, 115, 1286–1290.

Ou, S. Y., Jackson GM, Jiao X et al. (2007) *Journal of Agriculture and Food Chemistry*, 55, 3191.

Paz-Samaniego, R., Carvajal-Millan, E., Sotelo-Cruz, N., et al. *Sustainability* 2016, 8, 1104.

Ross-Murphy, S. B. (1984). Rheological Methods. In H. W. S. Chan, Ed.; *Biophysical Methods in Food Research*. Blackwell Scientific Publications: Oxford, UK, pp. 138–199.

Singh, V., Doner, L. W., Johnston, D. B., Hicks, K. B., Eckhoff, S.R. (2000). Comparison of Coarse and Fine Corn Fiber for Corn Fiber Gum Yields and Sugar Profiles. *Cereal Chemistry*, 77, 560–561.

Smith, M. M., Hartley, R. D. (1983). Occurrence and Nature of Ferulic Acid Substitution of Cell-wall Polysaccharides in Graminaceous Plants. *Carbohydrate Research*, 118, 65–80.

Van Laere, K. M. J, H.artemink, R., Bosveld M. et al. (2000). Fermentation of Plant Cell Wall Derived Polysaccharides and Their Corresponding Oligosaccharides by Intestinal Bacteria. *Journal of Agricultural Food Chemistry* 48, 1644–1652.

Vansteenkiste, E., Babot, C., Rouau, X. Micard, V. (2004). Oxidative Gelation of Feruloylated Arabinoxylan as Affected by Protein. Influence on Protein Enzymatic Hydrolysis. *Food Hydrocolloids*, 18, 557–564.

Index

S

T

V

Z